AF389381

Current Advances in Meat Nutritional, Sensory and Physical Quality Improvement

Current Advances in Meat Nutritional, Sensory and Physical Quality Improvement

Special Issue Editors

Mohammed Gagaoua
Brigitte Picard

MDPI • Basel • Beijing • Wuhan • Barcelona • Belgrade • Manchester • Tokyo • Cluj • Tianjin

Special Issue Editors
Mohammed Gagaoua
Teagasc Ashtown Food Research Centre
Ireland

Brigitte Picard
INRAE-Auvergne-Rhôn-Alpes
France

Editorial Office
MDPI
St. Alban-Anlage 66
4052 Basel, Switzerland

This is a reprint of articles from the Special Issue published online in the open access journal *Foods* (ISSN 2304-8158) (available at: https://www.mdpi.com/journal/foods/special_issues/Meat_Nutritional_Sensory_Physical_Quality).

For citation purposes, cite each article independently as indicated on the article page online and as indicated below:

LastName, A.A.; LastName, B.B.; LastName, C.C. Article Title. *Journal Name* **Year**, *Article Number*, Page Range.

ISBN 978-3-03928-690-4 (Pbk)
ISBN 978-3-03928-691-1 (PDF)

Contents

About the Special Issue Editors

Mohammed Gagaoua is a food/meat scientist who received his engineering degree at Constantine University, Algeria. In 2009, Mohammed started a high postgraduate and conducted a research on the main endogenous muscle proteolytic systems and serpins inhibitors before he receiving his Ph.D. in 2015 on meat and muscle biochemistry at INRA Research Centre, France, and Constantine University on biomarkers of sensory qualities of beef: understanding of the biological mechanisms and prediction. He worked as a lecturer at the Université de Béjaia from 2011 to 2014, as an associate professor at Université Mentouri Constantine from 2012 to 2017, before moving to INRA Auvergne-Rhône-Alpes to manage a project about the farmgate to fork continuum data to understand beef quality variability using rearing factors and carcass properties based on metadata. In February 2019, he was awarded a Marie Skłodowska-Curie Career-FIT program project at the Department of Food Quality and Sensory Science, Teagasc Food Research Centre, Ashtown, Dublin 15, Ireland titled "Omics-Based Biomarkers for Beef Carcass Quality Management". Mohammed's academic focus is on animal and food sciences, food biotechnology, meat and meat products, foodomics, and data integration. His main research interests are omics of meat quality, novel strategies to improve meat quality, biomarkers of meat qualities, and use of plant proteases recovered by three-phase partitioning as meat-tenderizing agents. As a career achiever, Mohammed was runner-up in 2015 for the International Meat Secretariat Award; he received the ICoMST2018 award by Robin Shorthouse for outstanding contribution to advancing Meat Science and the ICoMST2018 Meat Science journal award for best presentation in Australia. He has published more than 60 peer-reviewed papers in international journals, 7 book chapters, 24 proceedings, along with numerous contributions in national and international conferences.

Brigitte Picard is Director of Research at the National Research Institute for Agriculture, Food and Environment (INRAE) in France. After a bachelor at the University of Clermont-Ferrand and a thesis on the baking quality of hybrid wheat, she was recruited by INRA in 1991. Her research topic at the Joint Unit of Research on Herbivores concerns skeletal muscle development and ruminants meat quality. Her skills include: skeletal muscle physiology, analysis of proteins (electrophoresis, Western blot, ELISA, and immunohistochemistry), primary cells cultures (myoblasts and satellite cells), and proteomic analysis of skeletal muscle. She particularly works on the prediction of beef tenderness from proteomic biomarkers and the link between rearing practices and beef sensory qualities. Brigitte, with an h-index of 37, is the author of 172 original scientific articles, 46 reviews, and 558 papers in conferences. She has contributed to 21 book chapters and coordinated 5 collective editions. From 2004 to 2009, she was the French correspondent of the COST 925 action "The importance of prenatal events for postnatal muscle growth in relation to the quality of muscle-based foods". Then, she was a member of the COST FA1002 "Advances in Farm Animal Proteomics". From 1992 to 2016, she was the head of several INRA research teams. Since 2017, she has been co-director of the regional Ruminant Meat Consortium associating professionals of the meat sector and researchers.

Preface to "Current Advances in Meat Nutritional, Sensory and Physical Quality Improvement"

Meat is an important source of proteins, vitamins, minerals, and fat, and these nutrients are important for their beneficial effects on human health. In recent years, meat quality has become a more relevant topic for consumers with regard to health and sensory characteristics, and for beef industry stakeholders because it affects their profitability. Therefore, the control of meat quality, including technological, sensory, and nutritional quality traits, constitutes an important target for any farm animal production. What those qualities are and how we best evaluate them at the different levels of the supply chain from the farm to fork are critical to understanding meat production and consumption patterns. However, despite the efforts of industry to control the eating and nutritional quality traits of meat and meat products, a high level of variability remains, which is one reason for consumer dissatisfaction. This Special Issue focuses on the study of the various aspects from farm to fork that impact the control of the nutritional, sensory, and technological aspects of carcass, muscle, meat, and meat-product qualities. The Issue groups 14 original studies and one comprehensive review within five main topics: (i) production systems and rearing practices, (ii) prediction of meat quality, (iii) statistical approaches for meat quality prediction/management, (iv) muscle biochemistry and proteomics techniques, and (v) consumer acceptability, development, and characterization of meat products.

Mohammed Gagaoua, Brigitte Picard
Special Issue Editors

 *foods*

Editorial

Current Advances in Meat Nutritional, Sensory and Physical Quality Improvement

Mohammed Gagaoua [1,2,*] and **Brigitte Picard** [1,*]

[1] Université Clermont Auvergne, INRAE, VetAgro Sup, UMR Herbivores, 63122 Saint-Genès-Champanelle, France

[2] Food Quality and Sensory Science Department, Teagasc Ashtown Food Research Centre, Ashtown, Dublin 15, Ireland

* Correspondence: gmber2001@yahoo.fr (M.G.); brigitte.picard@inrae.fr (B.P.)

Received: 6 March 2020; Accepted: 9 March 2020; Published: 10 March 2020

Abstract: Meat is an important source of proteins, vitamins, minerals and fat, and these nutrients are important for their beneficial effects on human health. In recent years, meat quality has become a more relevant topic for consumers with regard to health and sensory characteristics, and for beef industry stakeholders because it affects their profitability. Therefore, the control of meat quality, including technological, sensory and nutritional quality traits, constitutes an important target for any farm animal production. What those qualities are and how we best evaluate them at the different levels of the continuum from the farm to fork are critical to understanding meat production and consumption patterns. However, despite the efforts of the industrials to control the eating and nutritional quality traits of meat and meat products, there remains a high level of variability, which is one reason for consumer dissatisfaction. This Special Issue focuses on the study of continuum aspects from farm to fork, which would have an impact on the control of the nutritional, sensory and technological aspects of carcass, muscle, meat and meat-product qualities. It groups fourteen original studies and one comprehensive review within five main topics that are (i) production systems and rearing practices, (ii) prediction of meat qualities, (iii) statistical approaches for meat quality prediction/management, (iv) muscle biochemistry and proteomics techniques and (v) consumer acceptability, development and characterisation of meat products.

Keywords: meat science; carcass and meat qualities; sensory and technological quality; muscle biochemistry; statistical tools for meat quality prediction; OMICs tools; production systems

---

Meat quality constitutes the main topic for both consumers with regard to health and sensory characteristics and beef industry stakeholders for economic considerations. Despite the efforts of beef sector actors to control the eating and nutritional quality of beef, there remains a high level of variability in these quality traits, which is one reason for consumer dissatisfaction. However, it is recognised that science and innovation will play a great role in helping the industry respond to consumer concerns and expectations. Accordingly, and within the idea and objective of bringing together original studies dealing with the continuum aspects, i.e., from farm to fork [1,2], having an impact on the nutritional, sensory and technological aspects of carcass, muscle, meat and meat-product qualities, we edit this Special Issue on "current advances in meat nutritional, sensory and physical quality improvement". From the fifteen published papers, five main research topics were covered: (i) production systems and rearing practices, (ii) prediction of meat qualities, (iii) statistical approaches for meat quality prediction/management, (iv) muscle biochemistry and proteomics techniques and (v) consumer acceptability, development and characterisation of meat products.

In the first topic dealing with production systems and rearing practices and their relationships with carcass and meat qualities, almost six papers can be categorized [3–8]. The first study by Moran et al. [3] compares the quality of beef from bulls slaughtered at 15 months of age and reared in typical Irish indoor production systems with those raised in novel grass-based systems. This study was conducted in response to some European markets seeking bulls to be slaughtered at less than 16 months of age with a carcass fatness score of six on the 15-scale. The context of the study by Moran and co-workers, therefore, was to compare cheaper alternative systems with those demonstrated to achieve the carcass weight and fat cover specifications. The authors found that the only grass-based system that reached the current market requirement was the ration based on grass silage and concentrates, offered indoors with associated costs of housing. Therefore, the production of late-maturing sired bulls for slaughter at less than 16 months of age from pasture does not seem to be an option for meeting current market requirements. It is worthwhile to note that the beef-eating quality of grass-fed animals was not detrimentally affected. The second study is by Couvreur et al. [4], which has been performed on the protected designation of origin (PDO) Maine–Anjou cull cows. The authors investigate the relationships among the characteristics of cull beef cows, rearing practices and physicochemical characteristics and sensory traits of two differing muscles, *Longissimus thoracis* (LT) and *Rectus abdominis* (RA). The novelty of this study is the consideration of the interaction between animal type and finishing practices at the farm scale on the final meat quality based on a clustering approach that distinguishes different rearing practices. The authors found that finishing practices, whatever the cluster, have less effect than animal type on RA and LT meat properties. Also, the effects observed on meat quality are directly related to farmers' practices and provide new advice and modifications in culled cows rearing practices that would help improve the meat quality of PDO Maine–Anjou. In the same context, two studies by Soulat et al. [5,6] investigate using the same experimental design on 96 heifers of the protected geographical indication (PGI) Fleur d'Aubrac (crossbreed Charolais × Aubrac); first, the effect of the rearing managements applied during the heifers' whole life on carcass and flank steak quality [5], and second, the effect on five muscles from ribs in the chuck sale section [6]. To achieve their goals, the authors conducted surveys with farmers to identify the main rearing managements applied, allowing them to select the animals with controlled slaughter and post-slaughter conditions to limit any effects on the final meat quality. In the first preliminary study, rearing management applied during the heifers' whole life seem to have more impact on carcass traits than on the flank steak properties. The authors conclude that carcass traits could be improved without altering the meat properties. In the second study, the authors investigate the impact of four different rearing managements (identified by a statistical clustering method) on the traits of the LT muscle (sensory, rheological, and colour) and the rheological traits of four other muscles, *complexus*, *infraspinatus*, *rhomboideus*, and *serratus ventralis*, in the ribs of the chuck sale section. It seems from the main results that the whole life period of the animals has no effect on tenderness (sensory or rheological analyses) of the rib muscles. The findings of these two studies by Soulat et al. [5,6] together with those of Couvreur et al. [4] demonstrate that it is possible to obtain similar meat qualities with different rearing managements. The fifth study is by Cafferky et al. [7] and it investigates, on a big cattle database at industrial scale of crossbred bull and steer progeny, the effect of breed (eight beef sire breeds representative of the Irish herd: Aberdeen Angus, Belgian Blue, Charolais, Hereford, Limousin, Parthenaise, Salers and Simmental) and gender on meat quality traits (Warner–Bratzler shear force (WBSF), intramuscular fat (IMF%), cook-loss%, drip-loss%, colour (L^*, a^*, b^*) and ultimate pH) of *Longissimus thoracis* et *lumborum* (LTL) muscle. The authors report on animals obtained and reared under the same feeding and environmental conditions that the sire breed had a significant effect on IMF%, cook-loss% and drip-loss%. With respect to breed, Aberdeen Angus-sired progeny had the highest IMF% and the lowest drip-loss%, Limousin-sired offspring had the lowest cook-loss%, while Belgian Blue- and Parthenaise-sired progeny scored the highest for drip-loss%. On another hand, the comparisons of bulls and steers highlight that castration significantly impacts WBSF, IMF% and cook-loss%. Steers in comparison to bulls had higher IMF% and reduced WBSF and cook-loss%, implying steer beef to be more tender and juicy, with more favourable IMF%. This study supports the hypothesis that breed and gender influence the final eating qualities of beef. Finally, the sixth study grouped in this first topic is by Chartrin et al. [8] on turkey meat. It evaluates the meat produced from breeder turkeys in comparison with that of standard turkeys. In line with the objectives of the Special Issue, the technological, nutritional, and sensorial quality of breasts and thighs with drumsticks of turkey male and female breeders was characterized in comparison to that of drumsticks of growing male and female turkeys from the Grademaker line. The authors prove that the differences that exist between males and females, and between standard and breeder turkeys, are mainly a consequence of differences in the age at slaughter and due to sexual dimorphism on body weight. The meat of female breeders had characteristics close to those of standard turkeys, whereas the meat of male breeders was clearly distinguishable, particularly by displaying lower tenderness and water-holding capacity.

In the second topic grouping of studies within the prediction of meat qualities, two papers by Berri et al. [9] and Sahar et al. [10] are published. The former is a review titled "Predicting the Quality of Meat: Myth or Reality?". The authors provide an overview of recent advances made in the field of meat quality prediction, particularly in Europe. The different approaches applied at the laboratory or at the production scale for the development of equations and tools using biological (genomic or phenotypic) or physical (spectroscopy) sources are extensively discussed. The authors develop the recent strategies and findings in Europe related to (i) the search of genes that control the quality of pork and chicken meat, (ii) the recent advances in the search and quantification of protein biomarkers that predict or explain beef tenderness, (iii) the potential of blood biomarkers to predict meat quality by giving the first encouraging results on the use of high-resolution nuclear magnetic resonance or NMR, followed by (iv) the potential of spectroscopic methods. For this last approach, the authors review that the most significant progress was achieved using near-infrared spectroscopy (NIRS) to predict the composition and nutritional value of meats. However, it is worthwhile to note that predicting the functional properties of meats using these methods, especially the sensory quality, is slightly difficult, and further studies are needed. Finally, the example of the Meat Standards Australia phenotypic model, which predicts the eating quality of beef based on a combination of upstream and downstream data, is briefly described. In frame of the objectives of this second topic and more specifically on the use of spectroscopic methods, the research paper by Sahar et al. [10] illustrate the potential of visible–near-infrared (Vis–NIR) spectroscopy to predict physicochemical quality traits (ultimate pH, colour, cook loss and drip loss) in a big dataset of 368 samples of bovine LTL muscle. The authors highlight that the application of Vis–NIR spectroscopy directly on the meat carcass is advantageous as it does not require the preparation of the sample before analysis, and it is applicable to the prediction of quality online using a fibre-optic probe.

In the third topic that groups papers dealing with statistical approaches for meat quality prediction/management, three original papers are published [11–13] and they are complimentary to the previous topic by addressing some of the objectives reported by Berri et al. [9] for the development of prediction/management tools of beef qualities. The first study by Ellies-Oury et al. [11] presenta a new methodology for the selection of protein biomarkers of tenderness in five different bovine muscles using a multi-block model: the data-driven sparse partial least square. In the same context, Gagaoua et al. [12] present an innovative approach for the prediction of beef tenderness by a combination of statistical methods that are "chemometrics" and "supervised learning" to manage the integrated data of the continuum from the farm to fork and select the potential predictors of beef tenderness. Among a total of 60 variables, including WBSF and belonging to 4 levels of the continuum that are farm–slaughterhouse–muscle–meat, partial least squares (PLS) and three decision tree methods (C&RT, classification and regression tree; QUEST, quick, unbiased, efficient regression tree; and CHAID, chi-squared automatic interaction detection) were tested to select the driving factors of WBSF and propose predictive decision tools using the selected variables. In the last study, Conanec et al. [13] propose a new method to manage the trade-off between four performance goals that are the nutritional and organoleptic properties of meat and cattle performances, including carcass properties. Among their findings, the authors show that there is no antagonism between organoleptic quality and nutritional quality. Moreover, the modelling approach is able to highlight the relation between the variables of different origins and the degree of their interconnectedness.

In the fourth topic dealing with muscle biochemistry and proteomics techniques, two original and independent papers are published [14,15]. The study by Zhu et al. [14] assess the usefulness of RNAlater®, regarded as a potential preservation method for proteins, to preserve *post-mortem* bovine muscle proteins compared with dry ice in a proteomic study. The protein profiles of muscle preserved in RNAlater® were found, whatever the sampling time, to be similar to those of dry ice. The results demonstrate that RNAlater® can be a simple and efficient way to preserve bovine muscle proteins for meat proteomics, where snap-freezing may not be a viable option for sample stabilization. The second study by Listrat et al. [15] deals with a deeper understanding of the ontogenesis of intramuscular connective tissue composition that would allow the control of muscle differentiation to improve beef quality. Therefore, the authors investigate the chronology of expression of ten extracellular matrix molecules in bovine *Semitendinosus* muscle using an immunohistology technique at five key stages of myogenesis. The data suggest that for the best controlling of the muscular differentiation to improve beef sensory quality, it would be necessary to intervene very early, i.e., at the beginning of the first-third of gestation.

In the fifth and last topic identified in this Special Issue, two original papers are categorised in the frame of consumer acceptability, development and characterisation of meat products. Čandek-Potokar et al. [16] aim to test the sensory acceptability of Slovenian consumers of a traditional product, the protected geographical indication dry-cured belly Kraška pancetta, produced from entire males, immunocastrates or surgical castrates. This preliminary study provides an overview of the sensory acceptability of dry-cured belly by Slovenian consumers in relation to gender, as characterized by leanness and boar taint level. The latest study by Burri et al. [17] tackles the subject of lipid oxidation known to affect the development and final qualities of meat and meat products [18]. Accordingly, the study by Burri et al. [17] had two main objectives: (i) develop a relevant oxidized processed meatball model to study the effects of supplemented antioxidants, and (ii) investigate lipid oxidation in meatballs without and with a range of eleven plant materials and extracts at different concentrations. The authors conclude that antioxidant-rich plant materials and extracts can efficiently prevent lipid oxidation in processed-meat products, such as meatballs.

In summary, the fifteen papers published in this Special Issue highlight a great part of the research activities in the field of meat science, aiming to characterize and improve the nutritional, sensory and physical quality of meat and meat products. This Special Issue feat, with the worldwide trend toward foods for nutrition and health, further states the importance of multidimensional and multidisciplinary approaches as exemplified in the papers described above. Finally, most authors who have contributed to this issue state that further research in their topic is required in every one of the presented papers and this assures an exciting time for future studies to improve and control the nutritional, sensory and physical quality aspects of meat and meat products.

Acknowledgments: We acknowledge the financial support from Project S3-23000846, funded by Région Auvergne-Rhône-Alpes and Fonds Européens de Développement Régional (FEDER). Mohammed Gagaoua is a Marie Skłodowska-Curie Career-FIT Fellow under the number MF20180029. He is grateful to the funding received from the Marie Skłodowska-Curie grant agreement No. 713654.

References

1. Gagaoua, M.; Monteils, V.; Couvreur, S.; Picard, B. Identification of Biomarkers Associated with the Rearing Practices, Carcass Characteristics, and Beef Quality: An Integrative Approach. *J. Agric. Food Chem.* **2017**, *65*, 8264–8278. [CrossRef] [PubMed]

2. Gagaoua, M.; Monteils, V.; Picard, B. Data from the farmgate-to-meat continuum including omics-based biomarkers to better understand the variability of beef tenderness: An integromics approach. *J. Agric. Food Chem.* **2018**, *66*, 13552–13563. [CrossRef] [PubMed]

3. Moran, L.; Wilson, S.S.; McElhinney, C.K.; Monahan, F.J.; McGee, M.; O'Sullivan, M.G.; O'Riordan, E.G.; Kerry, J.P.; Moloney, A.P. Suckler Bulls Slaughtered at 15 Months of Age: Effect of Different Production Systems on the Fatty Acid Profile and Selected Quality Characteristics of Longissimus Thoracis. *Foods* **2019**, *8*, 264. [CrossRef] [PubMed]

4. Couvreur, S.; Le Bec, G.; Micol, D.; Picard, B. Relationships Between Cull Beef Cow Characteristics, Finishing Practices and Meat Quality Traits of Longissimus thoracis and Rectus abdominis. *Foods* **2019**, *8*, 141. [CrossRef] [PubMed]

5. Soulat, J.; Picard, B.; Léger, S.; Ellies-Oury, M.-P.; Monteils, V. Preliminary Study to Determinate the Effect of the Rearing Managements Applied during Heifers' Whole Life on Carcass and Flank Steak Quality. *Foods* **2018**, *7*, 160. [CrossRef] [PubMed]

6. Soulat, J.; Monteils, V.; Picard, B. Effect of the Rearing Managements Applied during Heifers' Whole Life on Quality Traits of Five Muscles of the Beef Rib. *Foods* **2019**, *8*, 157. [CrossRef] [PubMed]

7. Cafferky, J.; Hamill, R.M.; Allen, P.; O'Doherty, J.V.; Cromie, A.; Sweeney, T. Effect of Breed and Gender on Meat Quality of M. longissimus thoracis et lumborum Muscle from Crossbred Beef Bulls and Steers. *Foods* **2019**, *8*, 173. [CrossRef] [PubMed]

8. Chartrin, P.; Bordeau, T.; Godet, E.; Méteau, K.; Gicquel, J.-C.; Drosnet, E.; Brière, S.; Bourin, M.; Baéza, E. Is Meat of Breeder Turkeys so Different from That of Standard Turkeys? *Foods* **2018**, *8*, 8. [CrossRef] [PubMed]

9. Berri, C.; Picard, B.; Lebret, B.; Andueza, D.; Lefevre, F.; Le Bihan-Duval, E.; Beauclercq, S.; Chartrin, P.; Vautier, A.; Legrand, I.; et al. Predicting the Quality of Meat: Myth or Reality? *Foods* **2019**, *8*, 436. [CrossRef] [PubMed]

10. Sahar, A.; Allen, P.; Sweeney, T.; Cafferky, J.; Downey, G.; Cromie, A.; Hamill, R.M. Online Prediction of Physico-Chemical Quality Attributes of Beef Using Visible—Near-Infrared Spectroscopy and Chemometrics. *Foods* **2019**, *8*, 525. [CrossRef] [PubMed]

11. Ellies-Oury, M.P.; Lorenzo, H.; Denoyelle, C.; Saracco, J.; Picard, B. An Original Methodology for the Selection of Biomarkers of Tenderness in Five Different Muscles. *Foods* **2019**, *8*, 206. [CrossRef] [PubMed]

12. Gagaoua, M.; Monteils, V.; Couvreur, S.; Picard, B. Beef Tenderness Prediction by a Combination of Statistical Methods: Chemometrics and Supervised Learning to Manage Integrative Farm-To-Meat Continuum Data. *Foods* **2019**, *8*, 274. [CrossRef] [PubMed]

13. Conanec, A.; Picard, B.; Durand, D.; Cantalapiedra-Hijar, G.; Chavent, M.; Denoyelle, C.; Gruffat, D.; Normand, J.; Saracco, J.; Ellies-Oury, M.-P. New Approach Studying Interactions Regarding Trade-Off between Beef Performances and Meat Qualities. *Foods* **2019**, *8*, 197. [CrossRef] [PubMed]

14. Zhu, Y.; Mullen, A.M.; Rai, D.K.; Kelly, A.L.; Sheehan, D.; Cafferky, J.; Hamill, R.M. Assessment of RNAlater® as a Potential Method to Preserve Bovine Muscle Proteins Compared with Dry Ice in a Proteomic Study. *Foods* **2019**, *8*, 60. [CrossRef] [PubMed]

15. Listrat, A.; Gagaoua, M.; Picard, B. Study of the Chronology of Expression of Ten Extracellular Matrix Molecules during the Myogenesis in Cattle to Better Understand Sensory Properties of Meat. *Foods* **2019**, *8*, 97. [CrossRef] [PubMed]

16. Čandek-Potokar, M.; Prevolnik-Povše, M.; Škrlep, M.; Font-i-Furnols, M.; Batorek-Lukač, N.; Kress, K.; Stefanski, V. Acceptability of Dry-Cured Belly (Pancetta) from Entire Males, Immunocastrates or Surgical Castrates: Study with Slovenian Consumers. *Foods* **2019**, *8*, 122. [CrossRef] [PubMed]

17. Burri, S.C.M.; Granheimer, K.; Remy, M.; Ekholm, A.; Hakansson, A.; Rumpunen, K.; Tornberg, E. Lipid Oxidation Inhibition Capacity of 11 Plant Materials and Extracts Evaluated in Highly Oxidised Cooked Meatballs. *Foods* **2019**, *8*, 406. [CrossRef] [PubMed]

18. Domínguez, R.; Pateiro, M.; Gagaoua, M.; Barba, F.J.; Zhang, W.; Lorenzo, J.M. A Comprehensive Review on Lipid Oxidation in Meat and Meat Products. *Antioxidants* **2019**, *8*, 429. [CrossRef] [PubMed]

Article

Suckler Bulls Slaughtered at 15 Months of Age: Effect of Different Production Systems on the Fatty Acid Profile and Selected Quality Characteristics of *Longissimus Thoracis*

Lara Moran [1,*,†] , Shannon S. Wilson [2], Cormac K. McElhinney [3], Frank J. Monahan [4], Mark McGee [3], Maurice G. O'Sullivan [2], Edward G. O'Riordan [3], Joseph P. Kerry [2] and Aidan P. Moloney [3,*]

1 Teagasc, Food Research Centre, Ashtown, Dublin 15, Ireland
2 School of Food and Nutritional Sciences, University College Cork, Cork T12 XF62, Ireland
3 Teagasc, Animal & Grassland Research and Innovation Centre, Grange, Dunsany, Co., Meath C15 PW93, Ireland
4 School of Agriculture and Food Science, University College Dublin, Dublin 4, Ireland
* Correspondence: Lara.Moran@ehu.eus (L.M.); Aidan.Moloney@teagasc.ie (A.P.M.)
† Current address: Lactiker Research Group, Department of Pharmacy and Food Science, University of the Basque Country (UPV/EHU), Edificio Lascaray, Avenida Miguel de Unamuno, 6, 01006 Vitoria-Gasteiz, Álava, Spain.

Received: 21 June 2019; Accepted: 15 July 2019; Published: 18 July 2019

Abstract: The objective was to compare the quality of beef from bulls reared in typical Irish indoor systems or in novel grass-based systems. Bulls were assigned to one of the following systems: (a) grass silage plus barley-based concentrate ad libitum (CON); (b) grass silage ad libitum plus 5 kg of concentrate (SC); (c) grazed grass without supplementation (G0); (d) grazed grass plus 0.5 kg of the dietary dry matter intake as concentrate (GC) for (100 days) until slaughter (14.99 months). Carcass characteristics and pH decline were recorded. *Longissimus thoracis* was collected for analytical and sensory analysis. Lower carcass weight, conformation and fatness scores were found for grazing compared to CON and SC groups. CON bulls had highest intramuscular fat and lighter meat colour compared with grazing bulls. The SC meat (14 days aged) was rated higher for tenderness, texture, flavour and acceptability compared with grazing groups. CON saturated and monounsaturated fatty acid (FA) concentration was highest, conversely, omega-3 FA concentration was higher for GC compared with CON, while no differences were found in polyunsaturated FA. In conclusion, while market fatness specification was not reached by grazed grass treatments, beef eating quality was not detrimentally affected and nutritional quality was improved.

Keywords: grass-fed; young bulls; nutritional quality; tenderness; carcass characteristics; pasture

---

1. Introduction

Some European markets require bulls to be slaughtered at less than 16 months of age and to have a carcass fatness score of six (1–15 scale) [1]. These market specifications are presumed to reflect an effect of age and carcass fatness on meat quality and consumer preferences and are generally achieved by the use of high energy rations [1]. However, these specifications limit the ability of farmers to explore the use of alternative lower-cost feed options and to increase the profitability of the bull beef enterprise.

Feedstuff costs are a major proportion of the total expenditure in cattle production [2]. In regions with a temperate climate, grazed grass is a cheaper feedstuff [3,4]. Conserved grass (silage) has been used in this bull production system to achieve the carcass specification required but few meat quality studies are available. As grazed grass is cheaper than conserved grass, there is interest in including grazed grass in bull production, while the inclusion of a grazing period would also decrease the costs associated with cattle housing for a prolonged period.

Grass-fed beef is increasingly appreciated due to its "green-healthy image" also related to an idyllic image of animal production [5], which, in turn, can play a crucial role in consumer acceptability [6]. Diet conscious consumers will also prefer the overall lower fat content and altered fatty acid profile of a grass-fed beef product [7]. However, tenderness is the most important quality characteristic for the consumer [8,9] and beef from grass-fed bulls has been reported as being less tender than beef from grain-fed bulls [10]. While market specifications reflect real and perceived market preferences, ideally, they should be supported by scientific evidence. There is, however, a poor relationship between carcass quality as assessed by the European classification system (EUROP) and beef quality characteristics, particularly eating quality [11]. Consequently, modified lower-cost production systems need to be studied not only for achievement of current market specifications but also for meat quality characteristics and consumer preferences.

Therefore, the aim of the present study was to compare two typical Irish bull production systems with novel grass-based systems to determine whether the carcasses comply with the actual market requirements and to determine whether carcasses that did not achieve the fat score specification were indeed inferior with respect to quality and nutritional characteristics.

2. Materials and Methods

2.1. Animals, Management and Feed Analyses

The study was carried out under license from the Irish Government Department of Health and Children (held by Edward G. O'Riordan B100/2483) and all procedures used complied with national regulations concerning experimentation on farm animals.

Spring-born late-maturing breed (Limousin and Charolais) sired suckler male cattle ($n = 60$; 386, s.d. 28.0 kg live weight; 337, s.d. 36.3 days of age) were housed in a slatted floor shed and offered grass silage to appetite, supplemented with 2 kg of a barley/soya-based concentrate for eight weeks prior to the start of the experiment. At the end of the winter (11 February), the animals were blocked by breed, weight and age and randomly assigned to one of four dietary groups ($n = 15$): Two indoor diets: (a) barley-based concentrate comprised of 862 g/kg rolled barley, 60 g/kg soya bean, 50 g/kg cane molasses and 28 g/kg vitamins and minerals offered ad libitum plus grass silage offered ad libitum (CON) or (b) grass silage ad libitum supplemented with the above concentrate (SC) and two outdoors diets: (a) rotationally grazed grass without concentrate supplementation (G0) or (b) grazed grass plus 0.5 of the dietary dry matter (DM) intake as the above concentrate (GC).

Indoor animals had a lying area of c. 2.72 m^2 per animal in a slatted, concrete floor shed. Beginning 23 February (considered as day 0 of the experiment), CON bulls were offered an increasing allowance of concentrates until the ad libitum level of consumption was achieved. The concentrate allowance for animals on SC was increased as the animals gained weight (4 kg, 5 kg, 6 kg and 6.5 kg per animal at days 0,10, 30, and 90 of the experiment, respectively). The grass silage (Table 1) was a first harvest from a predominantly perennial ryegrass sward, mowed and wilted for 24 h before ensiling. Grass silage and concentrates were offered separately.

Outdoors bulls remained on the pre-experimental ration until they were turned out to pasture (day 21 of the experiment) and rotationally grazed perennial ryegrass (*Lolium perenne* L.) dominant swards for 79 days. The total grazing area was a single block of 13.2 hectares (ha), split into three equal farmlets (4.4 ha). To ensure that the response to concentrates at pasture was not confounded with differences in herbage quality, the two grazing groups were allocated herbage of similar pre-grazing height and mass (2300 kg DM/ha) and the sub-paddock area was adjusted such that post-grazing sward height and herbage mass and, residency time were similar for each grazing treatment. The concentrate allowance, which averaged 5 kg per animal daily, was offered once daily and it was based on expected grass consumption as observed in previous studies with similar animals at this stage of the grass growing season. Based on previous research at this centre, initial grazing areas for GC were adjusted to be 0.73 of that of G0, and these proportions were altered during the experiment as necessary.

Samples of feeds were collected periodically throughout the study and analysed as previously described [12]. The chemical composition and fatty acid profile of the feedstuffs are shown in Table 1.

Table 1. Chemical composition of the dietary ingredients.

	Grazed Grass	Grass Silage	Concentrates
Dry matter (DM) (g/kg fresh matter)	180	247	804
pH	–	–	3.92
DM digestibility (g/kg)	761	680	–
Composition of DM (g/kg)			
Starch	–	–	544
Ash	112	92	55
Crude protein	163	130	144
Neutral detergent fibre	468	577	185
Acid detergent fibre	–	–	67
Water soluble carbohydrate	123	–	–
Acid hydrolysis ether extract	–	–	31
Fatty acid profile (%)			
14:0	0.11	0.36	0.21
16:0	18.18	15.59	22.12
16:1	0.00	0.05	0.12
17:0	0.09	0.04	0.08
18:0	2.32	1.86	1.61
18:1-c11	0.18	0.69	0.76
18:1-nc9	3.00	2.77	14.33
18:2-nc6	13.49	17.50	54.06
18:3-n3	55.67	56.52	4.44
20:0	0.19	0.29	0.17
20:1	0.00	0.00	0.68
20:5-n3	0.51	0.00	0.32
21:0	0.06	0.00	0.00
22:0	0.66	0.96	0.24
22:1-n9	0.07	0.06	0.08
24:0	0.89	0.84	0.13
Others	4.52	2.47	0.58

All animals were slaughtered on one day at 14.99 SD. 0.93 months of age with 53 of the 60 animals being under 16 months of age. The live weight was recorded on the farm on the morning of slaughter.

2.2. Slaughter, Sampling Procedures, PH and Colour Measurement

On the day of slaughter, the animals were transported approximately 30 km to a commercial slaughter plant and slaughtered immediately after arrival by bolt stunning followed by exsanguination from the jugular vein. Electrical stimulation was not applied, and the carcasses were hung by the Achilles tendon. The slaughter and dressing procedures were in accordance with European Union Regulations (EC) No. 1009/2009 and No. 853/2004 and carcass weight, conformation and fatness scores were recorded. Carcass grades for conformation and fatness were assessed using the numerical value within a 15-point scale [13]. Carcasses were then placed in a chiller set at 5 °C. Approximately 10 h later, the chiller temperature was set to 0 °C.

The pH and temperature of the *longissimus thoracis* (LT) at the 10th rib were recorded in the left side carcass at 1 h, 3 h, 5 h and 7 h post mortem, with a portable pH meter with temperature compensation (Model WP-80 (pH/ORP/T meter), TPS Pty, Ltd. Springwood, Queensland, Australia.) and a glass pH probe (Glass electrode: model EC-2010-06, Refex Sensors Ltd. Westport, Ireland.) using a scalpel incision for each measurement as described by Pearce et al. [14]. The pH meter was (re)calibrated at ambient temperature intermittently during the measurement period. The chill temperature was set to 0 °C approximately 12 h after the bulls were slaughtered.

After approximately 48 h in the chiller, carcasses were moved to the deboning hall (4 °C). The colour of the LT as Hunter lab values was measured at the 5th/6th rib interface, 1 h after cutting and exposure to air using a portable spectrophotometer (Miniscan EZ, HunterLab, Reston, VA, USA). The pH of LT was measured at this location as described above. The cube roll (commercial cut that begins between the 5th and 6th rib and ends between the 10th and 11th rib) was then removed, vacuum packed and transported to Teagasc, Food Research Centre, Ashtown, Dublin. Beginning at the 10th rib end, two steaks of 2.5 cm thick were stored at −20 °C for composition and fatty acid determination. The remainder of the cube roll was vacuum packed and wet-aged for 12 additional days (4 °C, in the dark) for a total of 14 days of ageing. Thereafter, it was sliced (2.5 cm thick steaks) for sensory evaluation, cook loss and instrumental texture analysis. All samples were then vacuum packed and frozen at −20 °C for subsequent analysis.

2.3. Meat Proximate Composition

Steaks for proximate analysis were thawed, trimmed of external fat and connective tissue, and the trimmed muscle was blended (R101, Robot Coupe SA, Vincennes Cedex, France). The moisture content of each sample was determined in duplicate using a microwave instrument (Microwave SMART Trac; CEM Corporation, Matthews, NC, USA) according to [15]. The intramuscular fat (IMF) content was measured as described by Folch et al. [16] (1957). Protein was determined in duplicate using a LECO protein analyser (Model FP-428, Leco Corporation, St. Joseph, MI, USA) based on the Dumas method [17].

2.4. Instrumental Texture and Sensory Evaluation

Frozen vacuum-packed steaks were thawed in circulating water at 20 °C. All external fat and connective tissue surrounding the muscle was removed, the steaks were conditioned for 15 min at 20 °C before cooking for sensory and instrumental texture analysis. After the excess moisture was removed, the weight of the steaks was recorded. The steaks were subsequently cooked in vacuum pack bags to an internal temperature of 70 °C, by immersing in a water bath (Model Y38, Grant Instruments Ltd. Royston, UK) at 72 °C. The internal temperature of the steaks was measured using a digital thermometer (HI 904, Hanna Foodcare Instruments, Bedfordshire, UK) [18]. After cooking, all the juices were poured out of the bag and the steaks were left to cool to room temperature, finally the weight of the cooked steak recorded. The cook loss (CL) was determined by the following formula:

$$CL\% = (\text{raw weight} - \text{cooked weight}) \div \text{raw weight} \times 100 \qquad (1)$$

All steaks were stored in a closed bag and tempered overnight at 4 °C for subsequent Warner–Bratzler shear force (WBSF) analysis [18]. Six cores (1.25 cm diameter) parallel to the direction of the muscle fibres were obtained and sheared using an Instron Universal testing machine (model 5543, Instron Corporation. Bucks, UK) equipped with a Warner–Bratzler shearing device. The crosshead speed was 5 cm/min. Instron Series IX Automated Materials Testing System software for Windows (Instron Corporation. Bucks, UK) was employed in the analysis. Three parameters were used to define the instrumental texture of meat: WBSF (N) or peak strength or force required to shear through a meat sample; modulus of deformability (Mpa) or the slope between the 20% to 80% segment of the total peak; and total energy (J) or total peak area.

Sensory testing was conducted using untrained assessors ($n = 15$) [19,20] who ranged in age from 20–50 and who consumed beef regularly. Sensory analysis was carried out in the sensory kitchen in University College Cork. The kitchen features sensory booths and conforms to the standards of the International Organization for Standardization [21]. The analysis was conducted under standard lighting

(LUX, 1000) in well-ventilated and portioned panel booths. Steaks were grilled to an internal temperature of 72 °C, assigned three-digit random codes and served to assessors as 1 cm^2 pieces, in randomised order [22]. Each assessor was asked to rate the sensory qualities of steak from each animal according to the methodology of the American Meat Science Association [18,23]. The assessors rated five sensory qualities on a scale (8-point hedonic) from 1–8 for tenderness (3–5 chews) where 1 = extremely tough and 8 = extremely tender, overall flavour where 1 = very poor and 8 = extremely good, overall firmness where 1 = extremely mushy and 8 = extremely firm, overall texture where 1 = very poor and 8 = extremely good and overall acceptability where 1 = not acceptable and 8 = extremely acceptable. Distilled water and unsalted soda crackers were provided to purge the palate of residual flavour notes between samples.

2.5. Meat Fatty Acid Analysis

Lipid extraction and fatty acid methylation was carried out as described by Noci et al. [12]. Fatty acid methlyesters (FAMEs) were analysed using a Varian 3500 GLC (Varian, Harbor City, CA, USA) fitted with a flame ionization detector. All samples were methylated in duplicate and each sample was injected, in splitless mode, twice onto the GLC column, using a Varian 8035 auto-sampler. Separation of the FAMEs was performed on a 100 m CP-Sil 88 column (100 m × 25 mm × 0.2 μm Supelco, Bellefonte, PA, USA) using H as the carrier gas. The GLC conditions have been described previously by Shingfield et al. [24]. Data were recorded and analysed on a Minichrom PC system (VG Data System, Manchester, UK). Individual FAMEs were identified by retention time with reference to the external standard (Supelco 37 component FAME Mix, Supelco Inc., Bellefonte, PA, USA) and quantified by using the internal standard C 23:0. The atherogenic index (AI) and thrombogenic index (TI) were calculated according to Ulbricht and Southgate [25].

2.6. Statistical Analysis

All data were subjected to analysis of variance (ANOVA) via the generalised linear mixed model, GLIMMIX procedure of SAS (SAS Inst. Inc., Cary, NC, USA) using block and production system as sources of variation. Animal was the experimental unit. The random tool of the GLIMMIX procedure was used to include "time" as a repeated measurement for the analysis of pH decline and assessor for the analysis of the sensory data. Data are presented as least squares means and when significant effects were detected, the post hoc Tukey test was used to separate the means. The level of significance used was $P < 0.05$.

The following models were used for instrumental and sensory measurements, respectively:

$$Y = \mu + P + B + P * ID + \sum$$

$$Y = \mu + P + B + C\,(ID) + \sum$$

μ = intercept; $\sum$ error. Fixed factors P: production system; B: animal block. Random effect: P * ID: Production System * Animal identity, C (ID): consumer within ID of the animal.

3. Results

Unless otherwise indicated, all stated differences are significant ($P < 0.05$).

3.1. Animal Performance, Meat Quality and Sensory Evaluation

Animal Performance, meat quality and sensory evaluation results are presented in Table 2.

Bulls fed on concentrates reached a higher live weight compared with SC which in turn was higher than GC and G0 which did not differ. Carcass weight and fat scores were higher for CON compared with SC which in turn were higher than GC and G0 which did not differ. Conformation scores were higher for CON compared to SC and GC which in turn were higher than G0.

Table 2. Animal performance and characteristics of *longissimus thoracis* muscle from bulls assigned to one of the following systems: grass silage plus barley-based concentrate ad libitum (CON); grass silage ad libitum plus 5 kg of concentrate (SC); grazed grass without supplementation (G0) or grazed grass plus 0.5 kg of the dietary dry matter intake as concentrate (GC) until slaughter at 15 months.

Variable	CON	SC	GC	G0	SEM	*P*-Value
BW (kg)	604 [a]	557 [b]	515 [c]	498 [c]	7.135	<0.001
Average daily weight (kg/day)	2.00 [a]	1.58 [b]	1.47 [b,c]	1.30 [c]	0.048	<0.001
Carcass weight (kg)	358 [a]	314 [b]	288 [c]	277 [c]	4.997	<0.001
Kill out proportion (%)	59.2 [a]	56.5 [b]	56.0 [b]	55.7 [b]	0.320	<0.001
Fat score (1–15)	7.27 [a]	6.00 [b]	4.27 [c]	3.67 [c]	0.231	<0.001
Conformation (1–15)	9.93 [a]	8.53 [b]	8.40 [b]	7.67 [c]	0.168	<0.001
Crude protein (%)	23.0 [a]	22.9 [a]	22.7 [a]	21.6 [b]	0.191	0.039
Moisture (%)	74.8 [c]	75.2 [b,c]	76.1 [a,b]	76.7 [a]	0.178	<0.001
Intramuscular fat (g/100 g meat)	2.20 [a]	1.85 [b]	1.48 [c]	1.59 [b,c]	0.063	<0.001
Ultimate pH	5.60	5.60	5.62	5.62	0.006	0.323
L	37.2 [a]	34.7 [b]	31.7 [c]	32.5 [c]	0.390	<0.001
a	19.3 [a]	19.2 [a]	18.3 [b]	18.4 [b]	0.139	0.022
b	12.7 [a,b]	12.8 [a]	11.8 [c]	12.2 [b,c]	0.096	0.001
Cook loss (%)	25.7	27.7	28.6	27.6	0.369	0.056
Warner–Bratzler shear force (N)	35.8	33.6	35.3	32.5	0.936	0.587
Slope (MPa)	0.785	0.770	0.845	0.733	0.024	0.452
Energy (J)	0.230	0.229	0.237	0.219	0.006	0.733
Tenderness (1–8)	4.70 [a,b]	5.22 [a]	4.24 [b]	4.60 [b]	0.109	<0.001
Overall flavour (1–8)	5.39 [a,b]	5.74 [a]	5.02 [b]	5.22 [b]	0.076	0.001
Overall firmness (1–8)	5.39 [a]	5.11 [a]	5.02 [a,b]	4.69 [b]	0.080	0.001
Overall texture (1–8)	5.02 [a,b]	5.42 [a]	4.71 [b]	4.81 [b]	0.087	0.001
Overall acceptability (1–8)	5.10 [b]	5.56 [a]	4.79 [b]	5.02 [b]	0.084	0.001

[a,b,c] Values within a row with different superscript differ significantly at $P < 0.05$.

The IMF concentration was higher for CON compared to SC which in turn was higher than GC and similar to G0 which did not differ. Protein concentration was lower for G0 compared with the other three diets. Moisture concentration was lowest for CON and highest for G0.

Figure 1 shows the evolution of pH/temperature of LT up to 7 h post-mortem. No significant differences were found between treatments at 1 h post-mortem. Thereafter, the decrease in pH was more pronounced for CON than for the grazing groups. The SC pH decline was similar to that from grazing animals from 1–3 h post-mortem, while the pH for SC after 5 h was lower than in grazing animals and similar to CON. At 7 h post-mortem the pH for SC did not differ from the other groups. There was no difference between groups in ultimate pH (pHu) (Table 2). Muscle temperature decline (Figure 1) was less pronounced in CON animals than in grazing animals with SC intermediate. Thus after 1 and 3 h, temperature for CON was higher than for grazing animals but similar to SC. At 5 and 7 h post-mortem temperature for CON was higher than for the other groups, while temperature for SC was higher than GC and G0 which did not differ.

With regard to colour (Table 2), CON meat was lighter (higher L value) than meat from SC, which in turn was lighter than that from GC and G0. No differences were found in lightness between grazing treatments. On the other hand, indoor treatments meat had higher redness (a value) than that from grazing treatment. Neither CON and SC, nor GC and G0 differ between them in a value. Finally, CON and SC meat showed similar yellowness in meat colour (b value), while meat form these two indoor treatments was more yellow than for GC treatment. G0 meat did not differ in yellowness from GC and SC treatments.

No differences ($P < 0.05$) were found after 14 days of ageing for any of the variables related to instrumental texture (WBSF, modulus and energy). On contrary, the sensory panel rated LT from SC higher than GC and G0 (which did not differ) but similar to CON, for tenderness and overall flavour. For overall firmness, LT from CON, SC and GC were rated similarly while LT from G0 was rated lower than CON and SC but similar to GC. For overall texture, LT from CON and SC were rated similarly, while SC was rated higher than GC and G0 which did not differ. For overall acceptability, meat from CON, GC and G0 were rated similarly, but lower than SC.

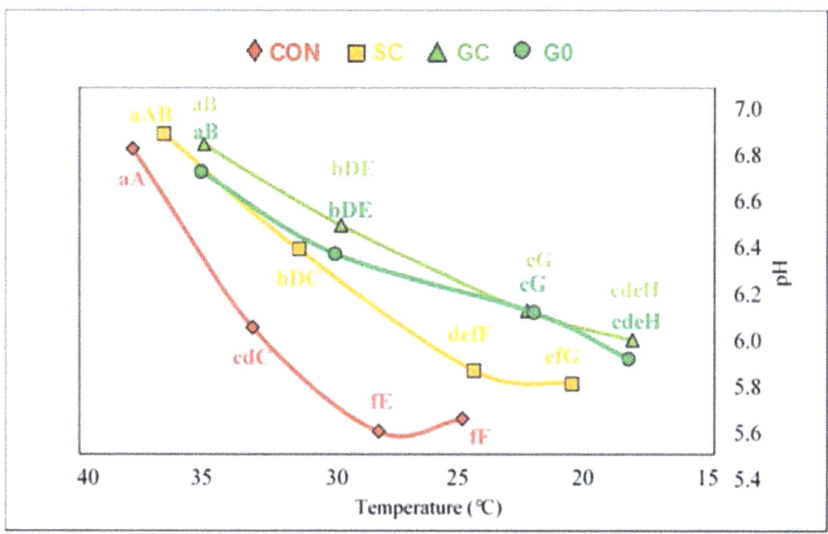

Figure 1. Pattern of pH/temperature decrease post-mortem (1, 3, 5 and 7 h) of longissimus muscle from bulls assigned to one of the following systems: grass silage plus barley-based concentrate ad libitum (CON); grass silage ad libitum plus 5 kg of concentrate (SC); grazed grass without supplementation (G0) or grazed grass plus 0.5 kg of the dietary dry matter intake as concentrate (GC) until slaughter at 15 months. [a,b,c,d,e,f] Different lowercase letters indicate statistical differences in pH ($P < 0.05$). [A,B,C,D,E,F,G,H] Different uppercase letters indicate statistical differences in temperature ($P < 0.05$).

3.2. Fatty Acid Profile

Data on fatty acid classes and relevant nutritional ratios are summarised in Table 3. The total fatty acid concentration and the concentration of total saturated fatty acids (SFA), total monounsaturated fatty acids (MUFA) and cis-MUFA were higher for CON than SC which in turn was higher than GC and G0 which did not differ. The concentration of trans-MUFA was higher for CON than for grazed animals while SC was intermediate

The concentration of total omega 3 polyunsaturated fatty acids (n-3) PUFA was similar for G0 and GC which was higher than SC which in turn was higher than CON. The concentration of total omega 6 (n-6) PUFA was similar for CON, SC and GC but higher than G0. The concentration of highly unsaturated fatty acids (HUFA) was higher for SC, GC and G0 (which did not differ) than CON. The n6 n3 PUFA ratio increased in the order G0 < GC < SC < CON. Grazing per se increased the 18:1-trans11: 18:1-trans10 ratio compared to SC and CON which did not differ. The PUFA:SFA ratio was similar for GC and G0 but higher than SC which in turn was higher than CON. The AI was similar for GC and G0 but lower than SC which in turn was lower than CON. The TI was similar for GC and G0 but lower than SC and CON which did not differ.

More detailed fatty acid profile analysis results are in Tables 4–6 provided in the discussion section.

The proportion of individual SFA and dimethyl acetals (DMA) is summarised in Table 4. Of the main SFA detected, the proportions of C14:0 and C16:0 were lower in GC and G0 than SC which in turn was lower than CON. The proportion of C18:0 was not affected by treatment. The proportions of C16:0 and C18:0 DMA were higher in GC and G0 than SC which in turn was higher than CON. The proportion of C18:1 DMA was higher in GC and G0 than SC and CON.

Table 3. Summary of total fatty acids (total FA mg/100 g muscle), and nutritional ratios of the *longissimus thoracis* muscle from bulls assigned to one of the following systems: grass silage plus barley-based concentrate ad libitum (CON); grass silage ad libitum plus 5 kg of concentrate (SC); grazed grass without supplementation (G0) or grazed grass plus 0.5 kg of the dietary dry matter intake as concentrate (GC) until slaughter at 15 months.

	CON	SC	GC	G0	SEM	P-Value
∑FA	1155 [a]	823 [b]	558 [c]	575 [c]	47.9	<0.001
SFA	517 [a]	340 [b]	196 [c]	208 [c]	24.9	<0.001
MUFA	478 [a]	308 [b]	169 [c]	180 [c]	25.3	<0.001
PUFA	125	126	132	123	2.15	0.409
trans-MUFA	25.6 [a]	17.4 [b]	14.1 [b]	16.2 [b]	1.24	0.007
cis-MUFA	451 [a]	291 [b]	154 [c]	164 [c]	23.01	<0.001
PUFA-n3	26.4 [c]	37.2 [b]	43.9 [a]	44.2 [a]	1.31	<0.001
PUFA-n6	85.7 [a]	77.4 [a]	78.8 [a]	67.7 [b]	1.68	0.003
HUFA	35.5 [b]	44.4 [a]	47.0 [a]	47.2 [a]	1.10	0.001
CLA	3.62	2.46	2.15	2.94	0.212	0.076
n6/n3 PUFA	3.24 [a]	2.16 [b]	1.83 [c]	1.55 [d]	0.095	<0.001
t11/t10 18:1	0.860 [b]	2.82 [b]	11.67 [a]	14.17 [a]	1.131	<0.001
PUFA:SFA	0.268 [c]	0.423 [b]	0.702 [a]	0.636 [a]	0.031	<0.001
Atherogenic index	0.643 [a]	0.553 [b]	0.377 [c]	0.408 [c]	0.019	<0.001
Thrombogenic index	1.65 [a]	1.63 [a]	1.36 [b]	1.45 [b]	0.027	<0.001

SFA = saturated fatty acid; MUFA = monounsaturated fatty acids; PUFA = polyunsaturated fatty acids; HUFA = Highly unsaturated fatty acids larger than 19 carbons; CLA = conjugated linoleic acid; n6 = omega 6 fatty acid; n3 = omega 3 fatty acids [a,b,c] Values within a row with different superscript differ significantly at $P < 0.05$.

Table 4. Individual saturated fatty acid (SFA) and dimethyl acetal (DMA) proportions (of total fatty acids [1]) in the *longissimus thoracis* muscle from bulls assigned to one of the following systems: grass silage plus barley-based concentrate ad libitum (CON); grass silage ad libitum plus 5 kg of concentrate (SC); grazed grass without supplementation (G0) or grazed grass plus 0.5 kg of the dietary dry matter intake as concentrate (GC) until slaughter at 15 months.

	CON	SC	GC	G0	SED	P-Value
SFA						
8:0	0.017 [c]	0.064 [b]	0.086 [a]	0.096 [a]	0.006	0.003
10:0	1.01 [c]	1.31 [b]	1.85 [a]	1.93 [a]	0.083	<0.001
11:0	0.125 [b]	0.138 [b]	0.242 [a]	0.230 [a]	0.013	<0.001
12:0	0.066 [b]	0.097 [b]	0.115 [a,b]	0.141 [a]	0.007	0.002
14:0	2.19 [a]	1.65 [b]	0.940 [c]	1.04 [c]	0.091	<0.001
iso-15:0	0.159 [b]	0.222 [a,b]	0.251 [a]	0.277 [a]	0.014	0.045
anteiso-15:0	0.162 [b]	0.175 [b]	0.218 [a]	0.246 [a]	0.007	<0.001
15:0	0.483	0.536	0.508	0.537	0.014	0.529
16:0	22.9 [a]	20.1 [b]	15.2 [c]	15.5 [c]	0.525	<0.001
iso-17:0 [2]	1.12 [a]	0.349 [c]	0.875 [b]	0.767 [b]	0.054	<0.001
17:0	1.09 [c]	0.905 [b]	0.670 [a]	0.697 [a]	0.030	<0.001
18:0	13.9	14.2	13.1	13.4	0.191	0.229
19:0 [3]	0.151	0.150	0.143	0.145	0.005	0.955
20:0	0.234 [a]	0.113 [c]	0.182 [b]	0.197 [b]	0.029	<0.001
24:0	0.107	0.131	0.100	0.098	0.030	0.975
DMA						
DMA 16:0	2.06 [c]	3.14 [b]	4.45 [a]	4.75 [a]	0.184	<0.001
DMA 18:0	1.20 [c]	2.04 [b]	2.90 [a]	2.91 [a]	0.122	<0.001
DMA 18:1	2.24 [b]	2.97 [b]	4.18 [a]	4.55 [a]	0.194	<0.001

[1] Only values superior to 0.05 have been reported [2] coeluted with t9-16:1. [3] coeluted with c15-18:1. [a,b,c] values within a row with different superscript differ significantly at $P < 005$.

The proportion of individual MUFA is summarised in Table 5. Of the main cis MUFA detected, the proportion of C16:1 and cis 9 C18:1 were lower in GC and G0 than SC which in turn was lower than CON. The proportion of cis 11 C18:1 was lower for G0 than GC, SC was similar to GC and G0 but lower than CON. Of the main trans MUFA detected, the proportion of trans 10 C18:1 was higher for CON than the other three groups which did not differ. The proportion of trans 11 C18:1 was similar for GC and G0 and higher than SC and CON which did not differ.

Table 5. Individual monounsaturated fatty acid (MUFA) proportion (of total fatty acids [1]) in the *longissimus thoracis* muscle from bulls assigned to one of the following systems: grass silage plus barley-based concentrate ad libitum (CON); grass silage ad libitum plus 5 kg of concentrate (SC); grazed grass without supplementation (G0) or grazed grass plus 0.5 kg of the dietary dry matter intake as concentrate (GC) until slaughter at 15 months.

	CON	SC	GC	G0	SEM	*P*-Value
MUFA						
14:1	0.345 [a]	0.230 [b]	0.088 [c]	0.103 [c]	0.019	<0.001
cis MUFA						
c7-16:1	0.179 [b]	1.30 [a]	1.54 [a]	1.80 [a]	0.115	<0.001
c9-16:1	2.77 [a]	2.21 [b]	1.39 [c]	1.47 [c]	0.100	<0.001
c5-17:1	0.094 [d]	0.226 [v]	0.363 [b]	0.504 [a]	0.025	<0.001
c7-17.1	0.110 [c]	0.633 [b]	0.518 [a]	0.518 [a]	0.029	<0.001
c9-17:1	0.700 [a]	0.383 [c]	0.602 [a,b]	0.534 [b]	0.024	<0.001
c10-17:1	0.303 [a]	0.001 [b]	0.093 [b]	<0.001 [b]	0.022	<0.001
c11-17:1	0.798 [b]	1.03 [b]	1.52 [a]	1.74 [a]	0.075	<0.001
c9-18:1	27.9 [a]	23.0 [b]	16.2 [c]	16.1 [c]	0.854	<0.001
c11-18:1	1.53 [a]	1.28 [b,c]	1.36 [b]	1.19 [c]	0.029	0.001
c12-18:1	0.314	0.389	0.376	0.275	0.023	0.276
c13-18:1	0.187 [a]	0.130 [b]	0.098 [b,c]	0.085 [c]	0.010	<0.001
c14-18:1	0.046	0.131	0.090	0.100	0.012	0.144
c16-18:1	0.106 [a]	0.105 [a]	0.072 [b]	0.077 [b]	0.005	0.010
trans MUFA						
t6+t8-18:1	0.145	0.112	0.105	0.090	0.006	0.064
t9-18:1	0.145	0.144	0.115	0.143	0.006	0.173
t10-18:1	0.726 [a]	0.397 [b]	0.230 [b]	0.253 [b]	0.062	0.023
t11-18:1	0.546 [b]	0.644 [b]	1.26 [a]	1.31 [a]	0.084	<0.001
t12+ t13-18:1	0.159 [a]	0.132 [a,b]	0.092 [b]	0.103 [b]	0.007	0.028
t15-18:1 [2]	0.024 [b]	0.084 [a]	0.197 [a]	0.343 [a]	0.039	0.017
t16-18:1	0.205 [a]	0.102 [b]	0.097 [b]	0.085 [b]	0.013	0.070
n-9						
20:1n-9	0.147	0.136	0.122	0.123	0.005	0.291
22:1n-9	0.053 [b]	0.114 [a]	0.131 [a]	0.107 [a]	0.008	0.013
24:1n-9	0.069	0.061	0.122	0.092	0.020	0.639

n-9: Monounsaturated fatty acid omega 9; t: trans isomer, c: cis isomer [1] Only values superior to 0.05 have been reported [2] coeluted with c10-18:1. [a,b,c] Values within a row with different superscript differ significantly at $P < 0.05$.

The proportion of individual PUFA is summarised in Table 6. Of the main n-6 PUFA detected, the proportion of C18:2 was similar in CON, SC and G0, while GC was higher than SC and CON. The proportion of C20:4 was higher in GC and G0, which did not differ, than in SC which in turn was higher than CON. Of the main n-3 PUFA detected, the proportions of C18:3, C22:5 and C20: 5 were higher in GC and G0, which did not differ, than in SC which in turn was higher than CON.

Table 6. Individual polyunsaturated fatty acid (PUFA) proportions (of total fatty acids [1]) in the *longissimus thoracis* muscle from bulls assigned to one of the following systems: grass silage plus barley-based concentrate ad libitum (CON); grass silage ad libitum plus 5 kg of concentrate (SC); grazed grass without supplementation (G0) or grazed grass plus 0.5 kg of the dietary dry matter intake as concentrate (GC) until slaughter at 15 months.

	CON	SC	GC	G0	SEM	*P*-Value
PUFA						
20:2	0.071 [c]	0.109 [a,b]	0.130 [a]	0.092 [b]	0.006	0.009
22:2	0.132 [b]	0.182 [b]	0.271 [a]	0.267 [a]	0.014	<0.001
Non conjugated 18:2						
9t,12t-18:2	0.128	0.110	0.107	0.094	0.006	0.444
t10,c15-18:2	0.133	0.131	0.123	0.148	0.006	0.612
t11,c15-18:2	0.127	0.159	0.149	0.132	0.009	0.555
c9,c15-18:2 [2]	0.214 [b]	0.254 [b]	0.341 [a]	0.348 [a]	0.013	<0.001
CLA						
c9,t11-CLA	0.286 [b]	0.279 [b]	0.364 [a,b]	0.471 [a]	0.021	0.003
t10,c12-CLA	0.025	0.008	0.010	0.009	0.003	0.167
n-6						
18:2n-c6	6.04 [b]	6.95 [b]	9.94 [a]	8.04 [a,b]	0.338	<0.001
18:3n-6	0.063 [b]	0.100 [b]	0.216 [a]	0.190 [a]	0.016	0.004
20:3n-6	0.421 [c]	0.558 [b]	0.804 [a]	0.714 [a]	0.029	<0.001
20:4n-6	1.58 [c]	2.32 [b]	3.30 [a]	3.30 [a]	0.131	<0.001
n-3						
18:3n-3	0.790 [c]	1.58 [b]	3.06 [a]	3.15 [a]	0.150	<0.001
20:3n-3	0.076	0.046	0.087	0.069	0.007	0.130
22:5n-3	0.718 [c]	1.30 [b]	2.06 [a]	2.15 [a]	0.127	<0.001
22:6n-3	0.106	0.461	0.444	0.365	0.068	0.315
20:5n-3	0.718 [c]	1.29 [b,a]	2.06 [a]	2.15 [a]	0.102	<0.001
n-9						
20:3n-9	0.135 [d]	0.229 [c]	0.340 [b]	0.398 [a]	0.016	<0.001

CLA = conjugated linoleic acid. n-6 = polyunsaturated fatty acids omega 6; n-3 = polyunsaturated fatty acids omega 3; n-9 = polyunsaturated fatty acids omega 9; t = trans isomer, c = cis isomer [1] Only values superior to 0.05 have been reported; [2] coeluted with c9-19: [a,b,c] Values within a row with different superscript differ significantly at $P < 0.05$.

4. Discussion

4.1. Carcass Characteristics

In this study, we considered the market specification that bulls must be younger than 16 months of age at slaughter to be the element of the production system that could not be changed. The context of the study therefore was to compare cheaper alternative systems with systems demonstrated to achieve the carcass weight and fat cover specifications [1]. The results indicate that the only grass-based system that reached the current market requirement was SC, i.e., the ration based on grass silage and concentrates, offered indoors with associated costs of housing. Therefore, the production of late-maturing sired bulls for slaughter at under 16 months of age from pasture seems not to be an option for meeting current market requirements. The age and fat score specification likely reflect a perception that meat from leaner and older animals is inferior in some quality characteristics. However, French et al. [26] reported a poor correlation between fatness scores and meat quality which was recently supported by Bonny et al. [11] from a much larger dataset. The primary objective of this study was to determine whether carcasses that did not achieve the fat score specification and were deemed too lean were indeed inferior with respect to quality for the consumer

4.2. Meat Quality

Mean carcass weights differed in this study and we acknowledge that carcass weight can influence meat characteristics [27]. In the current study, muscle from the grazing systems had lower IMF and lower protein in line with previous studies [10,28]. Banović et al. [29] reported that consumers pay more attention and choose more often meat products with lower fat content while Grunert [30,31] reported that visual fat has generally a negative effect on purchase decision. Therefore, while the leanness of beef cuts from grazing cattle may be appreciated by the consumer as a reduced fat product, low fat accretion in the animal will lead to a negative carcass classification based on EUROP scheme [32]. Thus, unlike the carcass grading system used in Australia [32], the lack of a relationship between carcass grading and eating quality in Europe is to the detriment of the primary producer and consumer.

The ultimate pH in the current investigation was within the normal range (5.4–5.8) described by Viljoen et al. [33] and agrees with previous studies [10,34]. Some authors have linked a higher pH in muscle from pasture animals with lower glycogen stores in muscle and/or increased stress during transport and slaughter management, since outdoor animals are not accustomed to human handling [27]. The latter is especially important in bulls, as male animals are more prone to suffer stress when forced into close social contact with their cohorts during transportation and lairage [35,36]. With correct management, normal ultimate pH can be achieved in under 16 months bulls independently of the dietary treatment.

The darker meat in the grazing groups is in line with previous studies [37,38]. Why SC was also darker than CON is not clear, since ultimate pH was similar, and the animals were managed similarly indoors. Darker colour may also be related to a higher myoglobin concentration in grass-fed animals. However, the redness (a) value indicates less red meat from grazed animals, which suggests that a higher pigment concentration is unlikely. However, myoglobin concentration measurement is needed to confirm this point.

In the present study, the steaks were aged for exactly 14 days as this is common industry practice. The lack of differences in the instrumental texture variables (WBSF, Modulus and Energy) agrees with those authors [39] who indicate that after 12 day of ageing differences in instrumental texture between factors such as gender or breed disappear [38,39].

Despite the lack of difference between treatments in instrumental texture, meat from SC was rated more highly by the sensory panel than meat from GC. Inconsistency between sensory and instrumental measurement results is common and is reflected in the modest correlation between both methods as reported by others [40,41]. Similar to the present study, Hedrick et al. [42] found that meat from cattle finished on silage was as tender or more tender than grain-finished cattle. The detrimental effect of concentrate supplementation outdoors compared to indoors (GC v SC) on tenderness was observed previously in 19-month-old bulls [34], albeit overall acceptability did not differ in that study.

The relationship of the rate and extent of post-mortem proteolysis with temperature and pH decline have been previously described [43]. Therefore, at the same chilling temperature, carcass fat cover impacts temperature and pH decline and sensory evaluation as previously highlighted [38,44,45].

The flavour of meat is dependent on the volatile profile [46] which in turn is greatly influenced by fat level and fatty acid composition [47,48]. Grass-produced beef can have a slightly less intense flavour than grain-produced beef [48], since higher levels of fat are associated with higher intensity of flavour. However, in the present experiment, flavour intensity was not assessed. In general, the effect of ration composition on flavour likeness results are inconsistent [38,49]. The higher rating in flavour for SC may be related to its higher fat content compared to outdoors animal and its lower SFA proportion compared to CON. This may give SC the best combination of IMF and fatty acid composition to meet consumer expectations. Overall, the lack of difference in sensory characteristics between CON and G0 is particularly noteworthy as G0 carcasses would be severely discounted relative to CON carcasses under the current EUROP grading system.

4.3. Fatty Acid Profile

Consumers are increasingly aware of the relationships between diet, health and well-being and this has resulted in a growing preference for foods which are healthier and more nutritious [50,51]. Some consumers prefer to purchase beef from grass fed cattle as meat and meat products from grass fed animals are often perceived as having higher amounts of nutritionally important compounds when compared with beef from non-grass-based production systems. In addition, 'grass fed' has been used to promote perception of animal health and well-being, and environmental sustainability [52]. Accordingly, the fatty acid composition was measured in beef from the contrasting production systems in this study. The higher concentration of fatty acids in the CON group per se reflects the higher energy consumption and associated fat deposition and is a consequence of this production system. The higher fatty acid concentration is the sum of higher concentrations of individual classes of fatty acids. When comparing studies in the literature with respect to the fatty acid profile of beef, it is important to be aware of differences in carcass weight or fatness between treatments. The higher concentration of SFA in CON animals compared to grass-fed animals in this study agrees with Aldai et al. [53] while other authors found no differences in SFA between grass and grain-fed meat [54].

An increase in IMF concentration on a common diet can alter the proportions of fatty acids, generally increasing MUFA and decreasing PUFA proportions [55,56]. Therefore, in this study the individual fatty acid data are presented as a proportion of total fatty acids to gain a better insight into any changes in the pattern or profile of fatty acids due to the different production systems examined.

The change in the individual SFA profile is generally in agreement with that reported by Alfaia et al. [54] and Aldai et al. [53]. The change in the SFA profile also reflects the fatty acid composition of the feedstuffs consumed, particularly the concentrate rations used. The lack of differences in total SFA concentration and in the SFA profile between both grazed grass-based diets (GC and G0) is in line with French et al. [28].

Three DMA derived from plasmalogen lipids were detected (16:0, 18:0 and 18:1). These DMA are generated from the vinyl chain linked at sn-1 position of plasmalogens, which are a particular class of glycerophospholipids present in cell membranes [57]. The lower proportions of DMA in muscle from indoor cattle (Table 4) agree with Aldai et al. [53]. Few studies have examined the nutritional importance of DMA in beef but DMA deficiency in humans has been associated with some diseases [57].

The higher concentration of MUFA in CON is consistent with previous investigations [53,54]. The alteration in the isomer profile of trans C18:1 observed in the present study has been previously reported [53,58]. A decrease in the trans-11 trans-10 18:1 ratio has been related with increased atherogenicity in animal models, while a higher ratio is related with increased rumenic acid (conjugated linoleic acid, see below) which has putative human health properties [59]. Changes in the proportions of these isomers has been suggested to reflect changes in rumen microbiota characteristics due to alteration in the rate of fermentation of the diet and associated changes in ruminal pH. C18:1-t11 rather than C18:1-t10 production in the rumen is associated with *Butyrivibrio fibrisolvens* which predominates in the rumen forage-finished animals [60]. While generally, a higher concentration of PUFA is observed in grass-fed beef compared to concentrate-fed beef [28,54,61], the lack of difference in the current study is likely related to the higher IMF concentration in CON. Also, many studies report fatty acid on a proportional basis. The proportions of PUFA (10.8, 15.3, 23.7 and 21.3 g/100 g fatty acids for CON, SC, GC and G0, respectively) in the present study support the literature but are high which likely reflects the very low total fatty acid concentration and very low neutral lipid proportion particularly in the grass-based production systems. In this regard, Nian et al. [62] reported a PUFA proportion of 23.5 g/100 g fatty acids for muscle from dairy origin bulls that had an IMF concentration of 0.5%, consistent with our data.

Linolenic acid (C18:3, omega 3) and linoleic acid (18:2n-6, omega 6) were the major PUFA of the omega series identified in this study. The omega series fatty acids cannot be synthesised by humans and are considered essential nutrients for humans [63]. The ratio is also considered important for human health since an excess of one family of omega 3 or omega 6 can interfere with the metabolism of the other [63]. It has been suggested that a healthy diet should have an n-6:n3 PUFA ratio higher than

4 [64], this ratio was only observed for outdoors animals suggesting that grass-fed beef has a higher nutritional value. For omega-3 fatty acids, EU (2010) states that a claim that a food is a source of omega-3 fatty acids may only be made where the product contains at least 300 mg α-linolenic acid per 100 g). The concentrations of linolenic acid in the present study were 9, 13, 17 and 18 mg/100 g muscle for CON, SC, GC and G0, respectively. None of the beef in this study meets this claim.

Higher linoleic acid in GC and G0 compared with CO animals is unusual based on the literature, but may reflect the extremely low IMF concentration in these groups. The higher proportion of linolenic acid its elongation products mainly C22-5n-3 and C20:5n-3 in GC and G0 is consistent with the literature [28].

Two CLA isomers were detected in the present study, the major isomer rumenic acid (C18:2-c9,t11) and the minor isomer, C18:2 t10c12. The higher rumenic acid proportion in GC and G0 is also consistent with the literature [28,53]. Animal studies demonstrated that CLA can reduce carcinogenesis, atherosclerosis and diabetes [65–67]. However, due to the higher IMF concentration in CON and SC compared to GC and G0 (Table 6), an individual consuming 100 g beef would consume a similar amount of rumenic acid from beef from all the production systems examined.

Since humans are able to transform trans vaccenic acid (C18:1-t11) into rumenic acid (9c11t-C18:2) at a rate between 5 to 12% [68], the higher concentration of trans vaccenic acid (6.3, 5.3, 7.0 and 7.5 mg/100 g muscle for CON, SC, GC and G0, respectively) would also contribute to a nutritional enhancement of beef from the grass-based production systems. We acknowledge that the proportion of CLA found in the present study was low compared with other studies, however the levels are similar to those reported by Aldai et al. [53], who explained the lower CLA concentration on the basis of the late-maturing breeds used.

The HUFA meat content is also an important nutritional factor, since the human efficiency of transforming α-linolenic acid to EPA (eicosapentaenoic acid C20:5n-3), DPA (docosapentaenoic acid C22:5n-3) and DHA (docosahexaenoic acid C22:6 n-3) is very low. On the other hand, HUFA (especially DHA and EPA) have been related with the prevention of atherosclerosis, heart attack, depression and cancer [69]. The higher concentration of HUFA and proportions of EPA and DPA in GC and G0 is similar to Alfaia et al. [54]. For longer carbon chain omega-3 fatty acids, EU (2010) states that a claim that a food is a source of omega-3 fatty acids may only be made where the product contains at least 40 mg of the sum of EPA and DHA per 100 g.). In the present study, EPA + DHA concentrations were 9.5, 14.4, 14.0 and 14.5 mg/100 g muscle for CON, SC, GC and G0, respectively. None of the beef in this study meets this claim.

In line with previous results, the nutritional indices, AI and TI, were better for grass-fed beef compared to CON indicating a general improvement of the nutritional quality. Similarly, the PUFA:SFA ratio is suggested to be above 0.4 [70] and only the grass-based production systems achieved this target.

5. Conclusions

In conclusion, while only the indoor production systems met the market carcass fat specifications, beef eating quality of grass-fed animals was not detrimentally affected. This, together with the fact that grass-beef systems result in leaner meat with a fatty acid profile better for the health of the consumer, makes these "grass-based" productions systems a feasible alternative especially for "health-concerned" consumers. It is clear that the carcass fat specifications required by the industry are not justified on an eating quality basis.

Author Contributions: Author's included in the present article have contributed substantially to the work reported. L.M. meat quality analysis writing—review and editing of the paper; A.P.M. Principal investigator of the project writing—review and editing of the paper; M.M. and E.G.O. animal production and carcass measurements; C.K.M. fatty acid analysis; F.J.M. meat quality writing—review; S.S.W., M.G.O. and J.P.K. sensory analysis.

Funding: This research was funded by the Irish Department of Agriculture, Food and the Marine's competitive research programmes, project ID (11/SF/322, "BullBeef").

Acknowledgments: The authors thank Kevin McMenamin for managing the production part of the study, Eugene Vesey for his help with sample management and Maeve Flood and Sibhekiso Siphambili for assistance with sample collection and analysis. We acknowledge the assistance of staff at Teagasc Animal & Grassland Research and Innovation Centre, Grange and at Kepak Group, Clonee, Co. Meath during sample collection and carcass measurement.

Conflicts of Interest: The authors declare no conflict of interest.

References

1. Teagasc. Beef Production System Guidelines. Available online: https://www.teagasc.ie (accessed on 21 June 2019).

2. Archer, J.A.; Richardson, E.C.; Herd, R.M.; Arthur, P.F. Potential for selection to improve efficiency of feed use in beef cattle: A review. *Aust. J. Agric. Res.* **1999**, *50*, 147. [CrossRef]

3. Finneran, E.; Crosson, P.; O'Kiely, P.; Shalloo, L.; Forristal, D.; Wallace, M. Stochastic modelling of the yield and input price risk affecting home produced ruminant feed cost. *J. Agric. Sci.* **2011**, *150*, 123–139. [CrossRef]

4. Bhandari, B.; Gillespie, J.; Scaglia, G. Labor use and profitability associated with pasture systems in grass-fed beef production. Sustain. *Agric. Res.* **2017**, *6*, 51–61. [CrossRef]

5. Boogaard, B.; Boekhorst, L.; Oosting, S.; Sørensen, J. Socio-cultural sustainability of pig production: Citizen perceptions in the Netherlands and Denmark. *Livest. Sci.* **2011**, *140*, 189–200. [CrossRef]

6. Musto, M.; Cardinale, D.; Lucia, P.; Faraone, D. Creating Public Awareness of How Goats Are Reared and Milk Produced May Affect Consumer Acceptability. *J. Appl. Anim. Welf. Sci.* **2016**, *19*, 2721–2734. [CrossRef] [PubMed]

7. Van Wezemael, L.; Caputo, V.; Nayga, R.M., Jr.; Chryssochoidis, G.; Verbeke, W. European consumer preferences for beef with nutrition and health claims: A multi-country investigation using discrete choice experiments. *Food Policy* **2014**, *44*, 167–176. [CrossRef]

8. Belew, J.; Brooks, J.; McKenna, D.; Savell, J.; Savell, J. Warner–Bratzler shear evaluations of 40 bovine muscles. *Meat Sci.* **2003**, *64*, 507–512. [CrossRef]

9. Guillemin, N.; Bonnet, M.; Jurie, C.; Picard, B. Functional analysis of beef tenderness. *J. Proteom.* **2011**, *75*, 352–365. [CrossRef]

10. Vestergaard, M.; Therkildsen, M.; Henckel, P.; Jensen, L.; Andersen, H.; Sejrsen, K. Influence of feeding intensity, grazing and finishing feeding on meat and eating quality of young bulls and the relationship between muscle fibre characteristics, fibre fragmentation and meat tenderness. *Meat Sci.* **2000**, *54*, 187–195. [CrossRef]

11. Bonny, S.P.F.; Pethick, D.W.; Legrand, I.; Wierzbicki, J.; Allen, P.; Farmer, L.J.; Polkinghorne, R.J.; Hocquette, J.-F.; Gardner, G.E. European conformation and fat scores have no relationship with eating quality. *Animal* **2016**, *10*, 996–1006. [CrossRef]

12. Noci, F.; Monahan, F.J.; French, P.; Moloney, A.P. The fatty acid composition of muscle fat and subcutaneous adipose tissue of pasture-fed beef heifers: Influence of the duration of grazing. *J. Anim. Sci.* **2005**, *83*, 1167–1178. [CrossRef] [PubMed]

13. Hickey, J.M.; Keane, M.G.; Kenny, D.A.; Cromie, A.R.; Veerkamp, R.F. Genetic parameters for EUROP carcass traits within different groups of cattle in Ireland1. *J. Anim. Sci.* **2007**, *85*, 314–321. [CrossRef] [PubMed]

14. Pearce, K.L.; Van De Ven, R.; Mudford, C.; Warner, R.D.; Hocking-Edwards, J.; Jacob, R.; Pethick, D.W.; Hopkins, D.L. Case studies demonstrating the benefits on pH and temperature decline of optimising medium-voltage electrical stimulation of lamb carcasses. *Anim. Prod. Sci.* **2010**, *50*, 1107–1114. [CrossRef]

15. *AOAC Official Methods of Analysis*; Method No. 985.14; Association of Official Analytical Chemists: Gaithersburg, MD, USA, 2000.

16. Folch, J.; Lees, M.; Stanley, G.H.S. A simple method for the isolation and purification of total lipides from animal tissues. *J. Boil. Chem.* **1957**, *226*, 497–509.

17. *AOAC Official Methods of Analysis*; Method No. 992.15; Association of Official Analytical Chemist: Gaithersburg, MD, USA, 2000.

18. *AMSA Research Guidelines for Cookery, Sensory Evaluation and Instrumental Tenderness Measurements of Fresh Meat*; American Meat Science Association: Chicago, IL, USA, 1995.

19. Stone, H.; Bleibaum, R.N.; Thomas, H.A. Affective testing. In *Sensory Evaluation Practices*; Stone, H., Bleibaum, R.N., Thomas, H.A., Eds.; Elsevier: San Diego, CA, USA, 2012; pp. 291–325.

20. Stone, H.; Sidel, J.L. Introduction to sensory evaluation. In *Sensory Evaluation Practices*; Stone, H., Sidel, J.L., Eds.; Elsevier Academic Press: San Diego, CA, USA, 2004; pp. 1–19.
21. ISO 67.240—Sensory Analysis. 1998. Available online: https://www.une.org/encuentra-tu-norma/busca-tu-norma/norma/?c=N0040176 (accessed on 8 July 2019).
22. Stone, H.; Bleibaum, R.N.; Thomas, H.A. Test Strategy and the Design of Experiments. In *Sensory Evaluation Practices*; Elsevier BV: Amsterdam, The Netherlands, 2012; pp. 117–165.
23. *AMSA Research Guidelines for Cookery, Sensory Evaluation and Instrumental Tenderness Measurements of Fresh Meat*; National Livestock and Meat Board: Campaign, IL, USA, 2005; pp. 4–8.
24. Shingfield, K.J.; Ahvenjärvi, S.; Toivonen, V.; Ärölä, A.; Nurmela, K.V.V.; Huhtanen, P.; Griinari, J.M. Effect of dietary fish oil on biohydrogenation of fatty acids and milk fatty acid content in cows. *Anim. Sci.* **2003**, *77*, 165–179. [CrossRef]
25. Ulbricht, T.; Southgate, D. Coronary heart disease: Seven dietary factors. *Lancet* **1991**, *338*, 985–992. [CrossRef]
26. French, P.; O'Riordan, E.; Monahan, F.; Caffrey, P.; Mooney, M.; Troy, D.; Moloney, A. The eating quality of meat of steers fed grass and/or concentrates. *Meat Sci.* **2001**, *57*, 379–386. [CrossRef]
27. Priolo, A.; Micol, D.; Agabriel, J. Effects of grass feeding systems on ruminant meat colour and flavour. A review. *Anim. Res.* **2001**, *50*, 185–200. [CrossRef]
28. French, P.; O'Riordan, E.; Monahan, F.; Caffrey, P.; Vidal, M.; Mooney, M.; Troy, D.; Moloney, A. Meat quality of steers finished on autumn grass, grass silage or concentrate-based diets. *Meat Sci.* **2000**, *56*, 173–180. [CrossRef]
29. Banovic, M.; Chrysochou, P.; Grunert, K.G.; Rosa, P.J.; Gamito, P. The effect of fat content on visual attention and choice of red meat and differences across gender. *Food Qual. Prefer.* **2016**, *52*, 42–51. [CrossRef]
30. Grunert, K.G. What's in a steak? A cross-cultural study on the quality perception of beef. *Food Qual. Prefer.* **1997**, *8*, 157–174. [CrossRef]
31. Grunert, K.G. Future trends and consumer lifestyles with regard to meat consumption. *Meat Sci.* **2006**, *74*, 149–160. [CrossRef] [PubMed]
32. Polkinghorne, R.J.; Thompson, J.M. Meat standards and grading: A world view. *Meat Sci.* **2010**, *86*, 227–235. [CrossRef] [PubMed]
33. Viljoen, H.; De Kock, H.; Webb, E.; De Kock, R. Consumer acceptability of dark, firm and dry (DFD) and normal pH beef steaks. *Meat Sci.* **2002**, *61*, 181–185. [CrossRef]
34. Moran, L.; O'Sullivan, M.; Kerry, J.; Picard, B.; McGee, M.; O'Riordan, E.; Moloney, A.; Lobato, L.M.; O'Sullivan, M. Effect of a grazing period prior to finishing on a high concentrate diet on meat quality from bulls and steers. *Meat Sci.* **2017**, *125*, 76–83. [CrossRef] [PubMed]
35. Field, R.A. Effect of Castration on Meat Quality and Quantity. *J. Anim. Sci.* **1971**, *32*, 849–858. [CrossRef]
36. Katz, L.S. Sexual behavior of domesticated ruminants. *Horm. Behav.* **2007**, *52*, 56–63. [CrossRef]
37. Vestergaard, M.; Oksbjerg, N.; Henckel, P. Influence of feeding intensity, grazing and finishing feeding on muscle fibre characteristics and meat colour of semitendinosus, longissimus dorsi and supraspinatus muscles of young bulls. *Meat Sci.* **2000**, *54*, 177–185. [CrossRef]
38. Muir, P.D.; Deaker, J.M.; Bown, M.D. Effects of forage- and grain-based feeding systems on beef quality: A review. *N. Z. J. Agric. Res.* **1998**, *41*, 623–635. [CrossRef]
39. Muchenje, V.; Dzama, K.; Chimonyo, M.; Strydom, P.; Hugo, A.; Raats, J. Some biochemical aspects pertaining to beef eating quality and consumer health: A review. *Food Chem.* **2009**, *112*, 279–289. [CrossRef]
40. Caine, W.; Aalhus, J.; Best, D.; Dugan, M.; Jeremiah, L. Relationship of texture profile analysis and Warner-Bratzler shear force with sensory characteristics of beef rib steaks. *Meat Sci.* **2003**, *64*, 333–339. [CrossRef]
41. Shackelford, S.D.; Wheeler, T.L.; Koohmaraie, M. Relationship between shear force and trained sensory panel tenderness ratings of 10 major muscles from *Bos indicus* and *Bos taurus* cattle1. *J. Anim. Sci.* **1995**, *73*, 3333–3340. [CrossRef] [PubMed]
42. Hedrick, H.B.; Paterson, J.A.; Matches, A.G.; Thomas, J.D.; Morrow, R.E.; Stringer, W.G.; Lipsey, R.J. Carcass and Palatability Characteristics of Beef Produced on Pasture, Corn Silage and Corn Grain. *J. Anim. Sci.* **1983**, *57*, 791–801. [CrossRef]
43. Koohmaraie, M. Effect of pH, temperature, and inhibitors on autolysis and catalytic activity of bovine skeletal muscle μ-calpain. *J. Anim. Sci.* **1992**, *70*, 3071–3080. [CrossRef] [PubMed]

44. Pearson, A.M. Desirability of Beef—Its Characteristics and Their Measurement. *J. Anim. Sci.* **1966**, *25*, 843–854. [CrossRef]

45. Smith, G.C.; Carpenter, Z.L. Eating Quality of Meat Animal Products and Their Fat Content, National Research Council (US). Available online: https://www.ncbi.nlm.nih.gov/books/NBK216525/ (accessed on 8 July 2019).

46. Hornstein, I.; Crowe, P.F. Meat flavor—A review. *J. Chromatogr. Sci.* **1964**, *2*, 128–131. [CrossRef]

47. Melton, S.L.; Amiri, M.; Davis, G.W.; Backus, W.R. Flavor and Chemical Characteristics of Ground Beef from Grass-, Forage-Grain- and Grain-Finished Steers. *J. Anim. Sci.* **1982**, *55*, 77–87. [CrossRef]

48. Melton, S.L. Effects of feeds on flavor of red meat: A review. *J. Anim. Sci.* **1990**, *68*, 4421–4435. [CrossRef]

49. Muir, P.D.; Smith, N.B.; Wallace, G.J.; Cruickshank, G.J.; Smith, D.R. The effect of short-term grain feeding on liveweight gain and beef quality. *N. Z. J. Agric. Res.* **1998**, *41*, 517–526. [CrossRef]

50. Verbeke, W.; Pérez-Cueto, F.J.; De Barcellos, M.D.; Krystallis, A.; Grunert, K.G. European citizen and consumer attitudes and preferences regarding beef and pork. *Meat Sci.* **2010**, *84*, 284–292. [CrossRef]

51. Hocquette, J.-F.; Botreau, R.; Picard, B.; Jacquet, A.; Pethick, D.W.; Scollan, N.D. Opportunities for predicting and manipulating beef quality. *Meat Sci.* **2012**, *92*, 197–209. [CrossRef]

52. Van Elswyk, M.E.; McNeill, S.H. Impact of grass/forage feeding versus grain finishing on beef nutrients and sensory quality: The U.S. experience. *Meat Sci.* **2014**, *96*, 535–540. [CrossRef] [PubMed]

53. Aldai, N.; Dugan, M.E.R.; Kramer, J.K.G.; Martinez, A.; López-Campos, Ó.; Mantecon, A.R.; Osoro, K. Length of concentrate finishing affects the fatty acid composition of grass-fed and genetically lean beef: An emphasis on trans-18:1 and conjugated linoleic acid profiles. *Animal* **2011**, *5*, 1643–1652. [CrossRef] [PubMed]

54. Alfaia, C.P.; Alves, S.P.; Martins, S.I.; Costa, A.S.; Fontes, C.M.; Lemos, J.P.; Bessa, R.J.; Prates, J.A.; Prates, J.A.M. Effect of the feeding system on intramuscular fatty acids and conjugated linoleic acid isomers of beef cattle, with emphasis on their nutritional value and discriminatory ability. *Food Chem.* **2009**, *114*, 939–946. [CrossRef]

55. Moreno, T.; Keane, M.; Noci, F.; Moloney, A. Fatty acid composition of M. Longissimus dorsi from Holstein–Friesian steers of New Zealand and European/American descent and from Belgian Blue×Holstein–Friesian steers, slaughtered at two weights/ages. *Meat Sci.* **2008**, *78*, 157–169. [CrossRef] [PubMed]

56. Warren, H.; Scollan, N.; Enser, M.; Hughes, S.; Richardson, R.; Wood, J. Effects of breed and a concentrate or grass silage diet on beef quality in cattle of 3 ages. I: Animal performance, carcass quality and muscle fatty acid composition. *Meat Sci.* **2008**, *78*, 256–269. [CrossRef]

57. Webb, E.; O'Neill, H. The animal fat paradox and meat quality. *Meat Sci.* **2008**, *80*, 28–36. [CrossRef]

58. Braverman, N.E.; Moser, A.B. Functions of plasmalogen lipids in health and disease. *Biochim. Biophys. Acta Mol. Basis Dis.* **2012**, *1822*, 1442–1452. [CrossRef]

59. Bessa, R.; Portugal, P.; Mendes, I.; Santos-Silva, J. Effect of lipid supplementation on growth performance, carcass and meat quality and fatty acid composition of intramuscular lipids of lambs fed dehydrated lucerne or concentrate. *Livest. Prod. Sci.* **2005**, *96*, 185–194. [CrossRef]

60. Bauchart, D.; Roy, A.; Lorenz, S.; Chardigny, J.-M.; Ferlay, A.; Gruffat, D.; Sébédio, J.-L.; Chilliard, Y.; Durand, D. Butters Varying in trans 18:1 and cis-9, trans-11 Conjugated Linoleic Acid Modify Plasma Lipoproteins in the Hypercholesterolemic Rabbit. *Lipids* **2007**, *42*, 123–133. [CrossRef]

61. Dugan, M.E.R.; Kramer, J.K.G.; Robertson, W.M.; Meadus, W.J.; Aldai, N.; Rolland, D.C. Comparing Subcutaneous Adipose Tissue in Beef and Muskox with Emphasis on trans 18:1 and Conjugated Linoleic Acids. *Lipids* **2007**, *42*, 509–518. [CrossRef] [PubMed]

62. Nuernberg, K.; Dannenberger, D.; Nuernberg, G.; Ender, K.; Voigt, J.; Scollan, N.; Wood, J.; Nute, G.; Richardson, R. Effect of a grass-based and a concentrate feeding system on meat quality characteristics and fatty acid composition of longissimus muscle in different cattle breeds. *Livest. Prod. Sci.* **2005**, *94*, 137–147. [CrossRef]

63. Nian, Y.; Allen, P.; Harrison, S.M.; Kerry, J.P. Effect of castration and carcass suspension method on the quality and fatty acid profile of beef from male dairy cattle. *J. Sci. Food Agric.* **2018**, *98*, 4339–4350. [CrossRef] [PubMed]

64. Wood, J.; Richardson, R.; Nute, G.; Fisher, A.; Campo, M.M.; Kasapidou, E.; Sheard, P.; Enser, M. Effects of fatty acids on meat quality: A review. *Meat Sci.* **2004**, *66*, 21–32. [CrossRef]

65. Moloney, F.; Yeow, T.-P.; Mullen, A.; Nolan, J.J.; Roche, H.M. Conjugated linoleic acid supplementation, insulin sensitivity, and lipoprotein metabolism in patients with type 2 diabetes mellitus. *Am. J. Clin. Nutr.* **2004**, *80*, 887–895. [CrossRef] [PubMed]

66. Ip, C.; Singh, M.; Thompson, H.J.; Scimeca, J.A. Conjugated linoleic acid suppresses mammary carcinogenesis and proliferative activity of the mammary gland in the rat. *Cancer Res.* **1994**, *54*, 1212–1215. [PubMed]

67. Park, Y.; Albright, K.J.; Liu, W.; Storkson, J.M.; Cook, M.E.; Pariza, M.W. Effect of conjugated linoleic acid on body composition in mice. *Lipids* **1997**, *32*, 853–858. [CrossRef] [PubMed]

68. Turpeinen, A.M.; Mutanen, M.; Aro, A.; Salminen, I.; Basu, S.; Palmquist, D.L.; Griinari, J.M. Bioconversion of vaccenic acid to conjugated linoleic acid in humans. *Am. J. Clin. Nutr.* **2002**, *76*, 504–510. [CrossRef] [PubMed]

69. Connor, W.E. Importance of n−3 fatty acids in health and disease. *Am. J. Clin. Nutr.* **2000**, *71*, 171S–175S. [CrossRef] [PubMed]

70. Wood, J.; Enser, M.; Fisher, A.; Nute, G.; Sheard, P.; Richardson, R.; Hughes, S.; Whittington, F. Fat deposition, fatty acid composition and meat quality: A review. *Meat Sci.* **2008**, *78*, 343–358. [CrossRef]

 *foods*

Article

Relationships Between Cull Beef Cow Characteristics, Finishing Practices and Meat Quality Traits of *Longissimus thoracis* and *Rectus abdominis*

Sébastien Couvreur [1,*], Guillain Le Bec [1], Didier Micol [2] and Brigitte Picard [2]

[1] Unité de Recherche sur les Systèmes d'Elevage, Ecole Supérieure d'Agricultures, Université Bretagne-Loire, 55, rue Rabelais, BP 30748, F-49007 Angers, France; g.lebec@groupe-esa.com

[2] Université Clermont Auvergne, INRA, VetAgro Sup, UMR Herbivores, F-63122 Saint-Genès-Champanelle, France; didier.micol@inra.fr (D.M.); brigitte.picard@inra.fr (B.P.)

* Correspondence: s.couvreur@groupe-esa.com; Tel.: +33-241235638

Received: 25 March 2019; Accepted: 22 April 2019; Published: 25 April 2019

Abstract: The aim of study was to investigate the relationships between the characteristics of cull beef cows in the Rouge des Prés breed, finishing practices and physicochemical characteristics and sensory traits of *Longissimus thoracis* (LT) and *Rectus abdominis* (RA) muscles from 111 cows. On the basis of our surveys, which qualify at cow level the animal characteristics and finishing diet, clusters of cull cows and finishing practices are created and their effects tested on LT and RA meat quality. Old and heavy cows with good suckling ability (95 months, 466 kg and 7.1/10) are characterized by LT with larger fibers, and higher intramuscular fat content and fat-to-muscle ratio. Young and heavy cows with low suckling ability (54 months, 474 kg and 4.4/10) are characterized by LT and RA with lower MyHC IIx and higher MyHC IIa and MyHC I proportions. MyHC IIx and IIa proportions are lower and a* and b* color indices higher when cows are finished on pasture, probably related to grass diet and physical activity. The fat-to-muscle ratio is higher without any effect on the intramuscular fat content when cows are finished over a short period (107 days) with a high level of concentrate (9.7 kg/day). The opposite effect is observed over a long period (142 days) with a low level of concentrate (5.8 kg/day), confirming the interaction effect between finishing duration and amount of energy concentrate on the allotment of adipose tissue deposit.

Keywords: suckling cattle; cull cow; meat quality; finishing practices; farm survey

---

1. Introduction

In France, more than 50% of the beef meat produced and consumed comes from cull dairy and beef cows. There is a high diversity of cows in terms of breed, age, and diets. Throughout their life, cows are fed with different diets (maize or grass silage, pasture, and hay associated with energy and protein supplementations) which led to different growth curves and fat deposition dynamics [1]. Before slaughtering, they can be fed with a finishing diet for several weeks which influences the muscle and fat contents of the carcass. This can induce a wide variability in carcass and meat quality traits such as tenderness, juiciness, or flavor [2,3]. Many experiments dealing with tenderness have shown that age can have a negative effect on tenderness from 24 months to more than 60 months [4,5]. Moreover, compensatory growth, high energy level, growth rates, and duration of finishing periods can have positive effects on tenderness that vary according to animal type (steer, heifer, young bull, or cull cow) and muscle [3]. Finally, morphological animal type has few effects on tenderness [6]. The effects of diet and animal type on flavor and juiciness have been assessed in a few studies. Duarte et al. [7] have shown, in Nellore cattle, that juiciness and flavor are less affected by the finishing diet and have a negative correlation to growth rate. Finishing diets leading to high fat deposition (and low growth rate), such as a long fattening period with a low energy level diet, may increase juiciness and flavor intensity [8,9].

Nevertheless, most studies dealing with the effects of diet and animal type on meat quality have mainly been conducted on young animals such as heifers, young bulls, and steers. In cull cows, few experiments have been conducted dealing with the effect of the breed [4–6], finishing diet [9,10], and age [5]. The main part of these studies has been conducted on cull dairy cows. They have shown that breed (dairy versus suckling), animal morphology in a breed (selection on body conformation or not), and interaction with finishing diet, can modify the fat deposition in the carcass and meat, the muscle fiber type, the collagen content and composition, and in consequence the meat quality in terms of tenderness, flavor, and juiciness. Moreover, the effects of finishing factors are different according to the animal category (young bulls or cull cows) [11]. Nevertheless, there are few experiments with cull cows because it is difficult to create homogenous experimental groups in terms of body condition score, weight, age, and raising practices before the experimentation (e.g., diets, sanitary events, and compensatory growth).

Moreover, most of the results have been produced in experimental conditions (trials dealing with one or two factors) and are difficult to apply to real multifactorial conditions such as farm management practices [12]. At farm scale, few experiments have studied the effects of rearing practices (diet) on meat quality of *Rectus abdominis* (RA) and *Longissimus thoracis* (LT) [3,13,14]. They have focused their work on the relationships of different types of finishing diets on beef meat quality traits (mainly tenderness) of heifers [3], young bulls [13], and cull beef cows [14], but they did not study factors such as age or animal type and their interaction with finishing diet. For these reasons, our objective was to assess the effects of the age, animal type (suckling ability, morphology) of cull beef cows and the finishing practices (forage and level of concentrate) on meat quality traits of RA (tenderness, color, juiciness, and flavor) and LT (tenderness and color) muscles in relation to their composition (fiber type, enzyme activities, fat content, and collagen content).

This work has been performed on protected denomination of origin (PDO) Maine-Anjou (MA) cull cows. All the animals raised in this PDO belong to the local suckling breed, Rouge des Prés (RdP), which was a dual-purpose breed until the late 1980s. We chose the population of PDO MA cull cows slaughtered in a year (around 1300 cows) for selection of our experimental animals because it is recognized for having the following diverse study factors at farm scale: cull cow morphology (age, body conformation, carcass weight, and fat deposition) and finishing practices [15].

2. Materials and Methods

2.1. Sampled Animals

Our objective was to sample 10% of the total population of PDO MA cows slaughtered throughout a year, in other words 110–120 cull cows. These cows represented the diverse finishing practices observed in PDO MA farms. A preliminary study on the whole PDO MA farm population conducted by Schmitt et al. [15] identified a representative selection of 45 farms based on their finishing practices (forage type, amount of concentrate, and the finishing period). In 2010, 111 cull cows from these farms and slaughtered throughout the year were selected and sampled.

The cows were collected in a commercial slaughterhouse (Elivia, Le Lion d'Angers, France) following standardized slaughtering procedures (last feed on the day before slaughter, transport duration less than 15 h, maximum 12-h resting period after arrival with available water, stunning procedure with a captive-bolt stunner, and vertical bleeding), as well as standardized chilling and storing procedures (deep muscle temperature of 6–7 °C achieved in 24 h). At 24 h post-mortem, cold carcass weight and fatness scores were collected. The conformation was judged by a trained classifier according to the EUROP classification with three levels per class (+, =, −) (CE1249/2008 regulation). Scores (from 1 to 5) related to the PDO MA agreement and assessing meat color (visual score), intramuscular fat content (IMF) (visual score), and tenderness (palpation of the fifth rib). For each carcass, the fifth rib from the left half of the carcass and the two RA muscles were removed at 24 h post-mortem. The LT muscle was sampled from the fifth rib the same day. Finally, the LT and the two RA muscles were vacuum packaged. The RA muscle was chosen because it is easily sampled, without any economic depletion of the carcass, and because it is more reactive to the variations of rearing practices than LT muscle [16]. In addition, the LT muscle sample was used as a reference for comparison with other studies using LT muscle and with the RA muscle results.

A 3 cm thick steak and samples of 110 g were removed from the LT muscle section and the left RA muscle for further analysis. The steaks were vacuum packaged, chilled for 14 days at 4 °C and then stored at −20 °C until shear force measurements were performed. From the 110 g, 10 g were used for the following fiber characterizations: fiber cross-section area, myosin heavy chains (MyHC) proportions, and enzyme activities. Samples intended for fiber area measurement were cut into 2 duplicates of 2 cm x 3 mm side cubes and stacked on cork in order to position the muscular fibers, progressively frozen in isopentane and then in liquid nitrogen, and then stored at −80 °C until analysis. Samples intended for MyHC proportions and for enzyme activities were cut into small cubes (down to 1 mm side), frozen in liquid nitrogen, and stored at −80 °C. The remaining muscles, 100 g, were used for the IMF, total, and soluble collagen content measurements. The samples were cut into 5 mm side cubes, freeze dried for 72 h and ground to obtain a meat powder. Then, the powder of each sample was vacuum packaged and stored at 4 °C until analysis.

The right RA muscles were vacuum packaged and chilled for 14 days at 4 °C for aging. After aging, they were frozen and stored at −20 °C until sensory analysis.

2.2. Physicochemical Measurements

2.2.1. Intermuscular Fat Content and Carcass Composition

The fifth rib was dissected and intermuscular fat, meat, and bones were separated from each other. Each tissue was weighted. Thereby, the fat to meat ratio of the fifth rib (i.e., the intermuscular fat content) and the carcass composition were calculated according to the equations developed by Robelin and Geay (1975) for Salers breed [17]. The equations used were: (i) Muscle in the carcass (kg) = −11.21 + 0.7449 × carcass weight (kg) − 72.52 × adipose tissue in the rib (kg) + 12.2 × muscular tissue in the rib (kg) and (ii) fat in the carcass (kg) = −5.22 + 0.1489 × carcass weight (kg) + 67.48 × adipose tissue in the rib (kg) − 10.39 × muscular tissue in the rib (kg).

2.2.2. Muscle Fiber Cross-Sectional Area

The fiber cross-sectional area was determined by computerized image analysis on 10-μm thick sections cut using a cryotome MICROM HM 500 M at a temperature of −25 °C [18]. The sections were stained with azorubine colorant to define the histological architecture of the muscle and to measure fiber proportion and diameter. Because the fibers did not have the same areas depending on animals and muscles, between 180 and 220 fibers from two different locations in the muscle were used to determine the mean fiber area using computerized image analysis. The surface area of each type of fiber and the mean fiber area were measured using the Visilog software program developed by Meunier et al. [18].

2.2.3. Myosin Heavy Chain Proportions

The MyHC IIx, IIa, and I isoforms were separated according to their molecular weight using the electrophoretic method developed by Picard et al. [19]. Around 100 mg of muscle were ground in 5 mL of a buffer solution containing 0.5 M NaCl, 20 mM Na pyrophosphate, 50 mM tris, 1 mM EDTA and 1 mM dithiothreitol. After centrifugation at 2500 × *g* for 10 min at 4 °C, 500 μL of supernatant were removed and diluted 1:1 v/v with glycerol. The protein content was measured by spectrometry according to Bradford [20]. Thereby, a volume of supernatant containing 5 μg of proteins was analyzed using a sodium dodecyl sulfate polyacrylamide gel electrophoresis (SDS-PAGE) according to Picard et al. [19]. After migration, the gels were stained in a solution of R250 Coomassie blue. The proteins were fixed in a solution made with ethanol (30%) and acetic acid (5%) for 20 min at room temperature. The gels were incubated in a colored solution containing propanol 2 (25%), acetic acid (10%) and Coomassie blue R250 (2 g/L) for 20 min. Then, the proportion (%) of the three MyHC was determined by densitometric analysis using an ImageQuant Software (Amersham Biosciences/GE Healthcare, Uppsala, Sweden).

2.2.4. Enzyme Activities

Muscle samples were homogenized with a polytron in a 5% (wt/v) solution with 10 mM tris (pH 8.0), 0.25 M sucrose, and 2 mM EDTA. One aliquot of homogenate was centrifuged at $10,000 \times g$ for 10 min at 4 °C for determination of lactate deshydrogenase (LDH) and isocitrate deshydrogenase (ICDH) activities. The LDH activity was measured by following the disappearance of nicotinamide adenine dinucleotide, reduced form (NADH) at 340 nm, and ICDH activity was measured by following the reduction of nicotinamide adenine dinucleotide phosphate (NADP) at 340 nm. LDH activity was determined according to Ansay [21], using 10 μL of supernatant diluted four-fold, in 2.9 ml of a reaction mixture that contained 50 mM triethanolamine (pH 7.5), 5 mM EDTA, 0.235 mM NADH and 2 mM pyruvate. Pyruvate concentration was determined for maximum LDH activity in bovine muscle. ICDH activity was determined according to Briand et al. [22], using 200 μL of supernatant in 2.7 mL of a reaction mixture that contained 36 mM Na_2HPO_4, 0.5 mM $MnCl_2$, 0.05% Triton X-100, 0.35 mM NADP and 1.3 mM isocitrate (pH 7.5). All enzyme activities were measured at 25 °C, performed in duplicate, and are expressed as micromoles of substrate concerted per minute and per gram of wet muscle (μmol/min/g).

2.2.5. Intramuscular Fat Content

An Accelerated Solvent Extractor 200 (Dionex Corporation, Sunnyvale, CA, USA) was used to measure out the IMF content of a large number of samples (111 cull cows × 2 muscles × 3 repetitions, $n = 666$). Exactly 1 ± 0.001 g of meat powder was placed in a 22 mL extraction cell previously prepared with a cellulose filter and silicon balls. The IMF was extracted with petroleum ether at a temperature of 125 °C and a pressure of 103 bar. The petroleum ether containing IMF was collected and transferred into an evaporation vial previously weighted (± 0.001 g). After 15 min of evaporation, the vial was placed in a drying oven at 105 °C for 17 h and then weighed (± 0.001 g) to determine the amount of IMF in the meat sample.

2.2.6. Total and Soluble Collagen Contents

The total collagen content of raw meat was derived from the hydroxyproline concentration (collagen = $7.5 \times$ HyPro) determined on five replicates using the method of Bergman and Loxley [23] adapted by Bonnet and Kopp (1984) [24]. The insoluble collagen content was determined following the same procedure on the residue obtained after heating the meat sample in a tris–HCl 0.02M, NaCl 0.23M, pH 7.4 buffer solution (1:5 w/v) at 90 °C for 2 h and subsequently discarding the heat-soluble fraction as described by Bonnet and Kopp [25].

2.2.7. Warner-Bratzler Shear Force

The shear force was measured according to the method developed by Honikel [26] using a Warner-Bratzler shear device (Synergie200 texturometer). After thawing 48 h at 4 °C, the RA and LT steaks were placed for 4 h in a thermostated bath at 18 °C. Then, they were cooked using an Infragrill E (Sofraca, France) set at 300 °C until the temperature at the heart of the steak reached 55 °C. From 3 to 5 test pieces (1 × 1 × 4 cm) were taken from the heart of the steak in the direction of the fibers and 3 to 4 repetitions per test tube were carried out. A 1 kN load cell and a 60 mm/min crosshead speed were used (universal testing machine, MTS, Synergie 200H). The peak load (N) and energy to rupture (J) of the muscle sample were determined.

2.2.8. Color

Meat color was monitored using a portable spectrocolorimeter (Minolta 508i, Minolta Konica, Japan) on LT and RA after a 30-min blooming period (24 h post-mortem, the day of cutting). The illuminant, D65, was the chosen because it closely approximates daylight [27]. The spectrocolorimeter was calibrated before measurement using a standard white calibration tile (Y = 93.58, x = 0.3150, y = 0.3217). Color coordinates were calculated in the CIELAB system: L* (lightness), a* (green to red color components), and b* (blue to yellow color components). Measurements were taken at nine locations on each muscle. Three consecutive measurements were averaged to give one value. Thereby, three values per muscle were obtained.

2.3. Sensory Analysis

The right RA muscles were thawed 48 h before sensory analysis at 4 °C. Then, they were cut into 15 mm steaks and cooked for 1 min 45 s in an Infra grill Duo Sofraca set at a temperature of 300 °C (SOFRACA, Morangis, France) to reach a core temperature of 55 °C. After cooking, the steaks were cut into 20 mm cubes that were served on a plastic plate at an internal temperature of 55 °C. Sensory assessment was conducted by a 12-member test panel trained in meat sensory analyses. The panelists evaluated the cooked samples for initial tenderness, overall tenderness, initial juiciness, final juiciness, global flavor, bovine flavor, persistence, and overall appreciation. Initial tenderness was defined as the ease of rupture at the first bite, whereas overall tenderness indicated the ease of rupture along chewing. Initial juiciness was defined as the released juice at the first bite; the final juiciness indicated the released juice at the end of chewing. The persistence characterized the duration of the taste's persistence all along chewing. Each attribute was rated on a nongraduated scale from 0 to 10 points. At each session, a monadic presentation of 6 RA muscle samples was completed, each sample being selected in random order. The sessions were carried out in a sensory analysis room equipped with individual boxes, under artificial noncolored lighting.

2.4. Surveys

For each sampled cow, the rearing practices were recorded using a survey carried out by directly interviewing farmers. The questionnaire included both quantitative and qualitative information about the finishing period and the characteristics of the cow related with its genetic type, its age at slaughter and other indicators of its management. More precisely, the information collected in the survey that related to diet characteristics of the finishing period was:

- The season of the finishing period (seasons are defined as spring (April to June), summer (July to August), autumn (September to November) and winter (December to March));
- The part of hay, either haylage or grass in the finishing diet (% DM of offered forage);
- The amount of concentrate in the diet (kg/day). Concentrate was defined by the amount of raw material (cereals, soya meal, dried beet pulp, …) or balanced compound feed purchased by the farmer;
- The duration of the finishing period (days);
- An estimation of the activity during the finishing period (% finishing days spent outside in a paddock);
- The number of cows in the finishing fattening batch;
- Duration and of the period between the last weaning and the beginning of the finishing period (days).

The information collected in the survey that related to the cull cow characteristics was:

- Birth weight from birth notification (kg);
- Age at first calving (months);
- Age at slaughter (months);

- Parity (i.e., calving number);
- Ascendance characteristics (muscular conformation x muscular conformation, maternal skills x maternal skills, muscular conformation x maternal skills);
- Estimated suckling value by the farmer (i.e., milk production; 0 for no ability to 10 for high ability);
- Sanitary events during productive period (yes/no);
- Long day-open periods (yes/no);
- Potential gestation during the finishing period (yes/no);
- Suckling during the finishing period (yes/no);
- The reason of culling (sanitary/other).

2.5. Statistical Analysis

The purpose of our analysis was to create clusters of cull cows based on finishing practices and cull cow morphology, and thereby to compare the created clusters on their muscle composition and meat quality traits. Two multivariate analyses, according to finishing practices and cull cow morphology, were performed. First, two principal component analyses (PCA) were implemented using a ade4 package (acp.dudi procedure) in the R software on the following discriminating and active variables: (i) part of different forages, amount of concentrate, duration of the finishing period, and activity for the finishing clustering; (ii) birth weight, age at first calving, age at slaughter, calving number, estimated suckling value, and carcass weight for the cull cow type clustering. Data were automatically normalized in the analysis in order to give equal weight to each variable in the PCA analysis. The other variables were not used because they were redundant or not discriminant. Afterwards, the item coordinates on the PCA dimensions were included in two hierarchical cluster analyses (HCA, ade4 package, cah procedure based on Ward's method, and R software). The program identifies the cluster, which has a minor variance within groups and the greater variance between groups. The HCA was done to identify different classes of finishing practices and cull cow types. The classes were reported in the database; each cow was associated with a class of finishing practices and cull cow type.

Classes of finishing practices, of cull cow types and their interactions were compared on data used for PCA, muscle composition, and meat quality traits. Variance analysis and mean multiple comparisons with one factor were carried out using the general linear model (GLM) procedure from SAS. Multiple comparisons of the adjusted means (LSMEANS) were done using the PDIFF option of the GLM procedure.

3. Results

3.1. Cull Cow and Finishing Practices Clusters

3.1.1. Cull Cow Type Clusters

The PCA distinguished three main components (total proportion of variability explained = 75.2%). The proportions of variability explained were 33.0%, 23.7%, and 18.5% and the eigenvalues were 1.65, 1.19, and 0.92 for the first, second, and third components, respectively. The coordinates of each item on the three PCA components were used to perform the HCA. It led to three clusters which were named Ylight, Omilk, and Yheavy (Table 1). Fifteen cows could not be included in the clustering due to a lack of data for the variables used in the analyses.

Table 1. Characteristics of the cull cow clusters identified by the hierarchical cluster analysis (adjusted mean from anova ± standard error of the mean).

	Ylight	**Omilk**	**Yheavy**	*p*
N	51	32	13	
Cold carcass weight, kg	415 [a] ± 3.0	466 [b] ± 5.9	474 [b] ± 4.0	***
Estimated suckling value, /10	5.5 [b] ± 0.17	7.1 [c] ± 0.14	4.4 [a] ± 0.26	***
Parity	1.5 [a] ± 0.09	5.2 [b] ± 0.23	1.7 [a] ± 0.26	***
Slaughter age, months	50.4 [a] ± 1.41	95.3 [b] ± 3.07	54.0 [a] ± 2.99	***
Age at 1st calving, months	32.8 [a] ± 0.53	32.0 [a] ± 0.84	33.5 [b] ± 0.84	***

***: $p < 0.001$; Ylight = cluster composed by young and light cows with medium suckling value; Omilk = old and heavy cows with good suckling value; Yheavy = young and heavy cows with low suckling value; [a,b,c] values within a row with different superscripts differ significantly at $p < 0.05$.

The first cluster is composed by light carcasses (415 kg on average) from young cull cows (50 months on average), with a parity below 2 and an average estimated suckling ability (5.5/10) (Ylight, $n = 51$). These cows are not kept for the herd replacement (weak development, low suckling ability) and are quickly culled. The second cluster is composed by heavy carcasses (466 kg on average) from old cows (95 months on average) with a parity above 4 and a high estimated suckling ability (7.7/10) (Omilk, $n = 32$). These cows are kept in the herd because they fit with the following objectives of the farmers: good skeletal and muscular development, reproductive performance, and suckling ability. Finally, the third cluster is composed by heavy carcasses (474 kg on average) from young cows (54 months on average) with a parity below 2 and a low suckling ability (4.4/10) (Yheavy, $n = 12$). These cows have a good muscular development and are quickly culled because of their higher economic value (related to a better carcass yield than the other classes). We observed no difference among clusters in terms of finishing practices.

3.1.2. Finishing Practices Type Clusters

The PCA distinguished three main components (total proportion of variability explained = 74.9%). The proportions of variability explained were 30.5%, 24.7%, and 19.7% and the eigenvalues were 2.14, 1.73, and 1.38 for the first, second, and third components, respectively. The coordinates of each item on the three PCA components were used to perform the HCA. It led to four clusters which were named LongF, HayF, ConcF, and PastF (Table 2). Fourteen cows could not be included in the clustering due to a lack of data for the variables used in the analyses.

The first cluster is characterized by a long finishing period (142 days on average) and a mix of hay and haylage based diet supplemented with 5.8 kg/day of concentrate on average (LongF, $n = 17$). Even if the daily amount of concentrate is the lowest among all the classes, the total amount distributed throughout finishing period is close to the average of the total population (819 kg). The second cluster is characterized by a short finishing period (80 d on average) and a hay-based diet supplemented with 8.0 kg/d (total = 632 kg) of concentrate (HayF, $n = 41$). The third cluster is characterized by duration of the finishing period close to the average of the total population (107 d). The diet is composed by hay supplemented with a high level of concentrate (9.7 kg/d and 1029 kg in total) (ConcF, $n = 18$). Finally, the fourth cluster is characterized by a short finishing period (86 d on average) and a pasture diet supplemented by 7.6 kg/d (total = 665 kg) of concentrate (PastF, $n = 21$). These clusters are consistent with the diverse practices in the PDO MA farm population [15]. We observed no difference among clusters in terms of prefinishing practices (8 weeks between weaning and finishing leading to similar body condition among cows) and cull cow characteristics.

Table 2. Characteristics of the finishing practices clusters identified by the hierarchical cluster analysis (adjusted mean from anova ± standard error of the mean).

	LongF	HayF	ConcF	PastF	*p*
Nb	17	41	18	21	
Duration, days	142 [c] ± 6.3	80 [a] ± 1.9	106 [b] ± 3.2	86 [a] ± 6.0	***
% finishing days spent outside	23 [a] ± 8.0	47 [b] ± 7.8	40 [b] ± 10.1	94 [c] ± 4.9	***
Grazed grass, % forage DM	8 [a] ± 2.7	5 [a] ± 1.5	4 [a] ± 1.7	83 [b] ± 2.7	***
Haylage, % forage DM	46 [b] ± 9.8	39 [b] ± 6.6	16 [a] ± 5.7	2 [a] ± 1.2	***
Hay, % forage DM	46 [b] ± 8.6	55 [b] ± 6.4	80 [c] ± 5.5	14 [a] ± 2.5	***
Concentrate, kg/d	5.8 [a] ± 0.31	8.0 [b] ± 0.26	9.7 [c] ± 0.28	7.6 [b] ± 0.57	***
Concentrate throughout finishing, kg	819 [b] ± 58.1	632 [a] ± 20.0	1029 [c] ± 36.5	665 [a] ± 68.0	***

***: $p < 0.001$; LongF = long finishing period with a mix hay-haylage diet supplemented with a low amount of concentrate; HayF = short finishing period with a hay or haylage diet supplemented with a medium amount of concentrate; ConcF = medium finishing period with a hay diet supplemented with a high amount of concentrate; PastF = short finishing period with a grazed grass diet supplemented with a medium amount of concentrate; [a,b,c] values within a row with different superscripts differ significantly at $p < 0.05$.

3.2. Effect of Cull Cow Clusters on Muscles and Meat Quality Traits

3.2.1. Longissimus Thoracis

There was no difference between clusters on total and insoluble collagen contents and shear force measurements (Table 3). Omilk LT had higher IMF content and fat-to-muscle ratio, in comparison to Ylight and Yheavy LT (+5.8 and +5.2 for IMF content, $p < 0.01$; +7.1 and +6.5 for fat-to-muscle ratio, $p < 0.01$). Yheavy LT tended to have lower cross-sectional area of fibers than Omilk LT (-477 μm^2, $p \leq 0.1$). The proportion of MyHC IIa was higher in the LT of Yheavy LT in comparison to Ylight and Omilk LT (+9.5 and +8.8 percentage points (pp), respectively, $p < 0.05$). On the other hand, the proportion of MyHC IIx tended to be lower (-9.5 and -6.8 pp, respectively, $p \leq 0.1$) and the proportion of MyHC I did not differ between classes. Compared to Omilk LT, the LDH activity in Yheavy LT tended to be lower (-73 $\mu mol/min/g$, $p \leq 0.1$). Finally, the b* index was higher in Yheavy LT in comparison to Ylight and Omilk LT (+1.13 and +0.63, respectively, $p < 0.05$).

Table 3. Effects of the cull cow clusters on *Longissimus thoracis* characteristics (adjusted mean from anova ± standard error of the mean).

	Ylight	Omilk	Yheavy	*p*
Nb	51	32	13	
ICDH [1]	1.06 ± 0.05	1.00 ± 0.06	1.12 ± 0.07	NS
LDH [1]	709 [a] ± 15.2	691 [a] ± 17.6	764 [b] ± 32.0	†
Fiber size, μm^2	2878 [a,b] ± 94.4	3142 [b] ± 126.9	2665 [a] ± 102.3	†
IIx, %	15.3 [b] ± 2.27	12.3 [a,b] ± 2.02	5.8 [a] ± 3.10	†
IIa, %	54.3 [a] ± 1.85	55.0 [a] ± 2.26	63.8 [b] ± 3.15	*
I, %	30.4 ± 0.99	32.6 ± 1.66	30.4 ± 1.01	NS
Total collagen [2]	3.09 ± 0.06	3.09 ± 0.08	2.95 ± 0.12	NS
Insoluble collagen [2]	2.43 ± 0.05	2.47 ± 0.06	2.26 ± 0.08	NS
Shear force, N/cm^2	45.6 ± 1.92	45.3 ± 1.69	42.2 ± 2.28	NS
IMF[3], g/100g DM	15.2 [a] ± 0.87	20.0 [b] ± 1.14	15.8 [a] ± 1.11	**
Fat-to-muscle ratio	29.2 [a] ± 1.23	36.3 [b] ± 2.31	29.8 [a] ± 1.38	**
L*	39.5 ± 0.31	39.9 ± 0.37	39.2 ± 0.83	NS
a*	8.61 ± 0.16	8.79 ± 0.19	9.19 ± 0.33	NS
b*	7.07 [a] ± 0.16	7.57 [a] ± 0.27	8.20 [b] ± 0.29	*

**: $p < 0.01$; *: $p < 0.05$; †: $p \leq 0.1$; NS: nonsignificant; [1] isocitrate dehydrogenase (ICDH) and lactate dehydrogenase (LDH) activities, in $\mu mole/min/g$; [2] in μg OH-proline/mg dry matter; [3] intramuscular fat content; [a,b,c] values within a row with different superscripts differ significantly at $p < 0.05$.

3.2.2. Rectus Abdominis

There was no difference between clusters on ICDH and LDH activities, fiber size, total and insoluble collagen contents, shear force, and color (Table 4). In comparison to Ylight RA, Yheavy RA also had lower proportions of MyHC IIx (−7.2 pp, $p < 0.05$), but in favor of the proportion of MyHC I (+6.1 pp, $p < 0.05$). The Yheavy RA tended to have higher IMF content than Ylight RA (+ 4.8 g/100 g, $p < 0.1$). Sensory scores were always below 5/10. The panelists were initially trained for LT analysis, not RA, probably explaining why scores were so low. Initial tenderness tended to be lower in Ylight RA, in comparison to Yheavy and Omilk RA. Overall appreciation tended to be higher in Omilk RA (Table 4).

Table 4. Effects of the cull cows clusters on *Rectus abdominis* characteristics (adjusted mean from anova ± standard error of the mean).

	Ylight	Omilk	Yheavy	p
Nb	51	32	13	
ICDH [1]	1.25 ± 0.5	1.37 ± 0.05	1.35 ± 0.13	NS
LDH [1]	645 ± 15.0	612 ± 13.9	669 ± 37.6	NS
Fiber size, μm^2	3516 ± 118.0	3664 ± 170.3	3743 ± 232.9	NS
IIx, %	27.4 [b] ± 1.68	21.3 [a] ± 1.78	20.2 [a] ± 3.89	*
IIa, %	36.2 ± 1.18	40.0 ± 1.54	37.2 ± 2.84	NS
I, %	36.4 [a] ± 1.03	38.7 [a,b] ± 1.25	42.5 [b] ± 1.78	*
Total collagen [2]	3.64 ± 0.10	3.50 ± 0.08	3.51 ± 0.17	NS
Insoluble collagen [2]	2.84 ± 0.08	2.75 ± 0.07	2.70 ± 0.13	NS
Shear force, N/cm^2	49.8 ± 1.21	50.6 ± 1.69	49.2 ± 2.61	NS
IMF[3], g/100g DM	17.3 [a] ± 1.07	19.6 [a,b] ± 1.15	22.1 [b] ± 2.59	†
L*	38.9 ± 0.31	38.7 ± 0.35	38.4 ± 0.67	NS
a*	5.77 ± 0.15	5.88 ± 0.17	5.63 ± 0.23	NS
b*	4.50 ± 0.13	4.78 ± 0.18	4.44 ± 0.31	NS
Initial tenderness, /10	4.7 [a] ± 0.06	4.8 [b] ± 0.08	4.9 [b] ± 0.08	†
Initial juiciness, /10	4.2 ± 0.06	4.2 ± 0.07	4.2 ± 0.07	NS
Global flavour, /10	4.7 ± 0.04	4.8 ± 0.05	4.8 ± 0.09	NS
Persistance, /10	4.5 ± 0.04	4.5 ± 0.05	4.5 ± 0.06	NS
Overall appreciation, /10	2.9 [a] ± 0.08	3.2 [b] ± 0.11	2.9 [a] ± 0.14	†

*: $p < 0.05$; †: $p \leq 0.1$; NS: nonsignificant; [1] isocitrate dehydrogenase (ICDH) and lactate dehydrogenase (LDH) activities, in µmole/min/g; [2] in µg OH-proline/mg dry matter; [3] intramuscular fat content; [a,b,c] values within a row with different superscripts differ significantly at $p < 0.05$.

3.3. Effect of Finishing Practices Clusters on Muscles and Meat Quality Traits

3.3.1. Longissimus Thoracis

We observed no effect of finishing practices on ICDH and LDH activities, fiber size, total and insoluble collagen contents, shear force, and IMF content (Table 5). The fat-to-muscle ratio was higher in ConcF LT in comparison to LongF, HayF, and PastF LT (+7.1, +8.8 and +7.2, respectively; $p < 0.05$). The proportion of MyHC IIa in PastF LT, in comparison to LongF, HayF, and ConcF LT, were higher (+9.0 pp, +6.2 pp and +10.4 pp, respectively, $p < 0.01$) and the b* index tended to be higher (+1.2, +0.8, and +0.8 for b* index, respectively, $p \leq 0.1$). Finally, the a* index was lower in HayF LT in comparison to LongF, ConcF, and PastF LT (−0.8, −0.9, and −0.8, respectively; $p < 0.01$).

Table 5. Effects of the finishing practices clusters on *Longissimus thoracis* characteristics (adjusted mean from anova ± standard error of the mean).

	LongF	**HayF**	**ConcF**	**PastF**	*p*
Nb	17	41	18	21	
ICDH [1]	1.06 ± 0.12	1.06 ± 0.05	1.03 ± 0.06	1.02 ± 0.04	NS
LDH [1]	710 ± 27.5	698 ± 18.5	705 ± 22.8	706 ± 22.9	NS
Fiber size, μm^2	2933 ± 184.5	2808 ± 94.6	2986 ± 130.7	2991 ± 160.9	NS
IIx, %	14.5 ± 2.96	12.8 ± 2.60	14.6 ± 2.97	7.0 ± 2.53	NS
IIa, %	53.1 [a] ± 2.54	55.9 [a] ± 1.97	51.7 [a] ± 2.63	62.1 [b] ± 3.23	*
I, %	32.4 ± 1.78	31.2 ± 1.05	33.6 ± 1.24	30.9 ± 2.24	NS
Total collagen [2]	3.06 ± 0.11	3.09 ± 0.06	3.12 ± 0.11	3.14 ± 0.09	NS
Insoluble collagen [2]	2.43 ± 0.09	2.45 ± 0.05	2.47 ± 0.08	2.47 ± 0.07	NS
Shear force, N/cm^2	49.1 ± 3.61	44.2 ± 1.60	43.8 ± 2.91	42.9 ± 1.64	NS
IMF[3], g/100g DM	18.2 ± 2.03	16.1 ± 0.90	15.8 ± 1.15	15.7 ± 1.17	NS
Fat-to-muscle ratio	30.9 [a] ± 2.40	29.2 [a] ± 1.21	38.0 [b] ± 3.62	30.8 [a] ± 1.41	*
L*	40.0 ± 0.55	39.9 ± 0.41	39.6 ± 0.34	39.6 ± 0.53	NS
a*	9.2 [b] ± 0.35	8.4 [a] ± 0.16	9.3 [b] ± 0.26	9.2 [b] ± 0.21	**
b*	7.0 [a] ± 0.36	7.4 [a] ± 0.23	7.4 [a] ± 0.26	8.2 [b] ± 0.33	†

**: $p < 0.01$; *: $p < 0.05$; †: $p \leq 0.1$; NS: nonsignificant; [1] isocitrate dehydrogenase (ICDH) and lactate dehydrogenase (LDH) activities, in μmole/min/g; [2] in μg OH-proline/mg dry matter; [3] intramuscular fat content; [a,b,c] values within a row with different superscripts differ significantly at $p < 0.05$.

3.3.2. Rectus Abdominis

Finishing practices have only influenced the proportion of MyHC I in RA muscle. LongF RA were characterized by a lower proportion of MyHC I in comparison to HayF, ConcF, and PastF RA (–4.6 pp, –5.9 pp and –5.3 pp, respectively, $p < 0.05$) (Table 6). Sensory traits were weakly modified by finishing practices. Initial tenderness was lower in HayF RA in comparison to ConcF RA (–0.4 pp, $p < 0.05$). And juiciness was higher in LongF RA in comparison to HayF, ConcF, and PastF RA (+0.3 pp, +0.3 pp, and +0.3 pp, respectively, $p < 0.01$) (Table 6).

Table 6. Effects of the finishing practices clusters on *Rectus abdominis* characteristics (adjusted mean from anova ± standard error of the mean).

	LongF	**HayF**	**ConcF**	**PastF**	*p*
Nb	17	41	18	21	
ICDH [1]	1.25 ± 0.09	1.29 ± 0.05	1.42 ± 1.00	1.22 ± 0.06	NS
LDH [1]	633 ± 17.2	656 ± 17.8	616 ± 20.1	648 ± 25.4	NS
Fiber size, μm^2	3855 ± 206.6	3466 ± 104.7	3559 ± 195.6	3636 ± 214.9	NS
IIx, %	25.2 ± 2.32	25.5 ± 2.04	24.8 ± 2.34	23.2 ± 2.58	NS
IIa, %	40.8 ± 1.91	36.9 ± 1.36	35.4 ± 2.02	37.6 ± 2.02	NS
I, %	33.9 [a] ± 1.46	37.5 [b] ± 1.03	39.8 [b] ± 1.79	39.2 [b] ± 1.52	*
Total collagen [2]	3.51 ± 0.11	3.67 ± 0.11	3.33 ± 0.09	3.57 ± 0.16	NS
Insoluble collagen [2]	2.73 ± 0.11	2.87 ± 0.09	2.55 ± 0.07	2.81 ± 0.13	NS
Shear force, N/cm^2	55.9 ± 4.69	51.4 ± 1.53	48.7 ± 1.45	50.7 ± 2.34	NS
IMF[3], g/100g DM	19.8 ± 1.62	17.5 ± 1.07	17.7 ± 1.61	19.8 ± 2.02	NS
L*	39.8 ± 0.36	39.0 ± 0.38	38.8 ± 0.50	38.3 ± 0.35	NS
a*	6.1 ± 0.27	5.5 ± 0.16	5.8 ± 0.26	5.9 ± 0.23	NS
b*	4.5 ± 0.24	4.4 ± 0.16	4.5 ± 0.27	4.6 ± 0.22	NS
Initial tenderness, /10	4.6 [a,b] ± 0.11	4.5 [a] ± 0.08	4.9 [b] ± 0.07	4.8 [a,b] ± 0.09	*
Initial juiciness, /10	4.5 [b] ± 0.10	4.2 [a] ± 0.05	4.2 [a] ± 0.10	4.2 [a] ± 0.08	**
Global flavour, /10	4.9 ± 0.06	4.7 ± 0.04	4.7 ± 0.05	4.7 ± 0.07	NS
Persistance, /10	4.6 ± 0.06	4.5 ± 0.05	4.4 ± 0.05	4.5 ± 0.07	NS
Overall appreciation, /10	3.1 ± 0.14	2.9 ± 0.08	3.1 ± 0.11	2.9 ± 0.12	NS

**: $p < 0.01$; *: $p < 0.05$; †: $p \leq 0.1$; NS: nonsignificant; [1] isocitrate dehydrogenase (ICDH) and lactate dehydrogenase (LDH) activities, in μmole/min/g; [2] in μg OH-proline/mg dry matter; [3] intramuscular fat content; [a,b,c] values within a row with different superscripts differ significantly at $p < 0.05$.

3.4. Effect of the Interaction Between Cull Cow Clusters and Finishing Practices Clusters on Muscles and Meat Quality Traits

The effects of the interaction were assessed by considering four finishing practices modalities and only two of the cull cow clusters (Ylight and Omilk). The number of Yheavy cows was too low to include this modality in the statistical analysis. We only observed two trends of interaction effects on IMF content and fat-to-muscle ratio in LT ($p \leq 0.1$). Finishing practices had no effect on IMF content and fat-to-muscle ratio in LT in Ylight LT. However, in Omilk LT, the IMF content tended to be high when cows were finished according to LongF practices (IMF content = 25.9%, fat-to-muscle ratio = 33.6%); conversely, the fat-to-muscle ratio was high when cows were finished according to ConcF practices (IMF content = 17.3%, fat-to-muscle ratio = 44.2%). These results might suggest that the partition of lipid deposition between inter- and intramuscular fat depend on feeding practices in Omilk cull cows.

4. Discussion

In comparison to LT muscle, RA muscle has significantly larger fibers, higher proportion of MyHC I at the expense of MyHC IIa, higher ICDH activity and lower LDH activity, higher shear force, higher collagen and IMF contents, and darker but less red and yellow meat. These differences match the results of [16] and will lead to strong differences in meat quality traits between LT and RA muscles, especially tenderness.

4.1. Effect of Cull Cow Clusters on Muscle Characteristics

In our study, the differences in meat quality traits among cull cow clusters are stronger for LT than RA. This result is consistent with Soulat [28] but not Oury et al. [16] who observed stronger differences for RA than LT of Charolais heifers in relation to several finishing practices.

Omilk cows are older, heavier, and have a higher parity and suckling ability than Ylight and Yheavy cows. These cows are kept in the herd for their good maternal skills (milk production to suckle the calves). Moreover, Jurie et al. [5] found no relation between age and carcass weight of suckling cull cows, whereas Liénard et al. [29] observed an increase of carcass weight of suckling cull cows until five to six years of age and then a 10 kg decrease when cows are slaughtered at eight or nine years of age. In our study, at least we can assume that the carcass weight and conformation of Omilk and Yheavy cows are similar and significantly higher than those of Ylight cows.

Omilk LT are characterized by larger fibers, and higher IMF content and fat-to-muscle ratio than Yheavy and Ylight LT. The latter result also means that the proportion of adipose tissue in the carcass is higher for Omilk cows in comparison to Yheavy and Ylight cows (19.4% versus 18.3 and 17.4%, respectively, $p = 0.003$). These differences could be related to the age and the suckling ability of the Omilk cows. Jurie et al. [5] have observed no difference in LT composition (IMF content, MyHC, fiber size and proportion of fat tissue in the carcass between groups of suckling cull cows differing in age (4–5, 6–7 and 8–9 years). In this study, the body composition score at slaughter was similar for all cull cows and could explain why no difference was observed. Moreover, between 4 and 9 years of age (as it was observed for RdP cows in our study), the biological mechanisms involved in fiber size and metabolism modifications do not exist anymore (growth of young cows [5]) or (aging of old cows [5]). The suckling ability could also explain the differences observed between Omilk cows and the others. It is well known that dairy animals (dairy breeds or dairy line within a suckling breed) tend to deposit more fat [5] and to have higher proportions of MyHC I in their muscles [30] when aging. For these reasons, Omilk LT composition would probably be intermediate between that of dairy and suckling cull cows.

Yheavy LT, in comparison to Omilk and Ylight LT, have lower MyHC IIx proportion and higher MyHC IIa proportion and tend to have higher glycolytic activity. Moreover, Yheavy RA, in comparison to Omilk and Ylight RA, have higher MyHC I proportion, and lower MyHC IIx proportion, without any difference in enzyme activities. These cows probably belong to an intra-breed muscular line as farmers are known to select animals on maternal abilities (RDP was a dual-purpose breed until the 90s) or muscular development abilities (some bulls have the double-muscle gene). The animals are leaner, the suckling ability is low, and the LT composition is closer to that of suckling cull cows (Limousine, Charolaise, Belgium Blue), with a higher proportion of IIa and higher glycolytic activities [30,31].

For the two muscles, we observed no difference in total and insoluble collagen contents between cull cow clusters, and in accordance to these results, no difference as well in shear force values. Our results are consistent with previous studies that have shown no relation between age and total and insoluble collagen contents in LT muscle of suckling cull cows [3–5]. Gerhardy [12] has also shown similar total collagen content between dairy cull cows (62 months at slaughter on average) and dairy heifers (23 months at slaughter on average). However, they observed a higher insoluble collagen content in the LT of suckling cows, in accordance with Purslow [32]. The existence of a long finishing period (at least 60 days) in our study could explain why there is no difference in insoluble collagen contents between cull cow clusters. During finishing, the collagen frame (perimysium) is remodeled. As the collagen turnover is long (half-life around 45 days), the effect of the remodeling fades after several months [32]. Thus, as finishing lasted more than 45 days in our study, we could assume that the remodeling of the collagen frame has been similar causing similar insoluble collagen contents between cows.

4.2. Effect of Finishing Practices Clusters on Muscle Characteristics

In our study, the differences in meat quality traits among and finishing practices clusters are weak for the two studied muscles. The main differences observed, mainly in LT muscle, concerned MyHC fibers proportions and color indices. This shows that RA is less reactive than LT to finishing practices as also stated in the literature [16].

MyHC IIx and IIa proportions are lower in the PastF LT, and the a* and the b* color indices are higher in the PastF LT, as compared with other LT. Muscle pH, physical activity, IMF content and color, and age are the main factors explaining meat color differences [2]. In our study, physical activity, diet, and IMF color are the only factors that could explain our results, other factors being similar between clusters [14]. Priolo et al. [33] have shown that the L* index is lower (related to a darker meat) when cattle graze during finishing. The physical activity and grazing (in comparison to hay diet), by modifying the fiber metabolic activity (mainly oxidative) [34], would lead to higher myoglobin content in the muscle and thereby the higher L* index [35]. However, at least 100 days of grazing would be necessary to observe significant effects [33] and probably would be adequate to increase the a* index. On the other hand, Kerth et al. [8] have shown that the b* index of subcutaneous fat is higher when cattle graze during finishing. Assuming that the IMF b* index is also higher, this could explain the higher b* index in PastF LT [36].

In ConcF LT, the fat-to-muscle ratio was higher without any effect on IMF content, whereas, in LongF LT, the IMF content was higher without any effect on the fat-to-muscle ratio. Even if the body condition is not known at the beginning of the finishing period, it seems that the finishing duration and the amount of energy concentrates (per day) influence the allotment of adipose tissue deposit during finishing [11]. A short finishing period with a high proportion of concentrate in the diet, as observed in the ConcF cluster, would lead to a deposit of adipose tissue in either the internal or intermuscular fat [3]. As shown by Robelin et al. [1], first the fat deposit occurs in the subcutaneous and internal fats, then in the intermuscular fat, and finally in the IMF. A short finishing period with a low forage-to-concentrate ratio diet promotes fat deposition as subcutaneous and internal fat, whereas a long finishing period with a high forage-to-concentrate ratio diet deposits fat in the IMF. Thereby, these feeding practices influence the IMF content and the fat composition of the carcass [9,37].

Shear force was not different among finishing practices clusters. This is consistent with the extensive literature that shows no or very few effects of forage type and amount of concentrate during finishing on shear force [3,36,38]. Nevertheless, the tenderness assessed by sensory analysis tended to be higher in ConcF and PastF RA, in comparison to HayF and LongF RA. Even if the main factors influencing tenderness (total and insoluble collagen contents, IMF content, carcass conformation, and fibers) were similar between clusters, the finishing practices could explain this slight effect. Vestergaard et al. [35] have shown that a low forage-to-concentrate ratio in the finishing diet, as observed in ConcF, could improve meat tenderness. Jurie et al. [34] have shown that a finishing diet based on grazing, as observed in PastF, could improve meat tenderness by modifying the metabolic activity and fiber types in the muscles.

In our study, cull cows were selected and collected in farms. Their characteristics before the finishing period (body condition score) and their history (growth, reproductive performance, diets, fat lipomobilization, sanitary events, ...) were highly variable. Although this information was collected (survey), its variability was too high to take it into consideration in our statistical analysis (clustering). This could have interfered in our analyses. Indeed, Apple et al. [39] have observed a linear relationship between the body condition score and the fat composition in the carcass of cull cows. Furthermore, it is well known in younger animals (heifers and steers) that the characteristics of the growth period (compensatory growth for instance) impact the effects of the finishing periods on the carcass composition and the meat quality (subcutaneous fat, IMF content, and tenderness) [40,41]. For that reason, often the growth period and the body condition scores are considered when animals are allocated to the experimental treatments in trials dealing with the effect of finishing practices on meat quality of heifers and steers. Thus, it could be interesting to perform another experiment taking into account those factors in order to study the effect of the interaction between the history of the cow before culling and the finishing practices on carcass and meat quality traits.

By exploring the differences between muscular and dairy lines within a local breed, our study gives new insights into the effect of animal type on meat quality. Our study is original because it considered the interaction between animal type and finishing practices at farm scale on the meat quality of cull beef cows. Finishing practices have less effect than animal type on RA and LT meat properties. Their effects also differed according to muscle type (RA or LT) and cull cows types on muscle composition. The effects observed on meat quality are directly related to farmers' practices and provide new advice and modifications in culled cows finishing practices to improve meat quality. As we only performed sensory analyses on RA muscle, we can only suppose that those differences might have effects on sensory attributes among muscles. It could be interesting to study other muscles to assess whether the effects of the animal type and finishing practices are similar regardless of the muscles considered. Moreover, to enhance characterization of animal types, it could be interesting to create animal clusters including genetic indices (e.g., suckling ability index) as a replacement for qualitative information from farmer surveys. It could also be interesting to increase the number of animals using a multifactorial approach, to study the effect of the overall farm management of cows on carcass and meat quality traits. This means studying the interaction between the practices before culling (sanitary problems, compensatory growth, feeding system, etc.), the animal characteristics at the beginning of the finishing period (e.g., body condition score), the animal type (as observed in our study) and the finishing practices. This interaction could partially explain the lack of effects observed among clusters (IMF content for instance among finishing clusters).

Author Contributions: Conceptualization, S.C. and G.L.B.; methodology, S.C., G.L.B., D.M., and B.P.; validation, S.C. and B.P.; formal analysis, S.C. and G.L.B.; investigation, S.C. and G.L.B.; resources, S.C., G.L.B., and B.P.; data curation, S.C., G.L.B., and B.P.; writing—original draft preparation, S.C.; writing—review and editing, G.L.B. and B.P.; visualization, S.C.; supervision, S.C.; project administration, S.C.; funding acquisition, S.C.

Funding: This research was funded by the INRA and Pays de la Loire Region (PSDR grant).

Acknowledgments: Authors thank the PDO Maine-Anjou association, ADEMA and the Elivia slaughterhouse for their help in the surveys, the farm selection and the sample collection. The authors remember Didier Micol who died during the writing of this paper.

Conflicts of Interest: The authors declare no conflict of interest. The funders had no role in the design of the study; in the collection, analyses, or interpretation of data; in the writing of the manuscript; or in the decision to publish the results.

References

1. Robelin, J.; Agabriel, J.; Malterre, C.; Bonnemaire, J. Changes in body composition of mature dry cows of Holstein, Limousin and Charolais breeds during fattening. I. Skeleton, muscles, fatty tissues and offal. *Livest. Prod. Sci.* **1990**, *25*, 199–215. [CrossRef]
2. Geay, Y.; Bauchart, D.; Hocquette, J.F.; Culioli, J. Valeur diététique et qualités sensorielles des viandes de ruminants. Incidence de l'alimentation des animaux. *INRA Prod. Anim.* **2002**, *15*, 37–52.
3. Oury, M.P.; Agabriel, J.; Agabriel, C.; Micol, D.; Picard, B.; Blanquet, J.; Labouré, H.; Roux, M.; Dumont, R. Relationship between rearing practices and eating quality traits of the muscle rectus abdominis of Charolais heifers. *Livest. Sci.* **2007**, *111*, 242–254. [CrossRef]
4. Dransfield, E.; Martin, J.F.; Bauchart, D.; Abouelkaram, S.; Lepetit, J.; Culioli, J.; Jurie, C.; Picard, B. Meat quality and composition of three muscles from French cull cows and young bulls. *Anim. Sci.* **2003**, *76*, 387–399. [CrossRef]
5. Jurie, C.; Martin, J.F.; Listrat, A.; Jailler, R.; Culioli, J.; Picard, B. Carcass and muscle characteristics of beef cull cows between 4 and 9 years of age. *Anim. Sci.* **2006**, *82*, 415–421. [CrossRef]
6. Jurie, C.; Picard, B.; Hocquette, J.F.; Dransfield, E.; Micol, D.; Listrat, A. Muscle and meat quality characteristics of Holstein and Salers cull cows. *Meat Sci.* **2007**, *77*, 459–466. [CrossRef] [PubMed]
7. Duarte, M.S.; Gionbelli, M.P.; Paulino, P.V.R.; Serão, N.V.; Silva, L.H.; Mezzomo, R.; Dodson, M.V.; Du, M.; Busboom, J.R.; Guimarães, S.E.; et al. Effects of pregnancy and feeding level on carcass and meat quality traits of Nellore cows. *Meat Sci.* **2013**, *94*, 139–144. [CrossRef]
8. Kerth, C.R.; Braden, K.W.; Cox, R.; Kerth, L.K.; Rankins, D.L., Jr. Carcass, sensory, fat color, and consumer acceptance characteristics of Angus-cross steers finished on ryegrass (Lolium multiflorum) forage or on a high-concentrate diet. *Meat Sci.* **2006**, *75*, 324–331. [CrossRef] [PubMed]
9. Vestergaard, M.; Madsen, N.T.; Bligaard, H.B.; Bredahl, L.; Rasmussen, P.T.; Andersen, H.R. Consequences of two or four months of finishing feeding of culled dry dairy cows on carcass characteristics and technological and sensory meat quality. *Meat Sci.* **2007**, *76*, 635–643. [CrossRef]
10. Minchin, W.; Buckley, F.; Kenny, D.A.; Monahan, F.J.; Shalloo, L.; O'Donovan, M. An evaluation of over-wintering feeding strategies prior to finishing at pasture for cull dairy cows on live animal performance, carcass and meat quality characteristics. *Meat Sci.* **2010**, *85*, 385–393. [CrossRef]
11. Soulat, J.; Picard, B.; Léger, S.; Monteils, V. Prediction of beef carcass and meat traits from rearing factors in young bulls and cull cows. *J. Anim. Sci.* **2016**, *94*, 1712–1726. [CrossRef] [PubMed]
12. Gerhardy, H. Quality of beef from commercial fattening systems in northern Germany. *Meat Sci.* **1995**, *40*, 103–120. [CrossRef]
13. Soulat, J.; Léger, S.; Picard, B.; Monteils, V. Improving beef sensory quality through breeding practices management. In Proceedings of the 61stInternational Congress of Meat Science and Technology, Clermont-Ferrand, France, 23–28 August 2015.
14. Gagaoua, M.; Monteils, V.; Couvreur, S.; Bicard, P. Identification of Biomarkers Associated with the Rearing Practices, Carcass Characteristics, and Beef Quality: An Integrative Approach. *J. Agri. Food Chem.* **2017**, *65*, 8264–8278. [CrossRef] [PubMed]
15. Schmitt, T.; Laurent, C.; Lautrou, Y.; Couvreur, S. Impact of production specifications on the management of beef cattle systems: Controlled Maine-Anjou Origin Appellation (AOC Maine-Anjou). *Rencontres Rech. Rumin.* **2008**, *15*, 151–154.
16. Oury, M.P.; Dumont, R.; Jurie, C.; Hocquette, J.F.; Picard, B. Specific fibre composition and metabolism of the rectus abdominis muscle of bovine Charolais cattle. *BMC Biochem.* **2010**, *11*, 12. [CrossRef]
17. Robelin, J.; Geay, Y. Estimation de la composition de la carcasse des taurillons à partir de la composition de la 6ème côte. *Bull. Tech. CRZV Theix INRA* **1975**, *22*, 41–43.
18. Meunier, B.; Picard, B.; Astruc, T.; Labas, R. Development of image analysis tool for the classification of muscle fibre type using immunohistochemical staining stochem. *Cell Biol.* **2010**, *134*, 307–317.
19. Picard, B.; Barboiron, C.; Chadeyron, D.; Jurie, C. Protocol for high-resolution electrophoresis separation of myosin heavy chain isoforms in bovine skeletal muscle. *Electrophoresis* **2011**, *32*, 1804–1806. [CrossRef]
20. Bradford, M.M. A rapid and sensitive method for the quantitation of microgram quantities of protein utilizing the principle of protein-dye binding. *Anal. Biochem.* **1976**, *72*, 248–254. [CrossRef]

21. Ansay, M. Individualité musculaire chez le bovin: Étude de l'équipement enzymatique de quelques muscles. *Ann. Biol. Anim. Bioch. Biophys.* **1974**, *14*, 471–486. [CrossRef]

22. Briand, M.; Talmant, A.; Briand, Y.; Monin, G.; Durand, R. Metabolic types of muscle in the sheep: I. Myosin ATPase, glycolytic and mitochondrial enzyme activities. *Eur. J. Appl. Physiol.* **1981**, *46*, 347–358. [CrossRef]

23. Bergman, I.; Loxley, R. Two improved and simplified methods for the spectrophotometric determination of hydroxyproline. *Anal. Chem.* **1963**, *35*, 1961–1965. [CrossRef]

24. Bonnet, M.; Kopp, J. Dosage du collagène dans les tissus conjonctifs, la viande et les produits carnés. *Cah. Tech. INRA* **1984**, *5*, 19–30.

25. Bonnet, M.; Kopp, J. Préparation des échantillons pour le dosage et la caractérisation qualitative du collagène musculaire. *Viandes Prod. Carnés* **1992**, *13*, 87–91.

26. Honikel, K.O. Reference methods for the assessment of physical characteristics of meat. *Meat Sci.* **1998**, *49*, 447–457. [CrossRef]

27. Hunt, M.C.; Kropf, D.; Morgan, B. Color measurement of meat and meat products. *Recipr. Meat Conf. Proc.* **1993**, *46*, 59–60.

28. Soulat, J. Pilotage Conjoint, à Partir des Pratiques D'élevage, des Caractéristiques de la Carcasse et de la Viande des Bovins. Ph.D. Thesis, Clermont Auvergne University, Clermont-Ferrand, France, 2017.

29. Liénard, G.; Lherm, M.; Pizaine, M.C.; Le Maréchal, J.Y.; Boussange, B.; Barlet, D.; Esteve, P.; Bouchy, R. Productivité de trois races bovines françaises, Limousine, Charolaise et Salers. *INRA Prod. Anim.* **2002**, *15*, 293–312.

30. Schreurs, N.M.; Garcia, F.; Jurie, C.; Agabriel, J.; Micol, D.; Bauchart, D.; Listrat, A.; Picard, B. Meta-analysis of the effect of animal maturity on muscle characteristics in different muscles, breeds, and sexes of cattle. *J. Anim. Sci.* **2008**, *86*, 2872–2887. [CrossRef]

31. Chriki, S.; Gardner, G.E.; Jurie, C.; Picard, B.; Micol, D.; Brun, J.P.; Journaux, L.; Hocquette, J.F. Cluster analysis application identifies muscle characteristics of importance for beef tenderness. *BMC Biochem.* **2012**, *13*, 29. [CrossRef]

32. Purslow, P.P. Intramuscular connective tissue and its role in meat quality. *Meat Sci.* **2005**, *70*, 435–447. [CrossRef]

33. Priolo, A.; Micol, D.; Agabriel, J. Effects of grass feeding systems on ruminant meat color and flavour. A review. *Anim. Res.* **2001**, *50*, 185–200. [CrossRef]

34. Jurie, C.; Ortigues-Marty, I.; Picard, B.; Micol, D.; Hocquette, J.F. The separate effects of the nature of diet and grazing mobility on metabolic potential of muscles from Charolais steers. *Livest. Sci.* **2006**, *104*, 182–192. [CrossRef]

35. Vestergaard, M.; Oksbjerg, N.; Henckel, P. Influence of feeding intensity, grazing and finishing feeding on muscle fibre characteristics and meat colour of semitendinosus, longissimus dorsi and supraspinatus muscles of young bulls. *Meat Sci.* **2000**, *54*, 177–185. [CrossRef]

36. Pordomingo, A.J.; Grigioni, G.; Carduza, F.; Volpi Lagreca, G. Effect of feeding treatment during the backgrounding phase of beef production from pasture on: I. Animal performance, carcass and meat quality. *Meat Sci.* **2012**, *90*, 939–946. [CrossRef] [PubMed]

37. Bunmee, T.; Jaturasitha, S.; Wicke, M.; Kreuzer, M. Effects of different finishing feeding strategies for culled cows on lipid and organoleptic characteristics of the meat. Songklanakarin. *J. Sci. Technol.* **2015**, *37*, 109–117.

38. Duckett, S.K.; Neel, J.P.S.; Lewis, R.M.; Fontenot, J.P.; Clapham, W.M. Effects of forage species or concentrate finishing on animal performance, carcass and meat quality. *J. Anim. Sci.* **2013**, *91*, 1454–1467. [CrossRef] [PubMed]

39. Apple, J.K.; Davis, J.C.; Stephenson, J.; Hankins, J.E.; Davis, J.R.; Beaty, S.L. Influence of body condition score on carcass characteristics and subprimal yield from cull beef cows. *J. Anim. Sci.* **1999**, *77*, 2660–2669. [CrossRef]

40. Guerrero, A.; Sanudo, C.; Alberti, P.; Ripoll, G.; Campo, M.M.; Olleta, J.L.; Panea, B.; Khliji, S.; Santolaria, P. Effect of production system before the finishing period on carcass, meat and fat qualities of beef. *Animal* **2013**, *7*, 2063–2072. [CrossRef]
41. Keogh, K.; Waters, S.M.; Kelly, A.K.; Kenny, D.A. Feed restriction and subsequent realimentation in Holstein Friesian bulls: I. Effect on animal performance; muscle, fat, and linear body measurements; and slaughter characteristics. *J. Anim. Sci.* **2015**, *93*, 3578–3589. [CrossRef]

 *foods*

Article

Preliminary Study to Determinate the Effect of the Rearing Managements Applied during Heifers' Whole Life on Carcass and Flank Steak Quality

Julien Soulat [1], Brigitte Picard [1], Stéphanie Léger [2,3], Marie-Pierre Ellies-Oury [1,4] and Valérie Monteils [1,*]

[1] Université Clermont Auvergne, INRA, VetAgro Sup, UMR Herbivores, F-63122 Saint-Genès-Champanelle, France; julien.soulat@live.fr (J.S.); brigitte.picard@inra.fr (B.P.); marie-pierre.ellies@agro-bordeaux.fr (M.-P.E.-O.)

[2] Université de Clermont Auvergne, Université Blaise Pascal, Laboratoire de Mathématiques, BP 10448, F-63000 Clermont-Ferrand, France; Stephanie.Leger@math.univ-bpclermont.fr

[3] CNRS, UMR 6620, Laboratoire de Mathématiques, F-63171 Aubière, France

[4] Bordeaux Science Agro, 1 Cours du Général de Gaulle, CS 40201, F-33175 Gradignan, France

* Correspondence: valerie.monteils@vetagro-sup.fr; Tel.: +330-4-73-98-13-45

Received: 27 August 2018; Accepted: 25 September 2018; Published: 1 October 2018

Abstract: The aim of this study was to investigate the impact of rearing managements applied during a heifers' whole life on the carcass and flank steak (rectus abdominis) meat traits. For this study, rearing managements applied on 96 heifers were identified by conducting surveys in farms. A heifers' whole life was divided into three key periods: Pre-weaning, growth, and fattening. The combination of the rearing factors applied during the heifers' whole life allowed us to characterize several rearing managements. Among them, four have been studied in depth. The main results displayed that the carcass traits were more sensitive to the rearing managements than the flank steak traits. The different managements considered had an impact on the weight, the dressing percentage and the conformation score of the carcass. Whereas, they had no impact on the sensory descriptors, the sheer force and the color of the flank steak. This study showed that the variations observed for carcass and meat traits could not be explained by the variation of only one rearing factor but could be explained by many rearing factors characterizing the rearing management applied. Finally, this study demonstrated that it was possible to improve carcass traits without deteriorating meat traits.

Keywords: pre-weaning period; fattening period; growth period; meat sensory properties; rearing managements; rearing surveys

---

1. Introduction

Beef carcass and meat traits are impacted by the animal type (sex and breed) [1–3]. Rearing managements have also been shown to influence these characteristics, with the fattening period as the main studied period of the animal's life. In the literature, the fattening period was mainly studied using experimental devices by controlling one or two rearing factors. The factors that have been most frequently studied during this period are the slaughter age [2,4–6], the slaughter weight [7,8], the fattening duration [9–11], the fattening diet [12–14], and the fattening management, i.e., a pasture period during the fattening period or the whole fattening in housing [15–18]. The rearing managements applied during the animal's whole life period (i.e., from birth to slaughter, WLP) are a complex combination of many rearing factors to achieve carcass traits expected by the target market and to maximize their value. Very few studies have jointly studied the influence of the rearing managements applied at different periods of the animal's life (growth, fattening or whole life) on carcass and meat traits [19–22]. The aim of the present study was to illustrate the effect of different rearing managements during the animal's whole life period on carcass traits and meat sensory properties. This study concerned the protected geographical indication (PGI) Fleur d'Aubrac. This PGI was chosen in order to work only on one animal type (i.e., crossbreed Charolais × Aubrac heifers) bred exclusively for the meat production with a slaughter age between 26 and 42 months. In France, the meat consumed mainly comes from female cattle. The shortest life duration of the heifers allows for the improvement of the accuracy of the WLP rearing managements collected by the survey. This study was undertaken using fixed slaughter and post-slaughter conditions to limit the potential bias caused by these parameters on the meat quality [23–25].

2. Materials and Methods

2.1. Animals and Rearing Factors Data Collected by Surveys

The 96 crossbreed Charolais × Aubrac heifers used in this study were born in Occitanie (France region) between December 2012 and May 2013. Their slaughters were distributed between February 2015 and June 2016.

To characterize the WLP rearing management, the heifer's life was divided into three key periods (Figure 1). Each key period of the heifer's life was characterized by many quantitative or qualitative rearing factors. The three steps of the survey were distributed over time to allow for the collection of information regarding different rearing factors that were applied during the whole life. The surveys were carried out by interviewing the farmers using questionnaires and establishing batch management practices [26].

In total, 46 rearing factors were used to characterize the three key rearing periods. The rearing factors characterizing the pre-weaning period (PWP, $q = 16$), the growth period (GP, $q = 13$) and the fattening period (FP, $q = 17$) are presented in the Tables 1–3, respectively.

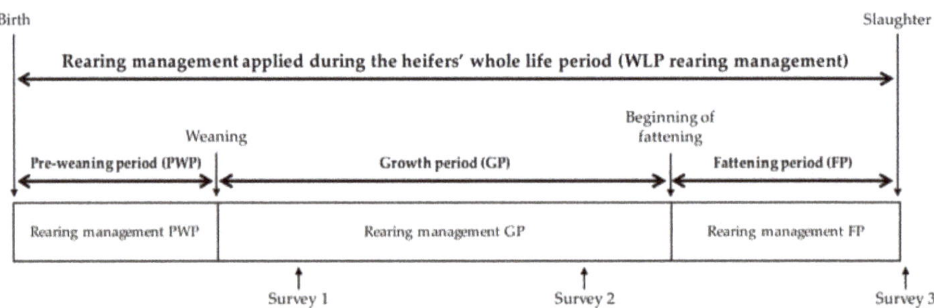

Figure 1. Description of the three key rearing periods during a heifers' whole life (WLP) and distribution of the three farm surveys conducted over time. Rearing managements characterizing the three key periods (pre-weaning, PWP; growth, GP and fattening, FP) are described in Tables 1–3, respectively.

Table 1. Description of the rearing factors characterizing the pre-weaning period (PWP) and the PWP rearing management clusters obtained for this period (PWP-clust).

Rearing Factors	Description of the Modalities	Overall ($n = 96$)			PWP Rearing Management Clusters									p
					PWP-Clust1 ($n = 37$)			PWP-Clust2 ($n = 39$)			PWP-Clust3 ($n = 20$)			
Quantitative Rearing Factors		Mean	SD	SE	Mean	SD	SE	Mean	SD	SE	Mean	SD	SE	
Birth weight (kg)	Calf weight at birth	42	4	0.4	42	4	0.7	41	4	0.6	41	5	1.1	0.80
ADG_PWP (kg/day)	Average daily gain of the calf during PWP	1.0	0.1	1.0	0.9 [b]	0.2	0.03	1.1 [a]	0.1	0.01	1.0 [a]	0.1	0.02	<0.001
Age of the cow (years)	Age of the heifer's mother at the heifer's mother	6.9	2.9	0.3	7.5	3.2	0.5	6.0	2.3	0.37	7.5	3.0	0.7	0.04
Age at the first calving (months)	Age of the heifer's mother at her first calving	34.5	3.0	0.3	33.7	3.3	0.5	34.9	2.8	0.4	35.2	2.4	0.4	0.09
Duration_day_CC (hour/day)	Time spent per day by the calf with her mother during the housing period	11.9	11.3	1.1	22.8 [a]	1.6	0.3	4.1 [b]	8.6	1.4	7.2 [b]	10.4	2.3	<0.001
Tot_duration_CC (days)	Total time spent by the calf with her mother between the birth and the weaning	192.0	67.8	6.9	252.6 [a]	38.2	6.3	144.3 [b]	44	7.0	173.0 [b]	64.5	14.4	<0.001
Conc_duration_PWP (days)	Number of days of offered concentrates in the diet during PWP	81.4	61.5	6.3	53.6 [b]	36.0	5.9	71.3 [b]	41.6	6.7	152.4 [a]	77.5	17.3	<0.001
Conc_CP_PWP (g/kg DM)	Calculated average of the concentrate's crude proteins content across the whole PWP	100	55	5.6	64 [c]	41	6.7	93 [b]	18	2.9	178 [a]	46	10.3	<0.001
Conc_E_PWP (Mcal/kg DM)	Calculated average of the concentrate's energy content across the whole PWP	0.8	0.5	0.05	0.5 [c]	0.3	0.05	0.8 [b]	0.2	0.03	1.5 [a]	0.3	0.07	<0.001
Pasture_duration_PWP (days)	Number of days spend on pasture during PWP	140.3	23.6	2.4	163.0 [a]	14.7	2.4	122.0 [c]	11.2	1.8	133.9 [b]	19.5	4.4	<0.001
PWP_duration (days)	Duration of PWP	253.9	34.9	3.6	257.6	39.0	6.4	251.3	28.4	4.5	252.1	39.4	8.8	0.72
Qualitative Rearing Factors														
Insemination type														0.11
Artificial	Artificial insemination using frozen semen	53.1%			45.9%			61.5%			50%			
Natural	Insemination realized by a bull	46.9%			54.1%			38.5%			50%			
Calving														<0.001
Easy	Natural calving	78.1%			86.5%			66.7%			85%			
Help	Farmer intervention during calving	21.9%			13.5%			33.3%			15%			

Table 1. *Cont.*

Rearing Factors	Description of the Modalities	Overall ($n = 96$)	PWP Rearing Management Clusters			p
			PWP-Clust1 ($n = 37$)	PWP-Clust2 ($n = 39$)	PWP-Clust3 ($n = 20$)	
Bull type						0.004
Bull-3 years	3-year-old bulls belonging to the farmer	34.4%	35.1%	41.0%	20%	
Bull->3 years	Bull older than 3 years belonging to the farmer	27.1%	21.6%	30.8%	30%	
Bull-IA-CE	Artificial insemination from frozen semen for calving ease	12.5%	10.8%	12.8%	15%	
Bull-IA-EM	Artificial insemination from frozen semen for early muscularity	11.4%	10.9%	7.7%	20%	
Bull-IA-CE & EM	Artificial insemination from frozen semen for calving ease and early muscularity	14.6%	21.6%	7.7%	15%	
Conc_housing_PWP						<0.001
Yes	Offered concentrates in housing calve diet during PWP	88.5%	81.1%	100%	80%	
No	No offered concentrates in housing calve diet during PWP	11.5%	18.9%	0%	20%	
Conc_pasture_PWP						<0.001
Yes	Offered concentrates in pasture during PWP	20.8%	0%	0%	100%	
No	No offered concentrates in pasture during PWP	79.2%	100%	100%	0%	

DM: dry matter. *n*: number of heifers. SD: standard deviation. SE: standard error. Values followed by different letters (a–c) are significantly different from each other at $p \leq 0.05$.

Table 2. Description of the rearing factors characterizing the growth period (GP) and the GP rearing management clusters obtained for this period (GP-clust).

Rearing Factors	Description of the Modalities	GP Rearing Management Clusters												p
		Overall (n = 96)			GP-Clust1 (n = 21)			GP-Clust2 (n = 46)			GP-Clust3 (n = 29)			
Quantitative Rearing Factors		Mean	SD	SE	Mean	SD	SE	Mean	SD	SE	Mean	SD	SE	
Age at weaning (months)	Age of heifer at the weaning	8.5	1.1	0.1	8.5	1.4	0.3	8.5	0.7	0.1	8.3	1.3	0.2	0.75
Weaning weight (kg)	Heifer weight at the weaning	296	47	4.8	260 [b]	53	11.6	311 [a]	29	4.3	299 [a]	52	9.7	<0.001
ADG_GP (kg/day)	Average daily gain of the heifer during GP	0.6	0.1	0.01	0.6	0.1	0.02	0.5	0.1	0.01	0.7	0.1	0.01	0.60
Conc_duration_GP (days)	Number of days of offered concentrates in the diet during the GP	206.4	118.1	12.0	133.0 [b]	0.0	0.0	150.9 [b]	80.4	11.8	347.6 [a]	84.7	15.7	<0.001
Conc_quanti_GP (kg)	Total concentrate quantity intake during GP	293.4	249.5	25.5	166.3 [b]	0.0	0.0	140.0 [b]	95.2	14.0	628.7 [a]	190.5	35.4	<0.001
Conc_CP_GP (g/kg DM)	Calculated average of the concentrate's crude proteins content across the whole GP	99	59	6.0	44 [c]	0.0	0.0	95 [b]	56	8.3	145 [a]	48	8.9	<0.001
Conc_E_GP (Mcal/kg DM)	Calculated average of the concentrate's energy content across the whole GP	0.9	0.5	0.05	0.5 [b]	0.0	0.0	0.6 [b]	0.3	0.04	1.5 [a]	0.4	0.07	<0.001
Pasture_duration_GP (days)	Number of days spend on pasture during GP	272.5	51.1	5.2	349.0 [a]	0.0	0.0	241.3 [c]	24.8	3.6	266.4 [b]	33.2	6.2	<0.001
Qualitative Rearing Factors														
Hay_GP (%)	Across the whole GP, the calculated average percentage of hay in the housing diet													<0.001
>80%		23.9%			0%			50%			0%			
(40%; 80%)		21.9%			0%			32.6%			20.7%			
(20%; 40%)		37.5%			100%			17.4%			24.1%			
<20%		16.7%			0%			0%			55.2%			
Grass_silage_GP (%)	Across the whole GP, the calculated average percentage of grass silage in the housing diet													<0.001
>50%		29.2%			100%			0%			24.1%			
<50%		15.6%			0%			13.0%			31.0%			
0%		55.2%			0%			87.0%			44.9%			

Table 2. *Cont.*

Rearing Factors	Description of the Modalities	GP Rearing Management Clusters				p
		Overall ($n = 96$)	GP-Clust1 ($n = 21$)	GP-Clust2 ($n = 46$)	GP-Clust3 ($n = 29$)	
Wrapped_haylage_GP (%)	Across the whole GP, the calculated average percentage of wrapped haylage in the housing diet					<0.001
>60%		15.6%	0%	4.4%	44.8%	
(40%; 60%)		15.6%	0%	32.6%	0%	
<40%		16.7%	0%	13.0%	34.5%	
0%		52.1%	100%	50%	20.7%	
GP_duration (days)	The duration of GP					<0.001
>500 days		76.0%	100%	50%	100%	
<500 days		24.0%	0%	50%	0%	
Nature of pasture						<0.001
Grass	During above 75% of the pasture period, the heifer diet was only grass	79.2%	4.8%	100%	100%	
Grass & Hay	During above 75% of the pasture period, the heifer diet was grass and a hay complement	20.8%	95.2%	0%	0%	

DM: dry matter. n: number of heifers. SD: standard deviation. SE: standard error. Values followed by different letters (a–c) are significantly different from each other at $p \leq 0.05$.

Table 3. Description of the rearing factors characterizing the fattening period (FP) and the FP rearing management clusters for this period (FP-clust).

Rearing Factors	Description of the Modalities	FP Rearing Management Clusters												p
		Overall (n = 96)			FP-Clust1 (n = 20)			FP-Clust2 (n = 21)			FP-Clust3 (n = 43)			
Quantitative rearing factors		Mean	SD	SE	Mean	SD	SE	Mean	SD	SE	Mean	SD	SE	
Age early fattening (months)	Age of the heifer at the beginning of FP	26.9	2.6	0.2	23.9 [b]	2.1	0.5	28.2 [a]	0.8	0.2	28.7 [a]	1.4	0.06	<0.001
Slaughter age (months)	Age of the heifer at the slaughter	33.0	3.0	0.3	30.7 [b]	2.4	0.5	34.5 [a]	2.9	0.6	33.3 [a]	2.0	0.3	<0.001
Initial weight (kg)	Live weight of the heifer at the beginning of FP	607	56	5.7	585 [b]	66	14.7	621 [a]	47	10.2	613 [ab]	46	7.0	0.02
Slaughter weight (kg)	Live weight of the heifer before the slaughter	734	62	6.3	745 [a]	41	9.2	747 [a]	68	14.8	691 [b]	58	8.8	<0.001
ADG_FP (kg/day)	Average daily gain of the heifer during FP	0.7	0.3	0.03	0.8 [a]	0.3	0.07	0.7 [ab]	0.4	0.02	0.5 [b]	0.2	0.03	0.01
Hay_FP (%)	Calculation of the hay percentage in the average diet across the whole FP	55.6	34.3	3.5	36.9 [b]	23.7	5.3	87.8 [a]	14.9	3.2	18.3 [c]	0.8	0.1	<0.001
Grass_silage_FP (%)	Calculation of the grass silage percentage in the average diet across the whole FP	22.5	23.0	2.3	20.4 [b]	12.9	2.9	9.3 [c]	13.2	2.9	52.7 [a]	23.6	3.6	<0.001
Wrapping_haylage_FP (%)	Calculation of the wrapped haylage percentage in the average diet across the whole FP	18.5	23.4	2.4	36.5 [a]	23.6	5.3	0.7 [b]	4.6	1.0	27.4 [a]	20.3	3.1	<0.001
Forage_CP_FP (g/kg DM)	Calculated average of the forage's crude proteins content across the whole FP	106	31	3.2	108 [b]	5	1.1	88 [c]	36	7.8	139 [a]	5	0.8	<0.001
Forage_E_FP (Mcal/kg DM)	Calculated average of the forage's energy content across the whole FP	1.2	0.1	0.01	1.2	0.1	0.02	1.2	0.1	0.02	1.2	0.1	0.6	0.10
Forage_NDF_FP (g/kg DM)	Calculated average of the forage's NDF content across the whole FP	577.0	43.5	4.4	579.5 [b]	28.7	6.4	606.6 [a]	26.1	5.7	512.5 [c]	1.5	0.2	<0.001
Conc_quanti_FP (kg)	Total concentrate quantity intake during FP	823.1	486	49.6	954.4 [a]	539.9	134.4	970.2 [a]	383.1	85.6	321.6 [b]	153.2	23.4	<0.001
Conc_E_FP (Mcal/kg DM)	Calculated average of the concentrate's energy content across the whole FP	1.9	0.1	0.01	1.9 [b]	0.1	0.02	1.9 [a]	0.04	0.01	1.9 [a]	0.001	0.0001	<0.001
Pasture_duration_FP (days)	Number of days spend on pasture during FP	50.4	74.0	7.5	0.0 [c]	0.0	0.0	57.3 [b]	88.5	19.3	113.3 [a]	34.6	5.3	<0.001
FP_duration (days)	Duration of FP	207.4	125.9	12.8	273.8 [a]	168.9	37.8	192.3 [b]	84.8	18.5	137.0 [b]	57.6	8.8	<0.001

Table 3. *Cont.*

Rearing Factors	Description of the Modalities	FP Rearing Management Clusters				*p*
		Overall (*n* = 96)	FP-Clust1 (*n* = 20)	FP-Clust2 (*n* = 21)	FP-Clust3 (*n* = 43)	
Qualitative rearing factors						
Conc_CP_FP (g/kg DM)						<0.001
>250 g/kg DM	Across the FP, the calculated average of the concentrate's crude proteins content was above 250 g/kg DM	26.0%	78.1%	0%	0%	
<250 g/kg DM	Across the FP, the calculated average of the concentrate's crude proteins content was below 250 g/kg DM	74.0%	21.9%	100%	100%	
Fattening management						<0.001
Pasture	The fattening was carried out on pasture	15.6%	0%	9.3%	52.4%	
Outside	The fattening was carried outside without grass	16.7%	0%	34.9%	4.8%	
Pasture & Housing	The fattening was started at pasture and then finished in housing	18.7%	0%	20.9%	42.8%	
Housing	The fattening was carried out in housing	49.0%	100%	34.9%	0%	

DM: dry matter. *n*: number of heifers. SD: standard deviation. SE: standard error. Values followed by different letters (a–c) are significantly different from each other at $p \leq 0.05$.

The energetic contents of the forages (hays, grass silages and wrapped haylages) in the fattening diet and all concentrates (crude proteins (CP), net energy and neutral detergent fiber (NDF) contents) were recorded. These data were used to calculate the chemical composition of the average forage (i.e., hays + grass silages + wrapped haylages) during the FP and the chemical composition of the average concentrate for each period.

2.2. Animals Slaughtering, Carcass Traits and Muscle Sampling

All heifers were slaughtered in the same industrial slaughterhouse (Abattoir du Gévaudan, Antrenas, France). The slaughter was done by exsanguination after stunning under the same conditions for all heifers. Carcasses were suspended vertically using the Achilles method without electrical stimulation. About 1 h (hot carcass weight) post-mortem, carcasses were weighted and graded visually by an official judge (conformation and fat scores) according to the EUROP grid system [27]. Until 24 h post-mortem, carcasses were chilled and stored at 2 °C in a cold room.

The EUROP system consists of a characterization of the carcass conformation using a scoring grid divided into 5 classes: E (extremely muscled), U, R, O and P (very poorly muscled). Each class of conformation was divided into 3 sub-categories using "+" (high), "=" (average), and "−" (low), so that the conformation was divided into 15 subclasses. A scale ranging from 1 (very poorly muscled) to 15 (extremely muscled), corresponding to each of the conformation sub-categories, was considered [27]. In the EUROP system, the fatness score of the carcass is divided into 5 classes where 1 = lean and 5 = very fat. In our study, the fatness score was not a discriminatory trait as all carcasses were scored 3.

Carcass quality was characterized using three measurements: Cold carcass weight (calculated from the hot carcass weight −2%, kg), dressing percentage (dressing% = ratio of cold carcass weight to live weight before slaughter, %) and conformation score. In our dataset, the distribution of each carcass trait was the following: Mean cold weight, 430 kg (standard deviation, SD = 42); mean dressing%, 58.5% (SD = 2.0) and mean conformation score, 11 (U=, SD = 1).

Twenty-four hours post-mortem, the full rectus abdominis (RA) muscle was collected from the right-hand side of the carcass. To conserve the same sampling conditions (24 h post-mortem), irrespective of the slaughtering day, only 77 RA muscles could be taken among the 96 samples. This muscle is an oxidative muscle with higher mean fibre areas [28,29] than the longissimus muscle (LM), which is considered as a reference muscle in most meat quality studies. According to the results of Cassar-Malek et al. [30], the RA muscle is more sensitive to the variation of rearing managements than the LM muscle. Moreover, Oury et al. [31] observed few differences between the sensory properties (tenderness, juiciness and flavor) of both muscles, RA and LM, when heifers were reared under the same conditions.

2.3. Meat Quality Evaluation

The color was measured 2 h after excision from the carcass on the fresh RA samples, using a spectrophotometer (Konica Minolta CM-600d, Osaka, Japan) and expressed in CIE L*a*b* units [32]. Before the color measures, meat samples were exposed to air for 20 min. The spectrophotometer was calibrated following the instructions of the manufacturer. A "black" and a "white" calibration were performed before each measurement session. To characterize the color of the whole RA muscle, the mean of 5 measurements (randomly distributed on the muscle) per RA muscle was used. The connective tissues were avoided.

The RA muscles were then weighted, vacuum-packaged and chilled for 14 days at 4 °C for ageing. After ageing, they were frozen at −20 °C until the analysis [33].

Sixteen members of a trained tasting panel conducted the sensory evaluation of the meat, using a monadic test. The members of the tasting panel had attended 20 training sessions before starting the sensory evaluation of the flank steak (RA muscle) samples [34]. In short, the assessors were selected and trained before the final studies in accordance with ISO 8586 [35] and were familiar with sensory assessment of the meat. After the training period of the panel composed of three sessions of different beef samples in which sensory descriptors were defined, assessors rated the descriptors on a structured scale extending from 0 to 10. Then, the concordance between the members of the trained tasting panel was checked using the Kendall concordance test [36].

The samples were randomly selected. Before the tasting session, the meat samples were thawed for 24 h at 4 °C and cut into 2 sub-samples: The first for the sensory evaluation and the second for the sheer force measurement. The sensory evaluation was carried out under conditions in accordance with AFNOR and SSHA [37,38].

The meat samples used for the sensory evaluation were cut into even 15 mm thick steaks. These steaks were cooked on a double-face grill at a temperature of 300 °C for 1 min 45 s in aluminum foil to remove the roasted taste (i.e., up to an internal temperature of 55 °C) [19]. After cooking and removing the meat borders, the steaks were cut into homogeneous pieces (size $15 \times 20 \times 20$ mm) that were served in batches of 3 or 4 on a plastic plate to the trained panel. During each tasting session, 5 samples were evaluated by the trained panel using a Latin square presentation.

The assessors evaluated five sensory descriptors: Initial tenderness, overall tenderness, overall juiciness, flavor intensity and fat presence. Initial tenderness was defined as the tenderness at the first bite, whereas overall tenderness was defined as an evaluation of the tenderness before swallowing the meat sample. Each sensory descriptor was rated on a 10-point non-graduated scale from a score of 0 (tough, dry, slight, and too lean) to a score of 10 (very tender, very juicy, strong, and highly fat).

The sheer force was measured according to the method described by Salé [39,40] on raw meat using material testing equipment (MTS Synergie 200). For each sample, 25 meat portions with different thicknesses (max 18 mm) were cut perpendicular to the fibers [41]. From the different measurements obtained, the force (daN) and the work (dJ) at 10 mm were determined. The sheer force was then calculated (ratio of work to force, dJ/daN).

In our dataset, the distribution of each flank steak trait was the following: Mean weight, 1.6 kg (SD = 0.2); mean L*, 26.2 (SD = 3.2); mean a*, 15.1 (SD = 2.5); mean b*, 10.9 (SD = 2.9); mean initial tenderness, 6.3/10 (SD = 0.8); mean overall tenderness, 6.1 (SD = 0.8); mean overall juiciness, 6.4 (SD = 0.5); mean flavor intensity, 4.5 (SD = 0.7); mean fat presence, 7.6 (SD = 0.6) and mean sheer force, 0.5 dJ/daN (SD = 0.07).

2.4. Statistical Analyses

The statistical analyses were performed using R 3.2.3 [42].

A descriptive analysis of the dataset to observe the normality of distribution was realized, using quantile-quantile plots [43]. After this descriptive analysis, the quantitative rearing factors that were discontinuous with few values were converted into qualitative parameters. The transformed rearing factors concerned: (i) For the growth period: The calculated average percentage of each forage (hay, grass silage, and wrapped haylage) in the housing diet and the growth period duration (Table 2); (ii) for the fattening period: The calculated average of the concentrate's crude protein content across the whole period (Table 3). These conversions allowed for the definition of homogeneous modality classes that could be processed statistically.

From the rearing factors characterizing the pre-weaning period ($q = 16$, Table 1), PWP rearing management clusters (PWP-clust) were constructed using a factor analysis for mixed data (FAMD), followed by a hierarchical clustering on the principal components (HCPC). The FAMD allowed for the consideration of quantitative and qualitative rearing factors simultaneously. The number of rearing management clusters was determined from the obtained dendrograms. For the growth and fattening periods, the same procedure was applied from the rearing factors characterizing these both periods (GP-clust and FP-clust), $q = 13$ (Table 2) and $q = 17$ (Table 3), respectively. The three FAMD were realized from the rearing data of the 96 heifers. The FAMD and HCPC were implemented using the "FactoMineR" package [44] in R.

Between the rearing management clusters defined for each key period of the heifer's life, analysis of variance (ANOVA) or χ^2 tests were conducted for each of the quantitative or qualitative rearing factors to evaluate their dependence on the defined rearing management clusters. If there was a significant difference in ANOVA, a Tukey test was performed to compare the average pairwise, using the "agricolae" package in R [45].

The WLP rearing management was determined from the combination of the obtained rearing management clusters (PWP-clust × GP-clust × FP-clust) in which the heifers were affected. Only the WLP rearing managements with at least 8 heifers for carcass and meat data were considered in the following study.

For each of the carcass and meat traits, an ANOVA was realized to evaluate their dependence on the considered WLP rearing managements. In the ANOVA, the effect of the farm was tested on all carcass and meat traits. If the effect of the farm was significant, the farm was considered as a random effect for the analysis. For each sensory parameter, the score given by each member of the trained tasting panel was used in the analysis. The effects of the member of the trained tasting panel and the effect of the animal were tested in the ANOVA. If they were significant, these factors were considered as random effects.

Finally, to investigate the effects of the WLP rearing managements on the carcass and meat traits, if there was not random effect, ANOVAs were performed. If not, mixed models were developed using the "lmerTest" package [46] in R. If the results of the ANOVA or mixed models were significant, a post-hoc Tukey test was conducted to compare the average pairwise. For the mixed models, the Tukey test was realized using the package "multcomp" [47] in R.

The effects were declared significant at $p \leq 0.05$ and tending toward significant was considered for $0.05 < p \leq 0.10$.

3. Results

After the HCPC analysis, three different rearing management clusters were defined for each key period of the heifer's life (PWP, GP and FP).

3.1. Description of the Three Rearing Management Clusters Obtained for the Pre-Weaning Period (PWP-Clust)

The PWP-clust1 contained 37 calves. The properties of this PWP rearing management cluster were an average daily gain significantly lower than the both other PWP clusters and the longest pasture duration (Table 1). The concentrate's crude protein and energy contents were lower than the other rearing managements applied during the pre-weaning period. During the housing, the time spent by the calf with its mother was longer than both of the other rearing managements applied during the pre-weaning period, as the calves were permanently with their mothers. Finally, in PWP-clust1, calves did not have concentrates during the pasture period.

The PWP-clust2 contained 39 calves. This PWP rearing management cluster was characterized by a shorter pasture duration than both other PWP rearing managements (Table 1). In this cluster, the concentrate's crude protein and energy contents had intermediate values compared to both of the other rearing managements applied during the pre-weaning period. In the PWP-clust2, the calves ingested concentrates only during the housing period of the pre-weaning period. Finally, in this rearing management, 66.7% of the calves were born without the intervention of the farmer.

The PWP-clust3 contained 20 calves. The properties of this cluster were a longer period with concentrate in the calves' diet and an intermediate pasture duration than the other rearing managements applied during the pre-weaning period. During the pasture, the calves received concentrate (Table 1). Moreover, in PWP-clust3, the concentrate's crude protein and energy contents were significantly higher than both of the other rearing managements applied during the pre-weaning period. The average daily gain of the calves from PWP-clust2 and PWP-clust3 was higher than that of the calves from PWP-clust1. In the PWP-clust2 and PWP-clust3, the time spent by the calf with her mother during the housing was the lowest. The calves were with their mother only during feeds.

3.2. Description of the Three Rearing Management Clusters Obtained for the Growth Period (GP-Clust)

The GP-clust1 contained 21 heifers. This GP rearing management cluster was characterized by the lowest weaning weight and the longest pasture duration (Table 2). In the GP-clust1 and GP-clust2, the period during which the heifers received concentrates in their diet was shorter than that in GP-clust3. In GP-clust1; the concentrate's crude protein content was lower compared to the other rearing managements applied during the growth period. During the housing period, the heifers received hay (between 20% and 40%) and grass silage (>50%). This management was applied for more than 500 days. In the GP-clust1, 95.2% of the heifers received a hay complement during the whole pasture duration of the growth period.

The GP-clust2 contained 46 heifers. This GP rearing management cluster had the shortest pasture duration (Table 2). In the GP-clust1 and GP-clust2, the heifers ingested a lower concentrate quantity during the growth period than those in the GP-clust3. Among the heifers in the GP-clust2, 82.6% ingested mostly hay (>80% or between 40 and 80% in diet) and 32.6% ingested also wrapped haylage (between 40 and 60% of the diet), during the housing period of the growth period. Then, 87% of the heifers had no grass silage in their diet. In the GP-clust2, the heifers were not supplemented with hay during the pasture of the growth period.

The GP-clust3 contained 29 heifers. This GP rearing management cluster was characterized by the longest period during which heifers received concentrate in their diet and the highest concentrate quantity intake (Table 2). In this cluster, the concentrate's crude protein and energy content values were the highest. This rearing management was applied for more than 500 days and during pasture of the growth period, the heifers were not supplemented with hay.

3.3. Description of the Three Rearing Management Clusters Obtained for the Fattening Period (FP-Clust)

The FP-clust1 contained 20 heifers. Properties of this FP rearing management cluster were a longer fattening period and an older age at slaughter than the other rearing managements applied during the fattening period (Table 3). In the FP-clust1, the heifers were younger and lighter at the beginning of the fattening period than those from the other fattening rearing managements. In the FP-clust1 and FP-clust2, the heifers' average daily weight gain was higher than that in the FP-clust3. In FP-clust1, the fattening diet offered contained on overage 36.9% hay, 36.5% wrapped haylage and 20.4% grass silage. Then, the forage's crude protein and NDF contents had intermediate values and the concentrate's energy content had lower values than those obtained in the other fattening rearing management clusters. In the FP-clust1, 78.1% of the heifers received a mean concentrate with more than 40 g/kg dry matter (DM) of crude proteins during the whole of the fattening period. In the FP-clust1 and FP-clust2, heifers had higher concentrate quantities during the whole of the fattening period than those from the FP-clust3. In FP-clust1, the heifers' fattening was carried out in housing.

The FP-clust2 contained 21 heifers. This rearing management cluster was characterized by heifers which had a higher initial weight and were slaughtered older than those from the FP-clust1 (Table 3). In this rearing management, the fattening diet was composed mostly of hay (87.8%) and the proportions of grass silage and wrapping haylage were the lowest. In the FP-clust2, the forage's crude protein and NDF content values were lower and higher respectively than those obtained in the other fattening rearing managements. Then, the crude protein content of the mean concentrate given, for the whole rearing period was below 250 g/kg DM. In FP-clust2, the main fattening managements were housing (34.9%) and outside (34.9%).

The FP-clust3 contained 49 heifers. The properties of this rearing management cluster were a lower average daily gain than FP-clust1 and the heifers were slaughtered with the lowest weight (Table 3). In this fattening rearing management, the highest proportion of grass silage (52.7%) and the lowest proportion of hay (18.3%) compounded the fattening diet. Then, the heifers ingested the lowest concentrate quantities and had the longest pasture duration. In the FP-clust3, the forage's crude protein and NDF content values were higher and lower respectively than those obtained in the other rearing management FP clusters. Then, the crude protein content of the mean concentrate given, for the whole fattening period was below 250 g/kg DM. In the FP-clust2 and FP-clust3, the fattening duration was shorter than that in the FP-clust1. In the FP-clust3, the main fattening managements were pasture (52.4%) and pasture and housing (42.8%).

3.4. Description of the WLP Rearing Managements Considered in This Study

From the rearing management (PWP, GP and FP) clusters, 12 different WLP rearing managements characterizing the whole life of the heifers are defined (Table 4). These rearing managements were the combination of rearing management clusters applied during the three key phases (pre-weaning, growth and fattening) of the heifers' life. For the rest of the study, only four WLP rearing managements: WLP-A, WLP-D, WLP-E, and WLP-F were considered to have at least eight heifers for the carcass and the meat data. The heifers from the WLP-A management received the rearing management clusters as following: PWP-clust1, GP-clust1, and FP-chust3, during their life. During the WLP-D management, the rearing management clusters applied during the heifers' life were: PWP-clust1, GP-clust3, and FP-clust2. The rearing management clusters characterizing the WLP-E were as following: PWP-clust2, GP-clust2, and FP-clust1. Finally, the rearing management cluster: PWP-clust2, GP-clust2, and FP-clust2 characterized the rearing management applied during the heifers' life, WLP-F.

Table 4. Description of the rearing managements applied during the heifers' whole life period (WLP).

WLP Rearing Managements	Rearing Management Clusters			Number of Heifers	
	PWP	**GP**	**FP**	**Carcass Data**	**Meat Data**
WLP-A	PWP-clust1	GP-clust1	FP-clust3	19	18
WLP-B	PWP-clust1	GP-clust2	FP-clust2	1	1
WLP-C	PWP-clust1	GP-clust3	FP-clust1	6	6
WLP-D	PWP-clust1	GP-clust3	FP-clust2	11	8
WLP-E	PWP-clust2	GP-clust2	FP-clust1	19	12
WLP-F	PWP-clust2	GP-clust2	FP-clust2	13	10
WLP-G	PWP-clust2	GP-clust3	FP-clust1	1	1
WLP-H	PWP-clust2	GP-clust3	FP-clust2	6	4
WLP-I	PWP-clust3	GP-clust1	FP-clust3	2	2
WLP-J	PWP-clust3	GP-clust2	FP-clust1	6	6
WLP-K	PWP-clust3	GP-clust2	FP-clust2	7	6
WLP-L	PWP-clust3	GP-clust3	FP-clust2	5	3

The WLP rearing managements were obtained from the rearing management clusters. The rearing management clusters were defined for each key period of heifers' life (pre-weaning, PWP; growth, GP and fattening, FP), described in Tables 1–3.

The WLP-A and WLP-D had only the pre-weaning rearing management in common. The WLP-E and WLP-F received the same rearing managements during the pre-weaning and growth period. WLP-D and WLP-F had only the fattening rearing management in common. The WLP-A and WLP-D were mostly characterized by a lower average daily weight gain than WLP-E and WLP-F during the pre-weaning period. Then, during the growth and fattening periods, the heifers ingested a higher quantity of concentrate in the WLP-A than WLP-D (Tables 2 and 3). In the WLP-E and WLP-F, the heifers ingested higher concentrate quantities than WLP-A only during the fattening period. In WLP-E, the fattening period duration was the longest and the heifers were slaughtered the youngest. During the fattening period, the heifers ingesting the highest hay percentage and the lowest grass silage and wrapping haylage percentage received the WLP-D and WLP-F managements. The WLP-A was characterized by the longest pasture duration during the heifers' life and the lightest heifers at slaughter. The heifers were slaughtered at the same age in the WLP-A, WLP-D and WLP-F managements.

3.5. Impact of the Four WLP Rearing Managements on Carcass and Flank Steak Traits

The WLP-E and WLP-F allowed for the production of a heifers' carcass that was heavier and had a higher conformation score than WLP-A (Table 5). The WLP-D, WLP-E, and WLP-F managements produced heifers' carcasses with similar weights and conformation scores.

Table 5. Impact of the four rearing managements applied during the heifers' whole life period (WLP) on carcass traits.

Carcass Traits	WLP Rearing Managements [1]												p
	WLP-A (n = 19)			WLP-D (n = 11)			WLP-E (n = 19)			WLP-F (n = 13)			
	Mean	SD	SE	Mean	SD	SE	Mean	SD	SE	Mean	SD	SE	
Carcass weight (kg)	393 [b]	36	8.3	422 [ab]	46	13.9	435 [a]	21	4.8	446 [a]	43	11.9	<0.001
Dressing (%)	57.3 [b]	1.6	0.4	59.1 [a]	2.2	0.7	58.6 [ab]	1.3	0.3	58.5 [ab]	1.3	0.4	0.01
Conformation score (scale 1–15)	10.3 [b]	0.9	0.2	10.8 [ab]	0.9	0.3	11.3 [a]	0.6	0.1	11.2 [a]	0.8	0.1	0.002
Number of carcasses per EUROP classes [2] (proportion of each conformation score)													
E−	0 (0%)			0 (0%)			1 (5%)			0 (0%)			
U+	1 (5%)			3 (27%)			4 (22%)			6 (46%)			
U=	8 (41%)			3 (27%)			13 (68%)			4 (31%)			
U−	6 (32%)			5 (46%)			1 (5%)			3 (23%)			
R+	4 (22%)			0 (0%)			0 (0%)			0 (0%)			

n: number of heifers. SD: standard deviation. SE: standard error. Values followed by different letters (a,b) are significantly different from each other at $p \leq 0.05$. [1], The WLP rearing management was the combinations of the rearing managements applied during the pre-weaning (PWP), growth (GP) and fattening (FP) periods of the heifers. The WLP rearing management A was the following combinations of PWP-clust1, GP-clust1 and FP-clust3. The WLP rearing management D was the following combinations of PWP-clust1, GP-clust3 and FP-clust2. The WLP rearing management E was the following combinations of PWP-clust2, GP-clust2 and FP-clust1. The WLP rearing management F was the following combinations of PWP-clust2, GP-clust2 and FP-clust2. The traits of the rearing management clusters are presented in the Tables 1–3. [2], The EUROP classes are E+ (extremely muscled), E=, […], P= and P− (very poorly muscled). In our dataset, conformation score of carcasses was only between E− and R+.

These difference in carcass traits, between these WLP rearing managements, could be explained by the distribution of the different conformation classes according to the EUROP grid. The conformation score of the carcasses from WLP-A was mainly U = (42%) but over half of the carcasses (52.6%) were conformed to U− and R+. In contrast, no carcasses with an R+ conformation were obtained and only one with a U− conformation score when heifers received the WLP-E. With the WLP-E, the conformation scores of carcasses produced by heifers were mainly U = (68.4%) or higher (26.3% of U+ and E− conformation scores). The heifers receiving the WLP-F produced carcasses with U+ and U= conformation scores (46.1% and 30.8%, respectively). The WLP-D allowed for the production of carcasses with a higher dressing% than WLP-A.

The four rearing managements applied during the heifers' whole life had no significant impact on the flank steak traits in terms of weight, all sensory descriptors, sheer force and color (Table 6).

Table 6. Impact of the four rearing managements applied during the heifers' whole life period (WLP) on flank steak (rectus abdominis, RA) traits.

Meat Traits	WLP Rearing Managements [1]												p
	WLP-A (n = 18)			WLP-D (n = 8)			WLP-E (n = 12)			WLP-F (n = 10)			
	Mean	SD	SE	Mean	SD	SE	Mean	SD	SE	Mean	SD	SE	
RA weight (kg)	1.4	0.2	0.04	1.5	0.2	0.07	1.6	0.2	0.06	1.7	0.2	0.06	0.35
Sensory descriptors of RA muscle (scale 0–10) [2]													
Initial tenderness	6.4	1.9	0.4	6.2	1.9	0.7	6.2	1.7	0.5	6.2	1.9	0.5	0.90
Overall tenderness	6.2	2.0	0.5	6.0	1.9	0.7	5.9	1.8	0.5	5.9	1.9	0.5	0.89
Overall juiciness	6.5	1.7	0.4	6.4	1.6	0.6	6.6	1.7	0.5	6.7	1.7	0.5	0.47
Flavour intensity	4.6	2.3	0.5	4.7	2.4	0.8	4.5	2.3	0.7	4.3	2.2	0.7	0.48
Fat presence	7.5	2.3	0.5	7.8	2.3	0.8	7.5	2.4	0.7	7.5	2.2	0.7	0.68
Sheer force (dJ/daN)	0.49	0.06	0.01	0.53	0.08	0.03	0.48	0.07	0.02	0.53	0.09	0.03	0.24
Colour													
L*	26.3	2.5	0.6	27.9	3.4	1.8	26.8	2.7	0.8	25.1	2.9	0.9	0.20
a*	15.0	2.4	0.6	13.9	2.0	0.7	15.4	2.1	0.6	14.4	3.1	1.0	0.50
b*	11.6	3.1	0.7	10.2	2.0	0.7	10.3	1.6	0.5	10.7	3.9	1.2	0.52

n: number of heifers. SD: standard deviation. SE: standard error. [1], The WLP rearing management was the combinations of the rearing managements applied during the pre-weaning (PWP), growth (GP) and fattening (FP) periods of the heifers. The WLP rearing management A was the following combinations of PWP-clust1, GP-clust1 and FP-clust3. The WLP rearing management D was the following combinations of PWP-clust1, GP-clust3 and FP-clust2. The WLP rearing management E was the following combinations of PWP-clust2, GP-clust2 and FP-clust1. The WLP rearing management F was the following combinations of PWP-clust2, GP-clust2 and FP-clust2. The traits of the rearing management clusters are presented in the Tables 1–3. [2], Scale for initial tenderness, overall tenderness, overall juiciness, flavour intensity, fat presence and overall appreciation: 0 = tough, dry, slight, too lean and highly disliked and 10 = very tender, very juicy, strong, highly fat and highly liked.

4. Discussion

4.1. Carcass Traits

The carcass results of this study match those of Hennessy et al. and Greenwood et al. [48,49] illustrating that animals with faster growth during the pre-weaning period were slaughtered with heavier carcass weights when the growth and fattening managements were similar. For the same slaughter weight, Cerdeño et al. [50] did not show significant differences on the conformation score and dressing% of calves when different rearing managements were applied during the pre-weaning period. However, if the calves were slaughtered higher, the conformation score and the dressing% of their carcass were improved [51]. The rearing management applied during the pre-weaning period could have an impact on the carcass traits. During the growth period, Guerrero et al. [52] did not observe any significant impact of rearing managements (intensive vs. extensive) on carcass traits (carcass weight, dressing% and conformation score) in young bulls with the same fattening management. The absence of significant differences in carcass traits between WLP-D and WLP-F is in accordance with these results. The effect of the rearing managements applied before weaning or fattening could interact with those applied during the next heifers' life period involving an attenuation or an amplification of the effects on carcass traits.

During the fattening period, the heifers from WLP-E were slaughtered at a younger age than those receiving the other rearing managements applied during the heifers' whole life. In the literature, the fattening period was the most frequently studied period and many rearing factors were known to influence carcass traits. Our results disagree with the results of many studies, which displayed that the heifers slaughtered at an older age were heavier [2,5]. The heavier carcass from WLP-E could be explained by the fattening period duration. This result agrees with many studies indicating an increase in the carcass weight of cull cows with an extension of the fattening period duration [10,11,53]. However, the WLP-D and WLP-F, with a shorter fattening period duration, obtained a similar carcass weight than WLP-E. In accordance with Soulat et al. [54], this result showed that is possible to attain the same carcass traits with different rearing managements applied during the fattening period. In our study, the heifers with the highest concentrate intake during the fattening period produced heavier carcasses confirming the results observed by Cook et al. [55] in heifers. However, our results displayed that it is possible to obtain the same carcass weight from a rearing management during the heifers' life using lower concentrate quantities in the diet during the growth and fattening periods. For the same concentrate quantity intake, in our study, the forage composition of the fattening diet had no impact on the carcass weight. This result is in accordance with many studies which did not find evidence for an impact of the fattening diet composition on carcass weight, in young bulls, steers, and cull cows [14,56,57]. In WLP-A, the heifers had the longest pasture duration during their life. This management produced lighter carcasses than the other rearing managements considered in this study. In accordance with these results, Keane and Allen and Neel et al. [15,58] observed a lighter carcass in steers from the fattening managements involving pasture. The longest pasture period of heifers during their whole life could explain part of the difference observed on the carcass weight between our four rearing managements applied during the heifers' whole.

Although the slaughter ages were different between WLP-E and WLP-D, the dressing% of the carcasses were not significantly different. This result is in accordance with the results of Bures and Barton [2] in heifers. Few studies observed the effect of this rearing factor on the dressing% in heifers. Many studies observed a higher dressing% when young bulls were slaughtered older [6,59]. However, our results showed that the slaughter age was not the only rearing factor that could have an impact on the dressing%. For a similar slaughter age, we observed a difference in the dressing% between carcasses from WLP-A and WLP-D. According to our results, a longer fattening period duration could have no impact on the dressing%. This result disagrees with the results of many studies indicating an increase in the dressing% of dairy cull cows when the fattening duration was longer [10,60]. The animal type and the breed, which were different could explain this difference. These parameters are known to influence the dressing%, which was higher for carcasses from WLP-D than WLP-A. In WLP-A, the heifers were slaughtered at a lighter weight. This result confirms the results of McEwen et al. [7] who observed an increase in the dressing% when steers were slaughtered at a heavier weight. However, our results showed that the dressing% was not systematically increased when heifers were slaughtered heavier. Moreover, these data illustrate that the variation observed for dressing% was not explained only by the live weight of the heifers. In contrast to the results of Price et al. [61] in young bulls and steers, in our study, the combination of a higher hay percentage in the fattening diet and heavier slaughter weight did not lead to a lower dressing%. As demonstrated in many studies, the fattening diet composition could have no impact on the dressing% in cattle [55,57,62,63]. It could explain the few differences observed between the four rearing managements applied during the heifers' whole life considered in our study. In the WLP-A, the heifers had the longest pasture duration during their life and their carcasses had a lower dressing%. In accordance with the results of Keane and Allen and Neel et al. [15,58], the pasture duration could explain in part the low dressing% values observed.

In our study, the heifers slaughtered younger had a similar conformation score than those slaughtered older. This result complies with the studies of Bures and Barton and Ahnstrom et al. [2,5]. However, with a similar slaughter age, the carcasses from the WLP-A had a lower conformation score than the carcasses from WLP-F. As for the others carcass traits, the variation of the conformation score could not be explained by only one rearing factor [54]. In accordance with different results obtained in cull cows [10,11,53], WLP-E with the longest fattening period allowed to produce carcasses with higher conformation scores. However, our results showed that it is possible to obtain conformation scores similar to WLP-E with a shorter fattening period. In our study, the conformation score did not seem to be impacted by the slaughter weight. This observation confirms the results obtained by Keane and Allen and Ellies-Oury et al. [15,64] in steers. According to many studies, the fattening diet composition would not seem to have any impact on the conformation score in heifers and steers [55,61,63]. On the other hand, WLP-A with the longest pasture duration produced carcasses with the lowest conformation score (R+). The pasture duration during the heifers' whole life could explain a part of the variation of the conformation score observed between our four rearing managements applied during the heifers' whole life. This observation confirms the results observed by Keane and Allen and Neel et al. [15,58] during the fattening period, in steers. However, the conformation scores were similar between WLP-A and WLP-D. Our study displayed that the rearing management applied during the heifers' whole life and not only one rearing factor and one heifers' life period explained the carcass traits variability. This study showed that an improvement of the carcass quality traits could be obtained from different rearing managements during the life of heifers. Moreover, our results showed that it is possible to modify the rearing managements at different period of heifers' life without having a negative impact on carcass traits.

4.2. Meat Traits

In accordance with the results of Hennessy and Morris [48], the variations in the growth rate during the pre-weaning period observed between the rearing managements considered in this study did not seem to have any impact on the meat's sensory traits. In the study conducted by Hennessy and Morris [48], steers and heifers were subjected to the same rearing managements after weaning in contrast to our study. Our results also confirm those of Picard et al. [65] showing that modifications of the rearing managements during the pre-weaning period had no impact on the fiber properties measured at slaughter on 18-month-old bulls However, Cerdeño et al. [50] showed that the rearing management applied during the pre-weaning period can have an impact on the tenderness of the calf meat. Considering the rearing management applied during the heifers' whole life, it is possible that the effect on the sensory properties of the flank steak could interact with the rearing managements applied after the weaning period. To the best of our knowledge, the impact of the rearing managements during the pre-weaning period on sheer force and meat color of the meat has never been studied before.

Our results also match the results of Durunna et al. [66] obtained on LM in steers. They showed that modifications of the rearing managements during the growth period did not seem to have an impact on sensory properties or on meat color. It is possible that the rearing managements during the fattening period could mitigate a possible negative impact on the meat, resulting from the rearing managements applied during the previous periods (pre-weaning and growth) of the heifers' life. Thanks to the plasticity of the muscle properties, this could explain the absence of differences between the rearing managements considered in our study with regard to meat traits.

In contrast to our results, Oury et al. [19] observed an impact of rearing managements applied during the heifers' whole life on flank steak tenderness. In cull cows, Couvreur et al. [20] also observed an impact of the rearing managements applied during the fattening period on the tenderness and juiciness of the flank steak. The different rearing managements and animal types (sex, breed, age) could explain the differences between these studies and our results. In the study of Oury et al. [19], the heifers ingested maize silage and higher concentrates quantities during the fattening period compared to our study. Other studies have demonstrated an effect of the animal type and breed on the LM meat traits in cattle [1,2,67]. Furthermore, Ellies-Oury et al. [64] demonstrated that an increase in the slaughter weight tended to increase tenderness and flavor intensity scores of flank steak in steers. In our study, the slaughter weight had no impact on the sensory properties of the flank steak. Moreover, as we worked at the scale of the combination of many rearing managements applied at different periods of the animal's life, it is difficult to identify the individual impact of the slaughter weight on the meat traits.

With regard to the RA color, Serrano et al. [68] did not demonstrate any effect of the fattening diet in young bulls. In their study, Oury et al. [69] observed few impacts of the rearing managements applied during the heifers' whole life on the L* parameter for flank steak. Our results are in accordance with these studies. However, there are few data on the impact of rearing factors on RA color in cattle. In our study, the different rearing managements considered allowed us to attain the same meat color.

Our study also showed that different combinations of rearing managements during the heifers' life could allow us to attain the same, or to improve the carcass and meat traits. That confirmed our previous results obtained for carcass and LM meat traits [54,70,71].

In our study, we observed that the carcass quality was weakly related to the meat quality of RA. Carcasses characterized by a low weight, conformation score, and dressing% can produce the same flank steak quality compared to carcasses which had high values for these three parameters. This result is in accordance with the observation of Gagaoua et al. [72] showing that a low contribution of carcass traits can explain meat traits of a cull cow during the fattening period. For other muscles, Bonny et al. [73] observed a very weak relationship between the carcass traits (conformation and fatness scores) and the sensory meat properties. However, a recent study showed that the carcass traits (fatness score, percentages of muscle and fat in the carcass) can have an impact on the sensory properties (tenderness, flavor and juiciness) of the meat [74]. The rearing management applied during the heifers' whole could improve carcass traits without decline RA meat traits.

A limitation of this study was the low number of animals for certain groups. To confirm the robustness and the repeatability of this approach it would be important to increase the number of heifers per rearing management. To complete this study, it would be interesting to expand to other rearing managements applied during the heifers' life such as those based on maize silage or corn. It would also be interesting to study other muscles to evaluate whether the effects of the rearing managements applied during the heifers' whole life are similar irrespective of the considered muscle. Finally, a consideration of the production costs and carcass valorization per rearing management would help meat sector stakeholders to reduce costs without declining the carcass traits and the eating quality of meat.

5. Conclusions

To conclude, the originality of this study was to consider the rearing managements applied during the heifers' whole life and to study their effects on carcass and meat traits. Our results showed that the rearing managements applied during the heifers' whole life seemed to have more impact on carcass traits than on the flank steak properties. According to the rearing managements considered in our study, the carcass traits could be improved without altering the meat properties. Different combinations of rearing managements during a heifers' life have been identified to improve the production at the farm level.

Author Contributions: J.S. realized the experiment, analyzed the data, prepared figures and/or tables, wrote the paper, and approved the final draft. B.P. co-conceived and designed the experiments, contributed to the analysis, co-wrote the paper, reviewed drafts of the paper and approved the final draft. S.L. participated in the statistical analysis of data and in the redaction of the statistical analysis section. M.-P.E.-O. supervised and performed sensory and mechanical analysis, reviewed drafts of the paper, approved the final draft. V.M. conceived and designed the experiments, contributed to the analysis, co-wrote the paper, reviewed drafts of the paper and approved the final draft.

Funding: This research was supported by the Agence Nationale de la Recherche of the French government through the program "Investissement d'Avenir" (16-IDEX-0001 CAP 20-25) and by the PGI "Fleur d'Aubrac".

Acknowledgments: The authors would first like to thank the members of the PGI "Fleur d'Aubrac" who participated in this study. The authors thank also the extension services of Aveyron and Lozère and the cereal firms concerned with the additional data provided. Finally, the authors would like to thank all the INRA staff and in particular David Chadeyron for sampling at slaughter and Sandrine Papillon (Bordeaux Sciences Agro) for sensorial and rheological analyses.

Conflicts of Interest: The author declares no conflicts of interest.

References

1. Chambaz, A.; Scheeder, M.R.L.; Kreuzer, M.; Dufey, P.A. Meat quality of Angus, Simmental, Charolais and Limousin steers compared at the same intramuscular fat content. *Meat Sci.* **2003**, *63*, 491–500. [CrossRef]
2. Bures, D.; Barton, L. Growth performance, carcass traits and meat quality of bulls and heifers slaughtered at different ages. *Czech J. Anim. Sci.* **2012**, *57*, 34–43. [CrossRef]
3. Pesonen, M.; Huuskonen, A.K. Production, carcass characteristics and valuable cuts of beef breed bulls and heifers in Finnish beef cattle population. *Agric. Food Sci.* **2015**, *24*, 164–172. [CrossRef]
4. Muller, L.; Perobelli, Z.; Feijo, G.L.D.; Grassi, C. Cull cow physiological maturity and its effect on carcass and meat quality. In Proceedings of the 38th International Congress of Meat Science and Technology, Clermont-Ferrand, France, 23–28 August 1992; pp. 101–104.
5. Ahnstrom, M.L.; Hessle, A.; Johansson, L.; Hunt, M.C.; Lundstrom, K. Influence of slaughter age and carcass suspension on meat quality in Angus heifers. *Animal* **2012**, *6*, 1554–1562. [CrossRef] [PubMed]
6. Marti, S.; Realini, C.E.; Bach, A.; Perez-Juan, M.; Devant, M. Effect of castration and slaughter age on performance, carcass, and meat quality traits of Holstein calves fed a high-concentrate diet. *J. Anim. Sci.* **2013**, *91*, 1129–1140. [CrossRef] [PubMed]
7. McEwen, P.L.; Mande, I.B.; Brien, G.; Campbell, C.P. Effects of grain source, silage level, and slaughter weight endpoint on growth performance, carcass characteristics, and meat quality in Angus and Charolais steers. *Can. J. Anim. Sci.* **2007**, *87*, 167–180. [CrossRef]
8. Lucero-Borja, J.; Pouzo, L.B.; de la Torre, M.S.; Langman, L.; Carduza, F.; Corva, P.M.; Santini, F.J.; Pavan, E. Slaughter weight, sex and age effects on beef sheer force and tenderness. *Livest. Sci.* **2014**, *163*, 140–149. [CrossRef]
9. Matulis, R.J.; McKeith, F.K.; Faulkner, D.B.; Berger, L.L.; George, P. Growth and Carcass Characteristics of Cull Cows after Different Times-on-Feed. *J. Anim. Sci.* **1987**, *65*, 669–674. [CrossRef] [PubMed]
10. Vestergaard, M.; Madsen, N.T.; Bligaard, H.B.; Bredahl, L.; Rasmussen, P.T.; Andersen, H.R. Consequences of two or four months of finishing feeding of culled dry dairy cows on carcass characteristics and technological and sensory meat quality. *Meat Sci.* **2007**, *76*, 635–643. [CrossRef] [PubMed]
11. Dumont, R.; Roux, M.; Touraille, C.; Agabriel, J.; Micol, D. The fattening of cull Charolais cows: Effect of linseed meal on fattening performance and meat quality. *Prod. Anim.* **1997**, *10*, 163–174.

12. Zahradkova, R.; Barton, L.; Bures, D.; Teslik, V.; Kudrna, V. Comparison of growth performance and slaughter characteristics of Limousin and Charolais heifers. *Arch. Anim. Breed.* **2010**, *53*, 520–528. [CrossRef]

13. Moloney, A.P.; Drennan, M.J. Characteristics of fat and muscle from beef heifers offered a grass silage or concentrate-based finishing ration. *Livest. Sci.* **2013**, *152*, 147–153. [CrossRef]

14. Aviles, C.; Martinez, A.L.; Domenech, V.; Pena, F. Effect of feeding system and breed on growth performance, and carcass and meat quality traits in two continental beef breeds. *Meat Sci.* **2015**, *107*, 94–103. [CrossRef] [PubMed]

15. Keane, M.G.; Allen, P. Effects of production system intensity on performance, carcass composition and meat quality of beef cattle. *Livest. Prod. Sci.* **1998**, *56*, 203–214. [CrossRef]

16. Cozzi, G.; Brscic, M.; Da Ronch, F.; Boukha, A.; Tenti, S.; Gottardo, F. Comparison of two feeding finishing treatments on production and quality of organic beef. *Ital. J. Anim. Sci.* **2010**, *9*, 404–409. [CrossRef]

17. Moloney, A.P.; Mooney, M.T.; Troy, D.J.; Keane, M.G. Finishing cattle at pasture at 30 months of age or indoors at 25 months of age: Effects on selected carcass and meat quality characteristics. *Livest. Sci.* **2011**, *141*, 17–23. [CrossRef]

18. Pordomingo, A.J.; Grigioni, G.; Carduza, F.; Volpi Lagreca, G. Effect of feeding treatment during the backgrounding phase of beef production from pasture on: I. Animal performance, carcass and meat quality. *Meat Sci.* **2012**, *90*, 939–946. [CrossRef] [PubMed]

19. Oury, M.P.; Agabriel, J.; Agabriel, C.; Micol, D.; Picard, B.; Blanquet, J.; Labouré, H.; Roux, M.; Dumont, R. Relationship between rearing practices and eating quality traits of the muscle rectus abdominis of Charolais heifers. *Livest. Sci.* **2007**, *111*, 242–254. [CrossRef]

20. Couvreur, S.; Le Bec, G.; Micol, D.; Aminot, G.; Picard, B. PDO Maine-Anjou culled cow characteristics and finishing practices influence meat quality. In Proceedings of the 20th Rencontres Recherches Ruminants, Paris, France, 5–6 December 2013; pp. 165–168.

21. Mezgebo, G.B.; Moloney, A.P.; O'Riordan, E.G.; McGee, M.; Richardson, R.I.; Monahan, F.J. Comparison of organoleptic quality and composition of beef from suckler bulls from different production systems. *Animal* **2017**, *11*, 538–546. [CrossRef] [PubMed]

22. Moran, L.; O'Sullivan, M.G.; Kerry, J.P.; Picard, B.; McGee, M.; O'Riordan, E.G.; Moloney, A.P. Effect of a grazing period prior to finishing on a high concentrate diet on meat quality from bulls and steers. *Meat Sci.* **2017**, *125*, 76–83. [CrossRef] [PubMed]

23. Vieira, C.; Garcia-Cachan, M.D.; Recio, M.D.; Dominguez, M.; Sanudo, C. Effect of ageing time on beef quality of rustic type and rustic × Charolais crossbreed cattle slaughtered at the same finishing grade. *Span. J. Agric. Res.* **2006**, *4*, 225–234. [CrossRef]

24. Ferguson, D.M.; Warner, R.D. Have we underestimated the impact of pre-slaughter stress on meat quality in ruminants? *Meat Sci.* **2008**, *80*, 12–19. [CrossRef] [PubMed]

25. Micol, D.; Jailler, R.; Jurie, C.; Meteau, K.; Juin, H.; Nute, G.R.; Richardson, R.I.; Hocquette, J.F. Sensory evaluation of beef eating quality in France and UK at two cooking temperatures. In Proceedings of the 18th Rencontres Recherches Ruminants, Paris, France, 1–2 December 2011; p. 208.

26. Ingrand, S.; Dedieu, B. Batch management diversity in suckling herds. *Prod. Anim.* **1996**, *9*, 189–199.

27. Concil Regulation (EC). No 1183/2006 of 24 July 2006 concerning the Community scale for the classification of carcasses of adult bovine animals. *Off. J. Eur. Union* **2006**, *214*, 1–6.

28. Oury, M.-P.; Dumont, R.; Jurie, C.; Hocquette, J.-F.; Picard, B. Specific fibre composition and metabolism of the rectus abdominis muscle of bovine Charolais cattle. *BMC Biochem.* **2010**, *11*, 12. [CrossRef] [PubMed]

29. Gagaoua, M.; Couvreur, S.; Le Bec, G.; Aminot, G.; Picard, B. Associations among protein biomarkers and ph and color traits in longissimus thoracis and rectus abdominis muscles in protected designation of origin maine-anjou cull cows. *J. Agric. Food Chem.* **2017**, *65*, 3569–3580. [CrossRef] [PubMed]

30. Cassar-Malek, I.; Listrat, A.; Jurie, C.; Jailler, R.; Hocquette, J.-F.; Bauchart, D.; Ouali, A.; Lamarre, M.; Picard, B. Influence of compensatory growth on muscular characteristics of steers. In Proceedings of the 8th Rencontres Recherches Ruminants, Paris, France, 6–7 December 2001; p. 111.

31. Oury, M.P.; Agabriel, J.; Picard, B.; Jailler, R.; Dubroeucq, H.; Egal, D.; Micol, D. Growth performance, carcass quality, muscular characteristics and meat quality traits of Charolais steers and heifers. In Proceedings of the 58th European Federation for Animal Science, Dublin, Irland, 26–29 August 2007; pp. 1–8.

32. Commission International de l'Eclairage. *Colorimetry*, 2nd ed.; Commission International de l'Eclairage: Vienna, Austria, 1986.

33. Dransfield, E.; Martin, J.F.; Bauchart, D.; Abouelkaram, S.; Lepetit, J.; Culioli, J.; Jurie, C.; Picard, B. Meat quality and composition of three muscles from French cull cows and young bulls. *Anim. Sci.* **2003**, *76*, 387–399. [CrossRef]

34. Gagaoua, M.; Micol, D.; Picard, B.; Terlouw, C.E.; Moloney, A.P.; Juin, H.; Meteau, K.; Scollan, N.; Richardson, I.; Hocquette, J.-F. Inter-laboratory assessment by trained panelists from France and the United Kingdom of beef cooked at two different end-point temperatures. *Meat Sci.* **2016**, *122*, 90–96. [CrossRef] [PubMed]

35. International Organization for Standardization (ISO). *ISO_8586: Sensory Analysis—General Guidelines for the Selection, Training and Monitoring of Selected Assessors and Expert Sensory Assessors*; ISO: Geneva, Switzerland, 2012; pp. 1–28.

36. Donnell, E.M.; Hulin-Bertaud, S.; Sheehan, E.M.; Delahunty, C.M. Development and learning process of a sensory vocabulary for the odor evaluation of selected distilled beverages using descriptive analysis. *J. Sens. Stud.* **2001**, *16*, 425–445. [CrossRef]

37. AFNOR. Analyse sensorielle. In *Contrôle de la Qualité des Produits Alimentaires*; AFNOR: Paris, France, 1991.

38. Social Science History Association (SSHA). *Evaluation Sensorielle—Manuel Méthodologique*; Technique & Documentation: Paris, France, 1990.

39. Salé, P. Evolution of some mechanical properties of muscle during aging. *CRZV* **1971**, *6*, 35–44.

40. Bouton, P.E.; Harris, P.V. A comparison of some objective methods used to assess meat tenderness. *J. Food Sci.* **1972**, *37*, 218–221. [CrossRef]

41. Oury, M.P.; Picard, B.; Briand, M.; Blanquet, J.P.; Dumont, R. Interrelationships between meat quality traits, texture measurements and physicochemical characteristics of M. rectus abdominis from Charolais heifers. *Meat Sci.* **2009**, *83*, 293–301. [CrossRef] [PubMed]

42. R Core Team. *A Language and Environment for Statistical Computing*; R Foundation for Statistical Computing: Vienna, Austria, 2016.

43. Azaïs, J.-M.; Bardet, J.-M. *Le Modèle Linéaire par l'Exemple Régression, Analyse de la Variance et Plans d'Expériences Illustrations Numériques avec les Logiciels R, SAS et Splus*; *Sciences Sup*; Dunod: Paris, France, 2006.

44. Le, S.; Josse, J.; Husson, F. FactoMineR: An R package for multivariate analysis. *J. Stat. Softw.* **2008**, *21*, 1–18.

45. Agricolae: Statistical Procedures for Agricultural Research. Available online: http://CRAN.R-project.org/package=agricolae (accessed on 27 September 2018).

46. Kuznetsova, A.; Brockhoff, P.B.; Christensen, R.H.B. lmerTest Package: Tests in linear mixed effects models. *J. Stat. Softw.* **2017**, *82*, 1–26. [CrossRef]

47. Hothorn, T.; Bretz, F.; Westfall, P. Simultaneous inference in general parametric models. *Biom. J.* **2008**, *50*, 346–363. [CrossRef] [PubMed]

48. Hennessy, D.W.; Morris, S.G. Effect of a preweaning growth restriction on the subsequent growth and meat quality of yearling steers and heifers. *Aust. J. Exp. Agric.* **2003**, *43*, 335–341. [CrossRef]

49. Greenwood, P.L.; Cafe, L.M.; Hearnshaw, H.; Hennessy, D.W.; Morris, S.G. Consequences of prenatal and preweaning growth for yield of beef primal cuts from 30-month-old Piedmontese- and Wagyu-sired cattle. *Anim. Prod. Sci.* **2009**, *49*, 468–478. [CrossRef]

50. Cerdeño, A.; Vieira, C.; Serrano, E.; Mantecón, A.R. Carcass and meat quality in Brown fattened young bulls: Effect of rearing method and slaughter weight. *Czech J. Anim. Sci.* **2006**, *4*, 143–150. [CrossRef]

51. Cerdeño, A.; Vieira, C.; Serrano, E.; Mantecón, A.R. Effect of production system on performance traits, carcass and meat quality in Brown Swiss young cattle. *J. Anim. Feed Sci.* **2006**, *15*, 17–24. [CrossRef]

52. Guerrero, A.; Sanudo, C.; Alberti, P.; Ripoll, G.; Campo, M.M.; Olleta, J.L.; Panea, B.; Khliji, S.; Santolaria, P. Effect of production system before the finishing period on carcass, meat and fat qualities of beef. *Animal* **2013**, *7*, 2063–2072. [CrossRef] [PubMed]

53. Sugimoto, M.; Saito, W.; Ooi, M.; Oikawa, M. Effects of days on feed, roughage sources and inclusion levels of grain in concentrate on finishing performance and carcass characteristics in cull beef cows. *Anim. Sci. J.* **2012**, *83*, 460–468. [CrossRef] [PubMed]

54. Soulat, J.; Monteils, V.; Léger, S.; Ellies-Oury, M.P.; Picard, B. Is it possible to simultaneously pilot carcass and flank steak qualities by breeding practices, among Charolais heifers during fattening? In Proceedings of the 22th Rencontres Recherches Ruminants, Paris, France, 2–3 December 2015; pp. 379–382.

55. Cooke, D.W.I.; Monahan, R.; Brophy, P.O.; Boland, M.P. Comparison of concentrates or concentrates plus forages in a total mixed ration or discrete ingredient format: Effects on beef production parameters and on beef composition, colour, texture and fatty acid profile. *Ir. J. Agric. Food Res.* **2004**, *43*, 201–216.

56. Agabriel, J.; Garel, J.P.; Lassalas, J.; Petit, M. The fattening of cull cows in montain conditions. *Prod. Anim.* **1991**, *4*, 389–397.

57. Barton, L.; Kudrna, V.; Bures, D.; Zahradkova, R.; Teslik, V. Performance and carcass quality of Czech Fleckvieh, Charolais and Charolais × Czech Fleckvieh bulls fed diets based on different types of silages. *Czech J. Anim. Sci.* **2007**, *52*, 269–276. [CrossRef]

58. Neel, J.P.S.; Fontenot, J.P.; Clapham, W.M.; Duckett, S.K.; Felton, E.E.D.; Scaglia, G.; Bryan, W.B. Effects of winter stocker growth rate and finishing system on: I. Animal performance and carcass characteristics. *J. Anim. Sci.* **2007**, *85*, 2012–2018. [CrossRef] [PubMed]

59. Aydin, R.; Yanar, M.; Diler, A.; Kocyigit, H.; Tuzemen, N. Effects of different slaughter ages on the fattening performance, slaughter and carcass traits of brown swiss and holstein friesian young bulls. *Indian J. Anim. Res.* **2013**, *47*, 10–16.

60. Franco, D.; Bispo, E.; González, L.; Vázquez, J.A.; Moreno, T. Effect of finishing and ageing time on quality attributes of loin from the meat of Holstein-Fresian cull cows. *Meat Sci.* **2009**, *83*, 484–491. [CrossRef] [PubMed]

61. Price, M.; Jones, S.; Mathison, G.; Berg, R. The effects of increasing dietary roughage level and slaughter weight on the feedlot performance and carcass characteristics of bulls and steers. *Can. J. Anim. Sci.* **1980**, *60*, 345–358. [CrossRef]

62. Hernandez-Calva, L.M.; He, M.; Juarez, M.; Aalhus, J.L.; Dugan, M.E.R.; McAllister, T.A. Effect of flaxseed and forage type on carcass and meat quality of finishing cull cows. *Can. J. Anim. Sci.* **2011**, *91*, 613–622. [CrossRef]

63. Keady, T.W.J.; Gordon, A.W.; Moss, B.W. Effects of replacing grass silage with maize silages differing in inclusion level and maturity on the performance, meat quality and concentrate-sparing effect of beef cattle. *Animal* **2013**, *7*, 768–777. [CrossRef] [PubMed]

64. Ellies-Oury, M.P.; Renand, G.; Perrier, G.; Krauss, D.; Dozias, D.; Jailler, R.; Dumont, R. Influence of selection for muscle growth capacity on meat quality traits and properties of the rectus abdominis muscle of Charolais steers. *Livest. Sci.* **2012**, *150*, 220–228. [CrossRef]

65. Picard, B.; Robelin, J.; Geay, Y. Influence of castration and postnatal energy restriction on the contractile and metabolic characteristics of bovine muscle. *Ann. Zootech.* **1995**, *44*, 347–357. [CrossRef]

66. Durunna, O.N.; Block, H.C.; Lwaasa, A.D.; Scott, S.L.; Robins, C.; Khakbazan, M.; Dugan, M.E.R.; Aalhus, J.L.; Aliani, M.; Lardner, H.A. Impact of calving seasons and feeding systems in western Canada. II. Meat composition and organoleptic quality of steaks. *Can. J. Anim. Sci.* **2014**, *94*, 583–593. [CrossRef]

67. Bures, D.; Barton, L.; Zahradkova, R.; Teslik, V.; Krejcova, M. Chemical composition, sensory characteristics, and fatty acid profile of muscle from Aberdeen Angus, Charolais, Simmental, and Hereford bulls. *Czech J. Anim. Sci.* **2006**, *51*, 279–284. [CrossRef]

68. Serrano, E.; Pradel, P.; Jailler, R.; Dubroeucq, H.; Bauchart, D.; Hocquette, J.-F.; Listrat, A.; Agabriel, J.; Micol, D. Young Salers suckled bull production: Effect of diet on performance, carcass and muscle characteristics and meat quality. *Animal* **2007**, *1*, 1068–1079. [CrossRef] [PubMed]

69. Oury, M.-P.; Dumont, R.; Agabriel, C.; Agabriel, J.; Blanquet, J.; Dransfield, E.; Istasse, L.; Micol, D.; Picard, B.; Roux, M. Differentiation of meat sensorial quality linked to farming practices among Charolais heifers. In Proceedings of the 13th Rencontres Recherches Ruminants, Paris, France, 7–8 December 2006; pp. 313–316.

70. Soulat, J.; Léger, S.; Picard, B.; Monteils, V. Improving beef sensory quality through breeding practices management. In Proceedings of the 61th International Congress of Meat Science and Technology, Clermont-Ferrand, France, 23–28 August 2015; pp. 1–4.

71. Soulat, J.; Picard, B.; Leger, S.; Monteils, V. Prediction of beef carcass and meat traits from rearing factors in young bulls and cull cows. *J. Anim. Sci.* **2016**, *94*, 1712–1726. [CrossRef] [PubMed]

72. Gagaoua, M.; Monteils, V.; Couvreur, S.; Picard, B. Identification of biomarkers associated with the rearing practices, carcass characteristics, and beef quality: An integrative approach. *J. Agric. Food Chem.* **2017**, *65*, 8264–8278. [CrossRef] [PubMed]

73. Bonny, S.P.F.; Pethick, D.W.; Legrand, I.; Wierzbicki, J.; Allen, P.; Farmer, L.J.; Polkinghorne, R.J.; Hocquette, J.-F.; Gardner, G.E. European conformation and fat scores have no relationship with eating quality. *Animal* **2016**, *10*, 996–1006. [CrossRef] [PubMed]
74. Gagaoua, M.; Picard, B.; Soulat, J.; Monteils, V. Clustering of sensory eating qualities of beef: Consistencies and differences within carcass, muscle, animal characteristics and rearing factors. *Livest. Sci.* **2018**, *214*, 245–258. [CrossRef]

 *foods* 

Article

Effect of the Rearing Managements Applied during Heifers' Whole Life on Quality Traits of Five Muscles of the Beef Rib

Julien Soulat, **Valérie Monteils** and **Brigitte Picard** *

Université Clermont Auvergne, INRA, VetAgro Sup, UMR Herbivores, F-63122 Saint-Genès-Champanelle, France; julien.soulat@inra.fr (J.S.); valerie.monteils@vetagro-sup.fr (V.M.)
* Correspondence: brigitte.picard@inra.fr; Tel.: +33(0)-4-73-62-40-56

Received: 9 April 2019; Accepted: 8 May 2019; Published: 10 May 2019

Abstract: The aim of this work was to study the effects of four different rearing managements applied during the heifers' whole life period (WLP) on muscles from ribs in the chuck sale section. The characteristics of meat studied were the sensory, rheological, and color of the longissimus muscle (LM) and the rheological traits of four other muscles: complexus, infraspinatus, rhomboideus, and serratus ventralis. The main results showed that WLP rearing managements did not significantly impact the tenderness (sensory or rheological analyses) of the rib muscles. The LM had high ($p \leq 0.05$) typical flavor and was appreciated when heifers received a WLP rearing management characterized by a short pasture duration during the heifers' whole life (WLP-E). The heifers' management characterized by a long pasture duration during their life (WLP-A) or by a diet composed mainly of hay during the growth and fattening periods (WLP-F), had lower typical flavor and were less appreciated than those with WLP-E management. Moreover, the LM color was redder for heifers of WLP-E than those of the WLP-A and WLP-F groups. This study confirmed that it is possible to obtain similar meat qualities with different rearing managements.

Keywords: rearing managements; chuck sale section; meat sensory properties; meat rheological properties; color attributes; longissimus muscle; complexus muscle; infraspinatus muscle; rhomboideus muscle; serratus ventralis muscle

---

1. Introduction

Beef carcasses are composed of many skeletal muscles [1] with different properties, e.g., structural, metabolic and contractile [2,3]. After beef carcass cuts, the wholesale cuts purchased by consumers could be composed of different muscles, e.g., ribs and short ribs of beef. It is well known that meat quality traits can be impacted by different factors, e.g., animal type (sex and breed) [4–6], stress (transport, slaughter condition) [7–9] and rearing managements [10–12]. Many studies showed that different rearing factors observed during the fattening period (e.g., slaughter age [6,13], fattening duration [14,15], and fattening diet [16–18]) had an effect on meat quality traits. Recent studies have shown that the rearing management (combining different rearing factors) applied during the animal's whole life period (i.e., from birth to slaughter, whole life period (WLP)) could have an impact on the carcass and/or meat traits [10–12]. These recent studies had observed the effect of the WLP rearing managements on the flank steak (rectus abdominis muscle, RA) [10,11]. The aim of the present work was to study the effects of the WLP rearing managements defined by Soulat et al. [10] on the longissimus muscle (LM) traits (sensory, rheological and color properties), in the ribs of the chuck sale section. Moreover, the effects of the WLP rearing managements were also studied on the toughness (rheological) of four other muscles composing these ribs: complexus, infraspinatus, rhomboideus, and serratus ventralis. In the literature, the LM is considered as a reference muscle and the other muscles composing the rib have been poorly studied. One originality of this work was to observe the effects of the same WLP rearing management on these different muscles.

2. Materials and Methods

2.1. Animals and Rearing Managements

The present study was realized from the same experiment presented by Soulat et al. [10] in partnership with the protected geographical indication (PGI) Fleur d'Aubrac. Briefly, this PGI produces exclusively crossbreed Charolais with Aubrac heifers for the meat production. The breeding of the heifers was realized respecting the design brief of this PGI, which is based mainly on a grass diet (conserved and pasture) [19]. The heifers ($n = 48$) considered in this work, were produced in eight commercial farms. For this study, the same four rearing managements described by Soulat et al. [10] applied during the animal's whole life period were considered: WLP-A, WLP-D, WLP-E and WLP-F (Figure 1). Briefly, these WLP were the combination of different rearing management clusters characterizing three periods of the heifers' life: pre-weaning, growth and fattening [10]. The clusters at each period of heifers' life were obtained statistically from the data collected during the surveys. The heifers receiving the rearing management WLP-A and WLP-D had the same pre-weaning management (PWP-clust1). The rearing managements applied during the growth and the fattening periods (GP-clust1 and FP-clust3) were specific to the WLP-A rearing management. The WLP-D had specific management during the growth period of the heifers (GP-clust3). The WLP-D and WLP-F had the same management during the fattening period (FP-clust2). The WLP-E and WLP-F rearing managements had the same rearing management during the pre-weaning period (PWP-clus2) and the growth period (GP-clust2). The rearing management applied during the fattening period of the heifers (FP-clust1) was specific to the WLP-E rearing management. Briefly, the WLP-A and WLP-E rearing managements were mainly characterized by a long and a short pasture duration during the heifers' whole life, respectively. The WLP-D rearing management was mainly characterized by a high concentrate quantity intake by the heifers during the growth and the fattening periods. The WLP-F rearing management was mainly characterized by a diet composed mainly of hay during the growth and the fattening periods.

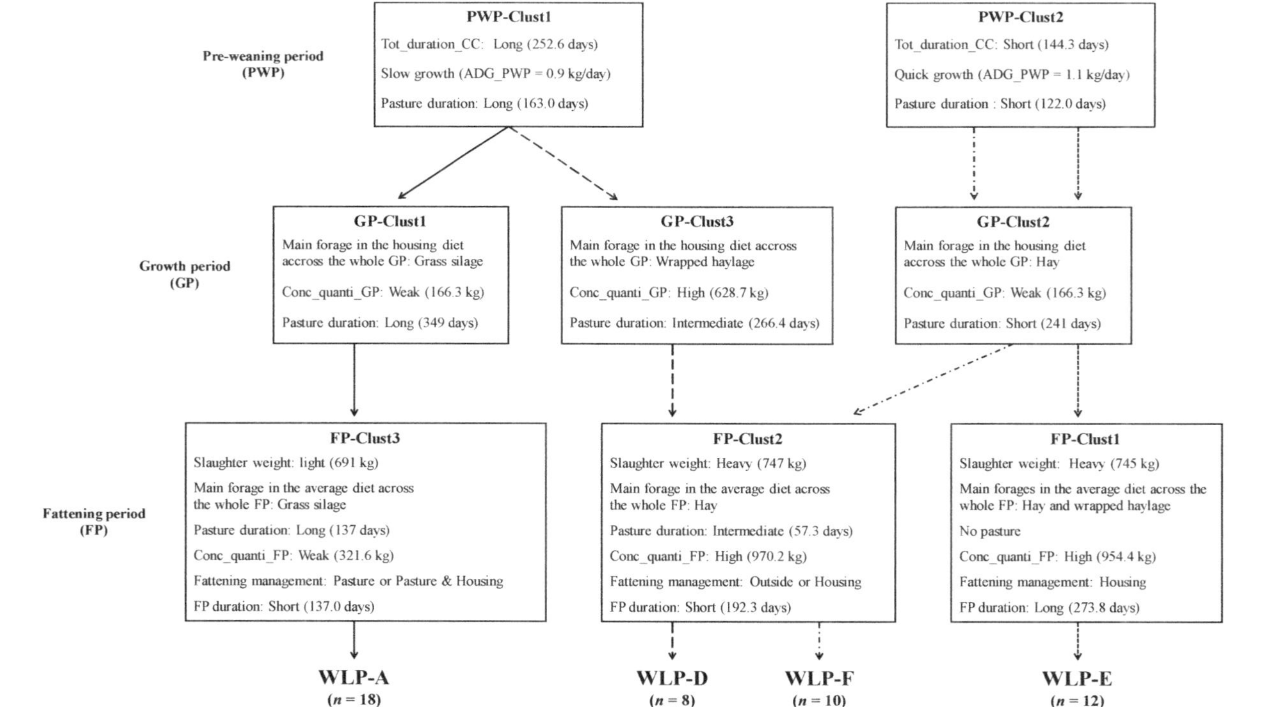

Figure 1. Description of the four rearing managements applied during the heifers' whole life period (WLP) defined by Soulat et al. [10] with a focus on the rearing factors with the most differences between groups. Tot_duration_CC: Total time spent by the calf with her mother between the birth and the weaning, ADG: Average daily gain, Conc_quanti: Total concentrate quantity intake during the period of heifers' life.

2.2. Animals Slaughtering, Muscle Sampling and Meat Quality Evaluation

The heifers were slaughtered by exsanguination after stunning in the same industrial slaughterhouse (Abattoir du Gévaudan, Antrenas, France) as described in Soulat et al. [10].

Twenty-four hours post-mortem, two beef ribs (the 5th and 4th ribs), localized in the chuck sale section, were collected from the right-hand side of the carcass. Two hours after excision from the carcass, the color was measured on the LM of the rib (side of the cutting section between the 6th and the 5th ribs). The color was expressed in CIE L*a*b* units [20], using a spectrophotometer (Konica Minolta CM-600d, Osaka, Japan) (light source D65, 8 mm diameter measurement area, and 0° standard observer). Before each measurement session, the spectrophotometer was calibrated by performing a black and a white calibration. Five measurements (randomly distributed on the muscle) per LM were realized to characterize the color of this muscle. The chroma ($C*$) and hue angle ($h*$) were calculated from the a* and b* values, as realized by Gagaoua et al. [21].

The rib samples of each animal were vacuum-packaged and chilled at 4 °C during 14 days for aging. At the end of the aging period, the samples were frozen at −20 °C until the analyses [22].

In this study, the sensory evaluation was only realized on the LM by a trained tasting panel (15 members), using a monadic test. The sensory evaluation was realized using the same process described by Soulat et al. [10]. Briefly, the members of the trained tasting panel had 20 training sessions (between 1 and 1.5 h per session) before starting the sensory evaluation of the LM, in accordance with ISO 8586 [23]. Before each tasting session, the ribs were thawed and dissected to separate the different muscles. The LM samples were cut into two sub-samples: The first for the sensory evaluation, and the second for the shear force measurement. For sensory evaluation, the LM samples were cut into steaks, and cooked on a double-face grill to reach an internal temperature of 55 °C. Then, samples were cut (size 15 × 20 × 20 mm) and 3 or 4 pieces were served to each member of the trained panel. At each tasting session, a Latin square presentation was used to evaluate the sensory traits of five samples.

The trained panel evaluated six sensory descriptors: The initial tenderness, overall tenderness, initial juiciness, overall juiciness, typical flavor, and overall acceptability. The initial tenderness and juiciness were defined as an evaluation at the first bite of the tenderness, and juiciness, respectively. In contrast, the overall tenderness and juiciness were an evaluation of the tenderness and the juiciness, respectively, before swallowing the meat sample. These six sensory descriptors were measured on a 10-point non-graduated scale from a score of 0 (tough, dry, slight, and highly disliked) to a score of 10 (very tender, very juicy, strong, and highly liked).

The shear force was measured on five muscles of the rib, located in the chuck sale section: The complexus (CP), infraspinatus (IF), longissimus (LM), rhomboideus (RH), and serratus ventralis (SV) (Figure 2). The shear force was evaluated using a Warner–Braztler apparatus (EZ-SX set assay EU RoHS, Shimadzu, Kyoto, Japan) on raw meat [24]. For each muscle from the rib, at least five meat portions (length: 1.5 to 3 cm, width: 1 cm and thickness: 0.5 to 1 cm) were cut perpendicular to the fibers [25]. From the different measurements per muscle, the shear force was calculated using the Trapezium X 1-5.1 software (Shimadzu, Kyoto, Japan).

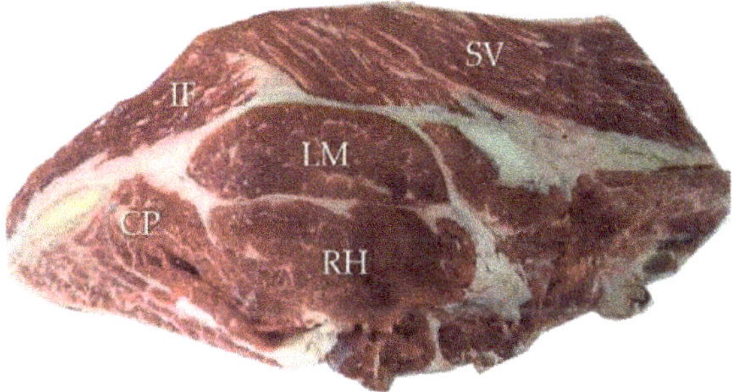

Figure 2. Localization of five rib muscles: complexus (CP), infraspinatus (IF), longissimus (LM), rhomboïdeus (RH), and serratus ventralis (SV).

The distribution of the meat rib traits considered in this study are presented in Table 1.

Table 1. Description of the rib muscle traits.

Meat Traits	Mean	SD	Min	Max
Sensory description of longissimus muscle (scale 0–10) [1]				
Initial tenderness	7.25	1.44	1.42	10.00
Overall tenderness	7.10	1.62	1.45	10.00
Initial juiciness	6.63	1.54	1.23	9.95
Overall juiciness	6.65	1.64	0.61	9.97
Typical flavor	6.60	1.57	0.84	9.99
Overall acceptability	6.37	1.68	0.27	9.96
Color of longissimus muscle				
L^*	32.90	2.80	27.70	41.32
a^*	18.23	2.53	12.67	26.52
b^*	17.82	2.47	8.64	21.39
C^*	25.10	4.39	3.06	33.03
h^*	43.50	6.93	3.98	50.42
Shear force (N/cm)				
Complexus	61.96	13.67	30.93	87.62
Infraspinatus	99.45	45.61	45.49	278.23
Longissimus	45.92	13.38	24.03	88.54
Rhomboideus	61.20	16.95	13.32	112.93
Serratus ventralis	56.23	18.52	31.18	125.29

SD: Standard deviation. Min: Minimum. Max: Maximum. [1], Scale for initial tenderness, overall tenderness, initial juiciness, overall juiciness, typical flavor, overall acceptability: 0 = tough, dry, slight, and highly disliked and 10 = very tender, very juicy, strong, and highly liked.

2.3. Statistical Analyses

The statistical analyses were realized using R 3.5.2 software (R Core Team, Vienna, Austria) [26].

A descriptive analysis of the dataset, using quantile-quantile plots, was performed to observe the normality of the distribution [27]. Then, for each meat trait, an analysis of variance (ANOVA) was performed to evaluate their dependence on the four considered WLP rearing managements. In all ANOVA, the farm effect was tested. If it was significant, it was considered as a random effect in the ANOVA. In the ANOVAs of the sensory parameters, the effect of the member of the trained tasting panel and the animal effect were considered as random effects. For the ANOVAs without a random effect, if the result was significant a post-hoc Tukey test was performed, using the agricolae package [28]. The ANOVAs containing random effects were developed using the lmerTest package [29].

If the results of these ANOVAs were significant, a post-hoc Tukey test was also performed, using the multcomp package [30].

An effect was considered significant at $p \leq 0.05$ and a tendency was considered for $0.05 < p \leq 0.10$.

3. Results and Discussion

The four WLP rearing managements considered, in this work, had no significant effect on the tenderness of the LM, as measured by a trained tasting panel or by shear force (Table 2). This result is in accordance with those of Soulat et al. [10] who observed no significant effect on the tenderness of the flank steak (RA muscle). However, tendencies ($p \leq 0.10$) were displayed for the initial tenderness and the shear force. The heifers receiving the WLP-E or WLP-F rearing managements (Figure 1) tended ($p \leq 0.10$) to produce LM with a higher initial tenderness than those receiving the WLP-A and WLP-D rearing managements (Table 2). The WLP-E and WLP-F rearing managements were characterized by a short pasture duration during the heifers' whole life and by a diet composed mainly of hay during the growth and fattening periods, respectively [10]. However, the WLP-A and WLP-F rearing managements were characterized by a long pasture duration during the heifers' whole life and by a high concentrate quantity intake during the growth and fattening periods. The raw LM also tended to be less tough when the heifers received the WLP-E or WLP-F rearing managements (Table 2). The importance of collagen in meat could explain this weak difference. However, this tendency was not found for the overall tenderness, as evaluated by the trained tasting panel. It is possible that the cooking mitigated this tendency. The four considered rearing managements did not also significantly ($p > 0.05$) affect the toughness (evaluated by shear force) of the other rib muscles uncooked: CP, IF, RH, and SV. Based on our overall results, the variation of rearing managements applied at the different key periods of the heifers' life seems to have no impact on the toughness of the five raw rib muscles. Our results show that it is possible to obtain the same overall tenderness of the LM with different rearing managements. The muscle traits could have more impact on the tenderness than the rearing management applied during the heifers' whole life.

The LM had a significant ($p < 0.05$) higher initial juiciness for heifers receiving the WLP-E rearing management than those receiving the WLP-A (Table 2). Nevertheless, the four WLP rearing managements had no significant effect on the overall juiciness. These results are in accordance with those of Soulat et al. [10] for the flank steak.

According to our results, the WLP-E rearing management produced an LM with a significantly ($p < 0.05$) higher typical flavor and higher appreciated meat than the WLP-A and WLP-F rearing managements (Table 2). The WLP-E and WLP-F rearing managements differ only by the rearing management applied during the fattening period, which was different (Figure 1). The fattening period duration was significantly ($p < 0.05$) longer in the WLP-E rearing management than the WLP-F and the WLP-A managements [10]. In cull cows, studies of the literature showed a significant increase in the LM flavor intensity when the fattening duration was longer [15,31]. Moreover, in the flavor prediction model developed by Soulat et al. [32], an increase of the fattening duration allowed to increase the LM flavor intensity in the cull cows. The main forages in the WLP-E and WLP-F rearing managements were different: Hay or hay and wrapped haylage, respectively. According to the results of different studies, the composition of the fattening diet could have no impact ($p > 0.05$) on the LM flavor intensity, in heifers and steers [33–35]. During the fattening period, the heifers from the WLP-F rearing management pastured, whereas those from the WLP-E were inside [10]. With less walking during the fattening period for the heifers from the WLP-E rearing management, it was possible that their LM was less oxidative and had more intramuscular lipid content than those from the WLP-F. In their study, Jurie et al. [36] showed that the RA muscle was more oxidative (an increase of isocitrate dehydrogenase concentration) when the steers moved during pasture. Moreover, studies showed that cattle with fattening management with pasture produced leaner carcasses [37,38]. In consequence, these carcasses could have less marbling. However, Soulat et al. [10] did not find a significant effect ($p > 0.05$) of the four WLP rearing managements on the flavor intensity of flank steak. The WLP-A and

WLP-E rearing managements had no rearing managements in common during the different key periods of the heifers' life (Figure 1). During the pre-weaning period, the heifers from the WLP-A rearing management had an average daily gain (ADG) lower than those from the WLP-E [10]. According to the results of Hennessy et al. [39], the meat flavor intensity was lower when cattle had a quick growth before weaning. Our results are not in accordance with these results. This difference could be explained by the fact that the heifers of our study did not have the same rearing management during the growth and the fattening periods. During the fattening period, the heifers from the WLP-A rearing management ingested a lower concentrate quantity than those from the WLP-E rearing management. Different studies showed that an increase of the concentrate quantity in the fattening diet increased the LM flavor intensity in steers [16,40]. However, other studies did not observe an effect of the concentrate quantity in the fattening diet on the LM flavor intensity [18,41,42]. As the WLP-F rearing management, the fattening period duration was shorter in the WLP-A than in the WLP-E rearing management [10]. The main fattening managements realized in the WLP-A was pasture or pasture and housing, whereas, it was housing in the WLP-E (Figure 1). In their study, Duckett et al. [43] observed that the LM flavor intensity was significantly higher when the fattening management was a concentrate ration compared to mixed pasture. In the WLP-E rearing management, the heifers were slaughtered heavier than those from the WLP-A. According to the results of different studies, the slaughter weight had no impact on the LM flavor intensity, in young bulls and steers [44,45]. In the study of Oury et al. [46] and Soulat et al. [10], in heifers, the flavor intensity of flank steak was not impacted ($p > 0.05$) by the rearing management applied during the heifers' whole life. Concerning the flavor, the LM seems to be more sensitive to the WLP rearing management variations compared to the flank steak.

In our study, the effect of the WLP rearing managements was the same on the overall acceptability and the typical flavor of the LM (Table 2). As the WLP rearing managements did not affect the tenderness and the juiciness of the LM, we suppose that the overall acceptability of the LM was strongly linked with the typical flavor. According to the results of different studies in cattle, the slaughter weight [45], the fattening diet composition or the fattening management [33,34,47] did not impact the LM overall acceptability.

For the LM color, the four rearing managements had no significant ($p > 0.05$) effect on the L*, b*, and h^* color parameters (Table 2). According to the a* values, the LM was significantly redder for heifers receiving the WLP-E than those receiving the WLP-A and WLP-D rearing managements. The heifers receiving the WLP-E management also produced LM with greater red color intensity, according to the C* values, than those receiving the WLP-A. To our knowledge, the impact of the pre-weaning and growth period on the LM color have not been studied. However, many studies showed that the fattening duration [48] and the composition of the fattening diet [18,49,50] did not affect ($p > 0.05$) the LM color. However, many studies showed that fattening management had a significant ($p < 0.05$) effect on the LM color [37,43,51]. In these studies, the cattle with a pasture period during their fattening produced LM meat with a lower a* than those fattened in housing. The L* and b* parameters were also impacted by the rearing managements applied during the fattening period. In our study, the heifers from the WLP-E rearing management had higher a* value than those receiving the WLP-A and WLP-F rearing managements (Table 2). This result is in accordance with the literature. In the WLP-E rearing management, the heifers were not pastured during their fattening compared to the heifers receiving the WLP-A and WLP-F rearing managements (Figure 1). In our study, in contrast to the results of Duckett et al. [43], Cozzi et al. [51], and Huuskonen et al. [37], the L* and b* color parameters were not impacted by the WLP rearing managements. As rearing management is multifactorial, it is possible that the effect of the fattening management was mitigated by this combination with the other rearing factors.

Table 2. Impact of the four rearing managements applied during the heifers' whole life period (WLP) on the traits of five rib muscles.

Meat Traits	WLP Rearing Managements [1]												p
	WLP-A ($n = 18$)			WLP-D ($n = 8$)			WLP-E ($n = 12$)			WLP-F ($n = 10$)			
	Mean	SD	SE	Mean	SD	SE	Mean	SD	SE	Mean	SD	SE	
Sensory description of longissimus muscle (scale 0–10) [2]													
Initial tenderness	7.10	1.55	0.24	7.04	1.43	0.10	7.53	1.30	0.10	7.34	1.34	0.11	0.06
Overall tenderness	6.91	1.74	0.26	6.98	1.51	0.11	7.36	1.47	0.11	7.22	1.61	0.13	0.21
Initial juiciness	6.46 [a]	1.58	0.25	6.61 [ab]	1.52	0.11	6.98 [b]	1.41	0.11	6.52 [ab]	1.59	0.13	0.04
Overall juiciness	6.53	1.75	0.28	6.53	1.51	0.11	6.97	1.50	0.11	6.58	1.67	0.14	0.12
Typical flavor	6.52 [a]	1.60	0.25	6.55 [ab]	1.32	0.09	7.00 [b]	1.47	0.11	6.32 [a]	1.75	0.14	0.01
Overall acceptability	6.14 [a]	1.68	0.26	6.29 [ab]	1.49	0.11	6.92 [b]	1.67	0.13	6.24 [a]	1.71	0.14	0.005
Color of longissimus muscle													
L^*	32.36	2.54	0.60	33.38	2.75	0.97	32.77	3.70	1.07	33.65	2.16	0.68	0.66
a^*	17.07 [a]	2.02	0.48	18.67 [ab]	1.67	0.59	20.51 [b]	2.78	0.80	17.23 [a]	1.79	0.57	<0.001
b^*	16.90	3.31	0.78	18.73	1.29	0.46	18.55	1.89	0.54	17.84	1.59	0.50	0.21
C^*	24.10 [a]	3.32	0.78	26.45 [ab]	1.99	0.70	27.70 [b]	2.93	0.85	24.84 [ab]	1.99	0.63	0.008
h^*	44.36	5.11	1.20	45.13	1.53	0.54	42.24	3.40	0.98	46.03	3.06	0.98	0.14
Shear force (N/cm)													
Complexus muscle	62.21	13.01	3.07	61.27	18.12	6.41	64.17	13.85	4.00	59.41	12.33	3.90	0.88
Infraspinatus muscle	115.92	48.42	11.41	97.00	69.56	24.59	80.64	21.54	6.22	94.33	32.50	10.28	0.21
Longissimus muscle	50.17	15.47	3.65	51.27	17.67	6.25	40.24	5.75	1.66	40.81	8.09	2.56	0.08
Rhomboideus muscle	61.74	13.12	3.09	64.38	28.49	10.07	58.90	14.16	4.09	60.45	16.82	5.32	0.92
Serratus ventralis muscle	59.60	15.95	3.76	46.01	16.20	5.73	62.34	23.29	6.72	51.00	15.74	4.98	0.16

n: number of heifers. SD: Standard deviation. SE: Standard error. Values followed by different letters (a, b) are significantly different from each other at $p \leq 0.05$. [1], The WLP rearing managements were the rearing managements described by Soulat et al. [10]. [2], Scale for initial tenderness, overall tenderness, initial juiciness, overall juiciness, typical flavor, overall acceptability: 0 = tough, dry, slight, and highly disliked and 10 = very tender, very juicy, strong, and highly liked.

Considering the WLP rearing managements, it is difficult to identify the rearing factors, which had the greatest impact on meat quality. In our study, the rearing factors varied independently, whereas, in the literature, many studies observed the effect of one or two rearing factors in an experimental condition. It is possible that some rearing factors have antagonist effects on meat quality during the life of the heifers. The difficulty of interpretation of our results was considering all the combinations of all the rearing factors. Moreover, the impact of rearing management applied during a period of the heifers' life can be mitigated or amplified by the rearing management applied during another period of life.

4. Conclusions

The originality of this study was to observe the impact of four different rearing managements applied during the heifers' whole life on the traits of the LM (sensory, rheological, and color) in the ribs of the chuck sale section. In accordance with the results of Soulat et al. [10], these new results showed that different rearing managements applied during the heifers' whole life obtained the same meat quality, particularly, the tenderness and juiciness. Finally, all the results of this experiment showed that the LM traits seem to be more sensitive to variations of the rearing managements than the flank steak, in particular, for the typical flavor, the overall acceptability, and the a* value. Our results also showed that these rearing managements did not significantly impact the toughness of the other four rib muscles uncooked: complexus, infraspinatus, rhomboideus, and serratus ventralis. These results demonstrate that farmers could adapt, to a certain extent, their rearing managements during the heifers' whole life, according to the hazards (e.g., drought, price of concentrates), with limited consequences on the meat quality.

Author Contributions: J.S. realized this experiment, the data analyses, wrote the article, prepared tables and/or figures and approved the final draft. V.M. and B.P. co-conceived and designed this experiment, co-wrote and reviewed this article and approved the final draft.

Funding: This research was supported by the PGI Fleur d'Aubrac through funding from the French regional council of Ocitanie (number 17011785) and by the Agence Nationale de la Recherche of French government through the program Investissement d'Avenir (16-IDEX-0001 CAP 20-25).

Acknowledgments: The authors thank the members of the PGI Fleur d'Aubrac who participated in this experiment. The authors thank also all staff of the Institut National de la Recherche Agronomique, in particular, David Chadeyron for sampling at slaughter. Finally, the authors would like to thank Marie-Pierre Elies-Oury and Sandrine Papillon (Bordeaux Sciences Agro) for the realization of meat sensory and rheological analyses.

Conflicts of Interest: The authors declare no conflicts of interest.

References

1.   University of Nebraska-Lincoln Bovine Myology. Available online: http://bovine.unl.edu/main/index.php (accessed on 20 March 2019).
2.   Totland, G.; Kryvi, H. Distribution Patterns of Muscle-Fiber Types in Major Muscles of the Bull (bos-Taurus). *Anat. Embryol. (Berl.)* **1991**, *184*, 441–450. [CrossRef] [PubMed]
3.   Schreurs, N.; Garcia, F.; Agabriel, J.; Jurie, C.; Micol, D.; Picard, B. Development of muscle characteristics with age in cattle according to muscle and animal type: meta-analysis. In Proceedings of the 14th Rencontres Recherches Ruminants, Paris, France, 6–7 December 2007; pp. 121–124.
4.   Chambaz, A.; Scheeder, M.R.L.; Kreuzer, M.; Dufey, P.A. Meat quality of Angus, Simmental, Charolais and Limousin steers compared at the same intramuscular fat content. *Meat Sci.* **2003**, *63*, 491–500. [CrossRef]
5.   Pesonen, M.; Huuskonen, A. Production, carcass characteristics and valuable cuts of beef breed bulls and heifers in Finnish beef cattle population. *Agric. Food Sci.* **2015**, *24*, 164–172. [CrossRef]
6.   Bures, D.; Barton, L. Growth performance, carcass traits and meat quality of bulls and heifers slaughtered at different ages. *Czech J. Anim. Sci.* **2012**, *57*, 34–43. [CrossRef]
7.   Gruber, S.L.; Tatum, J.D.; Engle, T.E.; Chapman, P.L.; Belk, K.E.; Smith, G.C. Relationships of behavioral and physiological symptoms of preslaughter stress to beef longissimus muscle tenderness. *J. Anim. Sci.* **2010**, *88*, 1148–1159. [CrossRef]

8. Velarde Calvo, A.; Dalmau, A. Preslaughter handling. In *Encyclopedia of Meat Sciences*, 2nd ed.; Dikeman, M., Devine, C., Eds.; Academic Press: Cambridge, MA, USA, 2014; Volume 1, pp. 95–101. ISBN 9780123847348.

9. Ferguson, D.M.; Warner, R.D. Have we underestimated the impact of pre-slaughter stress on meat quality in ruminants? *Meat Sci.* **2008**, *80*, 12–19. [CrossRef]

10. Soulat, J.; Picard, B.; Léger, S.; Ellies-Oury, M.-P.; Monteils, V. Preliminary Study to Determinate the Effect of the Rearing Managements Applied during Heifers' Whole Life on Carcass and Flank Steak Quality. *Foods* **2018**, *7*, 160. [CrossRef]

11. Soulat, J.; Picard, B.; Léger, S.; Monteils, V. Prediction of beef carcass and meat quality traits from factors characterising the rearing management system applied during the whole life of heifers. *Meat Sci.* **2018**, *140*, 88–100. [CrossRef] [PubMed]

12. Monteils, V.; Sibra, C. Rearing practices in each life period of beef heifers can be used to influence the carcass characteristics. *Ital. J. Anim. Sci.* **2019**, *18*, 734–745. [CrossRef]

13. Ahnstrom, M.L.; Hessle, A.; Johansson, L.; Hunt, M.C.; Lundstrom, K. Influence of slaughter age and carcass suspension on meat quality in Angus heifers. *Animal* **2012**, *6*, 1554–1562. [CrossRef]

14. Schnell, T.D.; Belk, K.E.; Tatum, J.D.; Miller, R.K.; Smith, G.C. Performance, carcass, and palatability traits for cull cows fed high-energy concentrate diets for 0, 14, 28, 42, or 56 days. *J. Anim. Sci.* **1997**, *75*, 1195–1202. [CrossRef]

15. Vestergaard, M.; Madsen, N.T.; Bligaard, H.B.; Bredahl, L.; Rasmussen, P.T.; Andersen, H.R. Consequences of two or four months of finishing feeding of culled dry dairy cows on carcass characteristics and technological and sensory meat quality. *Meat Sci.* **2007**, *76*, 635–643. [CrossRef] [PubMed]

16. Kerth, C.R.; Braden, K.W.; Cox, R.; Kerth, L.K.; Rankins, D.L., Jr. Carcass, sensory, fat color, and consumer acceptance characteristics of Angus-cross steers finished on ryegrass (Lolium multiflorum) forage or on a high-concentrate diet. *Meat Sci.* **2007**, *75*, 324–331. [CrossRef]

17. Geay, Y.; Picard, B.; Jailler, R.D.; Jailler, R.T.; Listrat, A.; Jurie, C.; Bayle, M.C.; Touraille, C. Effects of diet composition on growth performance, body composition, muscle characteristics and meat quality of young growing Salers bulls. In Proceedings of the 4th Rencontres Recherches Ruminants, Paris, France, 4–5 December 1997; pp. 307–310.

18. Faucitano, L.; Chouinard, P.Y.; Fortin, J.; Mandell, I.B.; Lafreniere, C.; Girard, C.L.; Berthiaume, R. Comparison of alternative beef production systems based on forage finishing or grain-forage diets with or without growth promotants: 2. Meat quality, fatty acid composition, and overall palatability. *J. Anim. Sci.* **2008**, *86*, 1678–1689. [CrossRef]

19. Legifrance.gouv.fr, Arrêté du 13 octobre 2008 portant homologation du cahier des charges de l'indication géographique protégée (IGP) «Génisse fleur d'Aubrac». Available online: https://www.legifrance.gouv.fr/eli/arrete/2008/10/13/AGRP0823338A/jo/texte (accessed on 8 April 2019).

20. Commission International de l'Eclairage. *Colorimetry*, 2nd ed.; Commission International de l'Eclairage: Vienna, Austria, 1986.

21. Gagaoua, M.; Picard, B.; Monteils, V. Associations among animal, carcass, muscle characteristics, and fresh meat color traits in Charolais cattle. *Meat Sci.* **2018**, *140*, 145–156. [CrossRef] [PubMed]

22. Dransfield, E.; Martin, J.F.; Bauchart, D.; Abouelkaram, S.; Lepetit, J.; Culioli, J.; Jurie, C.; Picard, B. Meat quality and composition of three muscles from French cull cows and young bulls. *Anim. Sci.* **2003**, *76*, 387–399. [CrossRef]

23. International Organization of Standardization (ISO). *ISO_8586: Sensory analysis-General Guidelines for the Selection, Training and Monitoring of Selected Assessors and Expert Sensory Assessors*; ISO: Geneva, Switzerland, 2012; pp. 1–28.

24. Wheeler, T.L.; Shackelford, S.D.; Johnson, L.P.; Miller, M.F.; Miller, R.K.; Koohmaraie, M. A comparison of Warner-Bratzler shear force assessment within and among institutions. *J. Anim. Sci.* **1997**, *75*, 2423–2432. [CrossRef]

25. Oury, M.P.; Picard, B.; Briand, M.; Blanquet, J.P.; Dumont, R. Interrelationships between meat quality traits, texture measurements and physicochemical characteristics of M. rectus abdominis from Charolais heifers. *Meat Sci.* **2009**, *83*, 293–301. [CrossRef] [PubMed]

26. R Core Team. *R: A Language and Environment for Statistical Computing*; R Foundation for Statistical Computing: Vienna, Austria, 2018.

27. Azaïs, J.-M.; Bardet, J.-M. *Le Modèle Linéaire par l'Exemple Régression, Analyse de la Variance et Plans d'Expériences Illustrations numériques avec les Logiciels R, SAS et Splus*; Sciences Sup; Dunod: Paris, France, 2006.

28. Agricolae: Statistical Procedures for Agricultural Research. Available online: http://CRAN.R-project.org/package=agricolae (accessed on 3 March 2019).

29. Kuznetsova, A.; Brockhoff, P.B.; Christensen, R.H.B. lmerTest Package: Tests in Linear Mixed Effects Models. *J. Stat. Softw.* **2017**, *82*, 1–26. [CrossRef]

30. Hothorn, T.; Bretz, F.; Westfall, P. Simultaneous Inference in General Parametric Models. *Biom. J.* **2008**, *50*, 346–363. [CrossRef] [PubMed]

31. Dumont, R.; Roux, M.; Agabriel, J.; Touraille, C.; Bonnemaire, J.; Malterre, C.; Robelin, J. Engraissement des vaches de réforme de race charolaise. Facteurs de variation des performances zootechniques, de la composition tissulaire des carcasses et de la qualité organoleptique de la viande. *INRA Prod. Anim.* **1991**, *4*, 271–286.

32. Soulat, J.; Picard, B.; Léger, S.; Monteils, V. Prediction of beef carcass and meat traits from rearing factors in young bulls and cull cows. *J. Anim. Sci.* **2016**, *94*, 1712–1726. [CrossRef] [PubMed]

33. Moloney, A.P.; Drennan, M.J. Characteristics of fat and muscle from beef heifers offered a grass silage or concentrate-based finishing ration. *Livest. Sci.* **2013**, *152*, 147–153. [CrossRef]

34. Moloney, A.P.; Mooney, M.T.; Kerry, J.P.; Stanton, C.; O'Kiely, P. Colour of fat, and colour, fatty acid composition and sensory characteristics of muscle from heifers offered alternative forages to grass silage in a finishing ration. *Meat Sci.* **2013**, *95*, 608–615. [CrossRef] [PubMed]

35. Juniper, D.T.; Browne, E.M.; Fisher, A.V.; Bryant, M.J.; Nute, G.R.; Beever, D.E. Intake, growth and meat quality of steers given diets based on varying proportions of maize silage and grass silage. *Anim. Sci.* **2005**, *81*, 159–170. [CrossRef]

36. Jurie, C.; Ortigues-Marty, I.; Picard, B.; Micol, D.; Hocquette, J.F. The separate effects of the nature of diet and grazing mobility on metabolic potential of muscles from Charolais steers. *Livest. Sci.* **2006**, *104*, 182–192. [CrossRef]

37. Huuskonen, A.; Jansson, S.; Honkavaara, M.; Tuomisto, L.; Kauppinen, R.; Joki-Tokola, E. Meat colour, fatty acid profile and carcass characteristics of Hereford bulls finished on grazed pasture or grass silage-based diets with similar concentrate allowance. *Livest. Sci.* **2010**, *131*, 125–129. [CrossRef]

38. Moloney, A.P.; Mooney, M.T.; Troy, D.J.; Keane, M.G. Finishing cattle at pasture at 30 months of age or indoors at 25 months of age: Effects on selected carcass and meat quality characteristics. *Livest. Sci.* **2011**, *141*, 17–23. [CrossRef]

39. Hennessy, D.W.; Morris, S.G.; Allingham, P.G. Improving the pre-weaning nutrition of calves by supplementation of the cow and/or the calf while grazing low quality pastures—2. Calf growth, carcass yield and eating quality. *Aust. J. Exp. Agric.* **2001**, *41*, 715–724. [CrossRef]

40. Roberts, S.D.; Kerth, C.R.; Braden, K.W.; Rankins, D.L.; Kriese-Anderson, L.; Prevatt, J.W. Finishing steers on winter annual ryegrass (Lolium multiflorum Lam.) with varied levels of corn supplementation I: Effects on animal performance, carcass traits, and forage quality. *J. Anim. Sci.* **2009**, *87*, 2690–2699. [CrossRef]

41. French, P.; O'Riordan, E.G.; Monahan, F.J.; Caffrey, P.J.; Vidal, M.; Mooney, M.T.; Troy, D.J.; Moloney, A.P. Meat quality of steers finished on autumn grass, grass silage or concentrate-based diets. *Meat Sci.* **2000**, *56*, 173–180. [CrossRef]

42. French, P.; O'Riordan, E.G.; Monahan, F.J.; Caffrey, P.J.; Mooney, M.T.; Troy, D.J.; Moloney, A.P. The eating duality of meat of steers fed grass and/or concentrates. *Meat Sci.* **2001**, *57*, 379–386. [CrossRef]

43. Duckett, S.K.; Neel, J.P.S.; Lewis, R.M.; Fontenot, J.P.; Clapham, W.M. Effects of forage species or concentrate finishing on animal performance, carcass and meat quality. *J. Anim. Sci.* **2013**, *91*, 1454–1467. [CrossRef]

44. McEwen, P.L.; Mande, I.B.; Brien, G.; Campbell, C.P. Effects of grain source, silage level, and slaughter weight endpoint on growth performance, carcass characteristics, and meat quality in Angus and Charolais steers. *Can. J. Anim. Sci.* **2007**, *87*, 167–180. [CrossRef]

45. Keane, M.G.; Allen, P. Effects of production system intensity on performance, carcass composition and meat quality of beef cattle. *Livest. Prod. Sci.* **1998**, *56*, 203–214. [CrossRef]

46. Oury, M.P.; Agabriel, J.; Agabriel, C.; Micol, D.; Picard, B.; Blanquet, J.; Labouré, H.; Roux, M.; Dumont, R. Relationship between rearing practices and eating quality traits of the muscle rectus abdominis of Charolais heifers. *Livest. Sci.* **2007**, *111*, 242–254. [CrossRef]

47. Cerdeño, A.; Vieira, C.; Serrano, E.; Lavín, P.; Mantecón, A.R. Effects of feeding strategy during a short finishing period on performance, carcass and meat quality in previously-grazed young bulls. *Meat Sci.* **2006**, *72*, 719–726. [CrossRef]

48. Sugimoto, M.; Saito, W.; Ooi, M.; Oikawa, M. Effects of days on feed, roughage sources and inclusion levels of grain in concentrate on finishing performance and carcass characteristics in cull beef cows. *Anim. Sci. J.* **2012**, *83*, 460–468. [CrossRef]

49. Holmer, S.F.; Homm, J.W.; Berger, L.L.; Stetzer, A.J.; Brewer, M.S.; Mckeith, F.K.; Killefer, J. Realimentation of Cull Beef Cows. II. Meat Quality of Muscles from the Chuck, Loin and Round in Response to Diet and Enhancement. *J. Muscle Foods* **2009**, *20*, 307–324. [CrossRef]

50. Moloney, A.P.; Keane, M.G.; Dunne, P.G.; Mooney, M.T.; Troy, D.J. Effect of concentrate feeding pattern in a grass silage/concentrate beef finishing system on performance, selected carcass and meat quality characteristics. *Meat Sci.* **2008**, *79*, 355–364. [CrossRef]

51. Cozzi, G.; Brscic, M.; Da Ronch, F.; Boukha, A.; Tenti, S.; Gottardo, F. Comparison of two feeding finishing treatments on production and quality of organic beef. *Ital. J. Anim. Sci.* **2010**, *9*, 404–409. [CrossRef]

Article

Effect of Breed and Gender on Meat Quality of *M. longissimus thoracis et lumborum* Muscle from Crossbred Beef Bulls and Steers

Jamie Cafferky [1,2], Ruth M. Hamill [1], Paul Allen [1], John V. O'Doherty [3], Andrew Cromie [4] and Torres Sweeney [2,*]

1   Department of Food Quality and Sensory Science, Teagasc Food Research Centre, Ashtown, 15 Dublin, Ireland; jamie.cafferky@teagasc.ie (J.C.); Ruth.Hamill@teagasc.ie (R.M.H.); paul.allen869@gmail.com (P.A.)
2   School of Veterinary Medicine, University College Dublin, Belfield, 4 Dublin, Ireland
3   School of Agriculture & Food Science, University College Dublin, Belfield, 4 Dublin, Ireland; john.vodoherty@ucd.ie
4   Irish Cattle Breeding Federation, Shinagh House, Bandon, P72 X050 Co. Cork, Ireland; acromie@icbf.com
*   Correspondence: torres.sweeney@ucd.ie; Tel.: +353-(0)17166244

Received: 12 April 2019; Accepted: 20 May 2019; Published: 21 May 2019

Abstract: The objective of this study was to determine whether sire breed and/or castration had an effect on meat quality of *M. longissimus thoracis et lumborum* (*LTL*) muscle from crossbred bulls and steers and to investigate the relationship amongst the traits examined. Warner–Bratzler shear force (WBSF), intramuscular fat (IMF)%, cook-loss%, drip-loss%, colour (*L**, *a**, *b**) and ultimate pH (upH) were determined in the *LTL* muscle from eight beef sire breeds representative of the Irish herd (Aberdeen Angus, Belgian Blue, Charolais, Hereford, Limousin, Parthenaise, Salers and Simmental). The results indicate that IMF%, cook-loss% and drip-loss% were associated with breed ($p < 0.05$); while WBSF, IMF% and cook-loss% differ between genders ($p < 0.05$). Steer *LTL* had a greater IMF% and exhibited reduced WBSF and cook-loss% in comparison to the bull *LTL* ($p < 0.05$). This study provides greater insight into how quality traits in beef are influenced by breed and gender and will support the industry to produce beef with consistent eating quality.

Keywords: beef quality; castration; breed; shear force; intramuscular fat

1. Introduction

Factors such as breed, gender, age of animal at slaughter, diet and feeding regime can influence muscle characteristics, which in turn affect meat quality [1–3]. Meat quality attributes such as tenderness, colour, flavour, juiciness, and water-holding capacity (WHC)-related traits—cook-loss% and drip-loss%—influence consumer satisfaction [4,5]. The relative importance and value of particular meat quality traits vary according to the type of meat product being produced and marketed and also the end target consumer of the product; e.g., tenderness is more important for beef meat than sheep meat [5]. As progress is made on the development of quality traits, their relative value is also altered, which has potential to impact on their prioritisation in animal breeding programmes [6]. In order to meet the demand for the high-quality product anticipated by consumers, beef producers must focus on improving the quality in addition to quantity [7]. This has increased the focus of both industry and academia on husbandry and breeding strategies aimed at improving meat eating quality traits [5].

Growth rate, carcass yield, feed efficiency and carbon efficiency are positively influenced in bulls (intact adult males) compared to steers (castrated adult males) [8,9]. Carcass fat is, however, a limiting factor for bull beef production [10]. Gender has been associated with many aspects of meat quality and has been proposed to favourably influence fat deposition and tenderness [11]. Steers are commonly used to produce the highest quality beef that is distinguished from beef attained from bulls and obtains a premium price in restaurants and markets in developed nations [11].

Breed also influences meat quality. Individual cattle breeds have been developed through extensive long-term selection for specific production attributes, such as increased growth rate, carcass conformation and intramuscular fat (IMF) [12,13]. Early maturing breeds such as Angus and Hereford have higher levels of IMF (and associated traits tenderness and flavour) in comparison to late maturing continental breeds [14–17] and this is reflected in the price [18].

Pre-slaughter handling and subsequent post-mortem processing play a major role in the final quality attributes of meat. Factors such as feed withdrawal [19], transport time [11] and stress during transport [20] can have a negative impact on subsequent meat quality. Bulls are more sensitive than steers to all these factors due to their sexual maturity and greater aggression in the lairage. This can lead to higher ultimate pH measurements and unfavourable meat quality [11,21,22].

Gaining insights into how meat quality traits are affected by animal breed and gender could inform pre-slaughter handling practises and post-mortem technologies aimed at maximising quality [22]. Furthermore, it could allow meat processors to optimise meat management systems based on specific quality traits (due to the animal's breed or gender) [23–25]. Therefore, the objective of this study was to determine whether sire breed and/or castration had an effect on the meat quality of *M. longissimus thoracis et lumborum* (*LTL*) muscle from crossbred bulls and steers and to investigate the relationship amongst the traits examined.

2. Materials and Methods

2.1. Animals and Sample Preparation

Crossbred bull and steer progeny were obtained and reared under the same feeding and environmental conditions by the Irish Cattle Breeders Federation Tully Progeny Test Centre (Tully, Kildare, Republic of Ireland). Bull and steer progeny examined in this study were bred from crossbred commercial suckler dams artificially inseminated by elite Irish beef breed bulls. Animals were acclimatised for approximately 30 days before starting a 90-day testing period. Bulls were offered an *ad-libitum* concentrate diet with 3 kilograms (kg) of fresh hay, while steers were offered 8 kg of concentrates and 5 kg of hay on a fresh-weight basis per head per day. Hay was offered to support the healthy functioning of the rumen and to reflect an Irish commercial high concentrate-based dietary regimen. All animals were finished to a specified carcass conformation and fat score range. For bulls this was U– to E+ conformation score, 3– to 5= fat score and 678 kg live weight (±58 kg); for steers this was R– to E= conformation score, 2+ to 5+ fat score and 637 kg live weight (±64 kg). Bulls were slaughtered at approximately 487 days old (±24 days), while steers were slaughtered at approximately 634 days old (±52 days). Eight beef breeds, with numbers representative of that in the Irish herd, were included as part of this study as follows: Aberdeen Angus (AA; bull $n = 36$, steer $n = 28$), Belgian Blue (BB; bull $n = 67$, steer $n = 10$), Charolais (CH; bull $n = 127$, steer $n = 41$), Hereford (HE; bull $n = 2$, steer $n = 11$), Limousin (LM; bull $n = 234$, steer $n = 62$), Parthenaise (PT; bull $n = 11$, steer = 4), Salers (SA; bull $n = 25$, steer $n = 16$) and Simmental (SI; bull $n = 63$, steer $n = 16$). Animals were slaughtered in batches of approximately 50, between February 2014 and May 2017 in a commercial plant by electrical stunning (50 Hz) followed by exsanguination from the jugular vein. Between 40–60 min post exsanguination, carcasses were split in half then chilled for 24 h at 2 °C. Twelve steaks with a thickness of 2.54 cm were removed sequentially from the right-side *LTL* 48 h post-mortem starting at the rump end and vacuum packaged. Steaks were labelled 1–12 according to the trait being measured in order to ensure the analysis was conducted in a consistent location within the *LTL* muscle. Steaks were frozen at −20 °C after 2 or 14 days of ageing at 4 °C, dependent on the trait being determined.

2.2. Warner–Bratzler Shear Force and Cook-Loss%

For the determination of cook-loss%, 14-day aged steaks were frozen at −20 °C until analysis then thawed within unsealed plastic vacuum bags in a circulating water bath at room temperature (20 °C). Steaks were trimmed of external fat, blotted lightly with tissue paper to remove moisture and weighed. Steaks were immersed in a water bath for cooking (Grant Instruments Ltd., Royston, England) at 72 °C until an internal core temperature of 70 °C was reached using a temperature probe (Eirelec Ltd., Dublin, Ireland). Samples were cooled to room temperature, blotted lightly with tissue paper and weight was recorded. Samples were then placed within new unsealed vacuum bags and left to temper overnight in a fridge at 4 °C. Cook-loss was expressed as a percentage of the raw weight of the steak as follows:

$$\text{Cook-loss (\%)} = (\text{raw weight} - \text{cooked weight})/\text{raw weight} \times 100 \tag{1}$$

Following cook-loss determination, the tempered steaks were used for Warner–Bratzler shear force analysis according to a modified version of American Meat Science Association guidelines [26]. Seven cores per steak were removed for analysis using a 1.27-cm core and sheared perpendicular to the fibre direction using the Instron 4464 Universal testing machine (Instron Ltd., Buckinghamshire, UK), with a load cell of 500 N and a cross head speed of 50 mm/min, and analysed using Bluehill®2 Software (Instron Ltd., Buckinghamshire, UK). The maximum peak force recorded during analysis was reported as Newton (N) shear force. The highest and lowest measurements were excluded with the average of the remaining 5 cores recorded as the result to reduce standard deviation.

2.3. Intramuscular Fat%

IMF% was determined on 2-day aged steaks using the Smart System-5 microwave moisture drying oven and NMR Smart-Trac rapid fat analyser (CEM Corporation, Matthews, NC, USA) using AOAC Official Method 985.14 [27]. In brief, steaks previously frozen at −20 °C were thawed within unsealed plastic vacuum bags in a circulating water bath at room temperature (20 °C). Once thawed, the steaks were trimmed of external fat, cut into cubes approximately 2.5 × 2.5 cm and placed into a RobotCoupe R2 blender and homogenised to a fine consistency. Two grams of homogenised meat free of connective tissue was then placed within the Smart-Trac for analysis.

2.4. Ultimate pH

Ultimate pH (upH) measurements were collected from carcasses 48 h post-mortem by placing a calibrated pH meter (Hanna HI 9125 pH meter, Woonsocket, RI, USA) within the loin between the 12th and 13th rib avoiding bone and connective tissue. Calibration of the pH electrode was performed with standardized buffers (pH 4.0 and 7.0).

2.5. Drip-Loss%

Drip-loss% was analysed according to the procedure of Honikel and Hamm [28]. From each steak (2-day aged samples, assigned for drip-loss analysis), a piece was removed (2.5 cm in thickness, 7.5 cm in length and 5 cm in width) avoiding connective tissue and large areas of fat not representative of the sample. Samples for drip-loss weighed approximately 100 ± 5 g and were lightly blotted with tissue paper, weighed, then suspended by string and an unfolded paperclip (formed into a hook shape) within an expanded clear plastic bag, with care taken to ensure the sample did not come into contact with the bag. The samples were suspended in a chill room at 4 °C for 96 h, after which, the surface was lightly blotted with a tissue and re-weighed. Drip-loss was expressed as a percentage of the original weight of the steak as follows:

$$\text{Drip-loss (\%)} = (\text{initial weight} - \text{final weight})/\text{initial weight} \times 100. \tag{2}$$

2.6. Colour

Colour measurements were performed using the HunterLab UltraScan Pro CIE *L*a*b** system with a dual beam xenon flash spectrophotometer (Hunter Associates Laboratory, Inc., Reston, VA, USA). CIE *L** (lightness), *a** (redness), *b** (yellowness) values were recorded. The illuminant (D65, 10°) consisted of an 8° viewing angle and a 9.9-mm port size. Calibration was carried out using a white standard tile ($L = 100$) and light trap ($L = 0$). The white tile was covered in cling film prior to calibration to prevent any effect on the colour reading. Steaks (2 d post-mortem) were wrapped in cling film and allowed to bloom for 1 h prior to measurement. Three measurements were taken in three separate locations on each steak, avoiding intramuscular fat and connective tissue.

2.7. Statistical Analysis

Statistical analysis was performed by two-way Analysis of Variance (ANOVA) using Tukey–Kramer adjusted generalised linear model (GLM) procedures of Statistical analysis Software (SAS) 9.4 (SAS Institute, Cary, NC, USA). Pearson Correlations between beef *LTL* quality attributes were calculated using the CORR procedure in SAS. Differences were considered significant at the $p < 0.05$ level.

3. Results

3.1. Correlations between Traits

Pearson correlations between eight beef quality traits examined as part of this study are presented in Table 1. Cook loss, drip loss, pH and colour traits showed a number of correlations with each other. WBSF and IMF were negatively correlated, with higher fat-content meat being associated with lower shear force. Cook loss was also linked to both fat content and shear force.

Table 1. Pearson correlation coefficients between quality traits of beef *LTL*.

	IMF (%)	upH	Cook-Loss (%)	Drip-Loss (%)	L*	a*	b*
WBSF	−0.26 ***	−0.05	0.19 ***	−0.16 ***	0.06	−0.15 ***	−0.05
IMF (%)		−0.015	−0.22 ***	−0.08	−0.07	0.1 **	0.05
upH			−0.15 ***	0.03	−0.01	−0.01	−0.01
Cook-loss (%)				0.06	0.18 ***	0.23 ***	0.16 ***
Drip-loss (%)					0.23 ***	0.04	0.13 **
L *						0.2 ***	0.42 ***
a *							0.83 ***

* $p < 0.05$, ** $p < 0.01$, *** $p < 0.001$; WBSF—Warner–Bratzler shear force; IMF—Intramuscular fat; upH—Ultimate pH.

3.2. Effect of Breed and Gender on Meat Quality

The effects of breed and gender on *LTL* muscle quality traits from eight beef breeds are presented in Table 2. Gender had a significant effect on WBSF values, with bull *LTL* samples having higher shear force than that of steer *LTL*. However, no effect was observed on WBSF values for breed.

IMF was associated with both breed and gender, with *LTL* from steers containing almost twice the IMF% of bull *LTL* (2.85 and 1.27%, respectively; Table 2). AA sired *LTL* samples had the highest levels of IMF in the current dataset (2.78%) with BB and PT (1.12%) sired progeny, the lowest. Cook-loss was significant for both breed and gender effects, with LM-sired progeny having the lowest cook-loss values (29.09%). Bull *LTL* cook-loss% was higher than that of steers (30.4 and 29.25%, respectively). Drip-loss values only tended to be higher in steers than in bulls (3.41 and 2.73%); however, sire breed had an effect on drip-loss% values. The AA sire progeny had the lowest drip-loss% measured (2.15%), with BB having the highest (4.11%).

Table 2. Effect of breed and gender on quality traits of *M. longissimus thoracis et lumborum* (least squares mean and standard error).

Trait	Breed								Gender		*p*-Value	
	AA	HE	LM	CH	SA	SI	BB	PT	Bull	Steer	Breed	Gender
n	64	13	296	168	41	79	77	15	565	188	-	-
WBSF	41.59 (1.2)	36.96 (3.8)	42.69 (0.7)	40.62 (0.9)	42.57 (1.6)	44.37 (1.4)	42.55 (1.7)	40.97 (2.9)	44.54 (1.2)	37.81 (1.1)	0.3122	0.0001
IMF (%)	2.78 (0.2) [b]	2.16 (0.5) [ab]	2.13 (0.1) [ab]	2.05 (0.1) [ab]	2.41 (0.2) [ab]	2.13 (0.3) [ab]	1.7 (0.2) [a]	1.12 (0.4) [a]	1.27 (0.1)	2.85 (0.1)	0.009	0.0001
upH	5.55 (0.02)	5.53 (0.05)	5.55 (0.01)	5.54 (0.01)	5.56 (0.02)	5.56 (0.02)	5.6 (0.03)	5.54 (0.04)	5.57 (0.01)	5.54 (0.01)	0.6594	0.286
Cook-loss (%)	30.15 (0.4) [ab]	29.09 (1.1) [ab]	29.09 (0.2) [a]	29.66 (0.2) [ab]	29.33 (0.5) [ab]	30.59 (0.4) [b]	29.63 (0.5) [ab]	31.07 (0.8) [b]	30.4 (0.3)	29.25 (0.3)	0.0032	0.005
Drip-loss (%)	2.15 (0.3) [a]	2.5 (0.5) [ab]	2.97 (0.1) [e]	3.22 (0.1) [f]	2.72 (0.2) [bcd]	2.52 (0.3) [abc]	4.37 (0.3) [g]	4.11 (0.5) [g]	2.73 (0.2)	3.41 (0.2)	0.0009	0.0950
*L**	41.89 (0.5)	41.68 (1.1)	42.66 (0.2)	42.52 (0.3)	41.75 (0.5)	42.27 (0.7)	42.14 (0.5)	42.11 (0.9)	42.33 (0.3)	41.93 (0.3)	0.5777	0.3769
*a**	14.37 (0.3)	14.15 (0.7)	14.31 (0.1)	14.56 (0.2)	14.63 (0.3)	14.27 (0.4)	13.61 (0.3)	14.17 (0.5)	14.01 (0.2)	14.52 (0.2)	0.2665	0.0635
*b**	11.15 (0.3)	10.67 (0.6)	11.41 (0.1)	11.6 (0.1)	11.49 (0.3)	11.12 (0.2)	10.82 (0.3)	11.31 (0.5)	11.33 (0.2)	11.06 (0.2)	0.2902	0.3130

n—number of animals in each breed; AA—Angus; BB—Belgium Blue; CH—Charolais; HE—Hereford; LM—Limousin; PT—Parthenaise; SA—Salers; SI—Simmental; ()—Standard error; WBSF—Warner–Bratzler shear force; IMF—Intramuscular fat; upH—Ultimate pH; [a,b,c,d,e,f,g] Within a row values not sharing a common superscript are significantly different for breed.

4. Discussion

This study illustrates the significant effect castration has on important technological beef quality traits, such as increasing IMF%, reducing WBSF and improving cook-loss%. Moreover, the quality attributes, IMF, cook-loss and drip-loss, varied significantly according to sire breed. However, very little variability exists for texture, as reflected by WBSF scores, and sire breed was not determined to be a significant factor for this trait.

Interesting correlations amongst the traits examined as part of our study were found. Notably, the three traits affected by gender, WBSF, IMF% and cook-loss%, were all significantly correlated to each other. In this study, traits relating to different aspects of meat quality show trends that were associated with each other, pointing to particular animals having an overall higher quality, relative to others within the study. For example, animals with higher IMF% and redness values had lower WBSF scores and reduced cook-loss%, which could be considered to be a combination of trait values associated with good overall meat quality. Monteiro et al. [29] also reported a positive relationship between WBSF and cook-loss%. They concluded that cook-loss% influenced WBSF more than any other physiochemical trait and was the reason for shear force variation. The multifactorial relationship amongst these traits suggests that selecting one of these correlated traits for improvement may have a beneficial effect on other meat quality attributes. The two water-holding capacity related traits (cook-loss% and drip-loss%) were not correlated with each other. The two traits capture different aspects of fluid loss in processing. Drip-loss is reflective of exudate, consisting mainly of water and proteins, whereas cook-loss may also be associated with glycogen potential and additionally melting fat during the thermal processing [30]. These traits have been well studied in pork [31]. Interestingly, a comparable lack of correlation between these traits is present in beef in the current study.

The highest scoring sire breed for IMF% was the early-maturing AA sire offspring, which was significantly different from the two lowest scoring sire breeds for this trait, i.e., the late maturing continental BB and PT sire progeny. In a comprehensive study by Gagaoua et al. [1], breed had a significant influence on IMF%, with AA bulls having twice the IMF% relative to continental breed bulls examined in their study. Differences in IMF% values between breeds is mostly attributed to genetics, with early maturing breeds (such as AA and HE) having higher fat deposition than continental breed animals [16]. Steers had greater than twice the IMF% relative to bulls, consistent with Moran et al. [10] and Nian et al. [32] who reported that gender was significant for IMF. The difference between IMF% values between bulls and steers is attributed to the removal of the testes in steers, which arrests sexual maturation leading to reductions in growth rate and muscular development and increases fat deposition and accelerates the fattening period [33]. Other hormones involved in muscle and adipose tissue metabolism include leptin, growth hormone, insulin, cortisol, insulin-like growth factor 1, thyroxin and triiodothyronine [34–36]. Testosterone binds to receptors within the muscle, increasing the incorporation of amino acids into protein and increasing the capacity for muscular development and growth rate [37]. This results in an increase in muscle mass without increases in IMF% [37]. Greater IMF values in steers are attributed to the diminished physiological effects of this androgen, reducing plasma lipids, increasing lipolysis by adipocytes and stimulating androgen receptors [38,39]. The castration of bulls is directly involved in the upregulation of the lipogenic gene expression of fatty acid synthase (FASN) and acetyl-CoA carboxylase (ACC); furthermore, it also downregulates the lipolytic gene expression of monoglyceride lipase (MGL) and adipose triglyceride lipase (ATGL) [40]. Hence, castration contributes to improved IMF deposition mediated through increased lipid uptake and lipogenesis and decreased lipolysis [40].

WBSF values exhibited a gender effect but not a breed effect, with steers exhibiting a WBSF value approximately 7 N lower than bulls, in agreement with Nian et al. [32]. A gender effect for WBSF between bulls and steers was also found by Mberema et al. [41]. However, they observed 14-day aged bull *LTL* to be more tender than steer and heifer *LTL*. The results of the current study contrast with Moran et al. [10] who found no effect on WBSF values between bulls and steers. Gerrard et al. [42] reported that any difference in WBSF values explained by gender is not observed following 13 days of ageing. However, the current study aged samples for 14 days and an effect was still observed. Other factors that contribute to the textural toughness of beef (namely sarcomere length and collagen) were not assessed but may play a role in the differences observed between genders, with bulls having shorter sarcomeres and higher levels of collagen within muscle [32]. Various studies on the use of late maturing beef bulls have also reported that breed has no significant effect on instrumental tenderness [13,43,44]. With regards to breed effect and sheer force, Marino et al. [45] reported significance when comparing three cattle breeds contrasting in traditional breed purpose (Friesian dairy bulls; Romagnola crossbred beef bulls; Podolian bulls—indigenous to southern Italy), with the indigenous Podolian bulls significantly tougher than their counterparts. This is in contrast to the findings of the current study as sire breed had no effect on shear force values.

In the current study, cook-loss% is associated with both breed and gender. The breed effect was observed between LM and SI sire breeds. This is in agreement with Chambaz et al. [14] who also observed a significant difference in cook-loss between these two breeds. IMF is one of the factors associated with cook-loss; as IMF% increases, cook-loss% values decrease [46]. Findings relating breed to cook-loss are sometimes conflicting. Mandell et al. [47] examined HE and SI breeds for cook-loss% and found no significance; however, when comparing AA, BB and LM bulls for cook-loss, Cuvelier et al. [2] found AA bulls to be significantly different. Steers had lower cook-loss% than the bulls examined which is in agreement with numerous studies [10,32,47]. The differences observed between bulls and steers may be attributed to the larger muscle fibre diameter of bulls, induced by androgens, with muscles of increased cross-sectional area exhibiting greater cook-loss [9,48]. In contrast, Knight et al. [49] found no gender effect between bulls and steers for cook-loss%. In our study, drip-loss% was significant for breed but not gender. Early maturing breeds (AA and HE) exhibited lower drip-loss% values in comparison to the larger, late maturing continental breeds such as Belgian Blue, Charolais and Parthenaise animals, indicating early maturing breeds have the potential to have juicier meat and less reduction in yield associated with hanging. Similar trends regarding drip-loss values and breed effect were shown by Cuvelier et al. [2] when studying BB, LM and AA breeds, and Chambaz et al. [14] when studying AA, LM, SI and CH breeds. Cuvelier et al. [2] suggested that higher drip-loss% values in larger animal breeds (such as BB and PT, the highest scoring for this trait) may be due to their higher meat:water content. The lower collagen content of double muscle animals may also contribute to increased drip-loss, as WHC is known to increase with increasing amounts of connective tissue [50]. Breeds exhibiting lower drip-loss values have the added economic benefit of less product weight loss, leading to higher financial gain on carcass and primal cuts.

Lightness (L^*), redness (a^*) and yellowness (b^*) values were not influenced by breed or gender in our study. This is consistent with Moran et al. [10] who did not see any difference when comparing colour values between bulls and steers. Nian et al. [32] also found no significant effect for gender on L^* values; however, in that study, steers had significantly higher a^* and b^* values in comparison to bulls, which is in contrast with the current findings. Papaleo Mazzucco et al. [16] found breed to have no effect on L^*, a^* or b^* values when comparing AA and HE steers. However, Cuvelier et al. [2] found breed to influence both L^* and a^* values when comparing AA, BB and LM bulls; with BB bulls having lighter and less red *LTL*, and AA bulls having the darkest and reddest *LTL* (b^* values were not reported).

5. Conclusions

Sire breed had a significant effect on three beef quality traits analysed, IMF%, cook-loss% and drip-loss%, which were also correlated to each other. With respect to breed, Aberdeen Angus sired progeny had the highest IMF% and the lowest drip-loss%, Limousin sired offspring had the lowest cook-loss%, while Belgian Blue and Parthenaise sired progeny scored the highest for drip-loss%. Castration significantly impacted three of the beef quality traits analysed: WBSF, IMF% and cook-loss%. In comparison to bulls, steers had higher IMF% and reduced WBSF and cook-loss%, implying steer beef to be more tender and juicy, with more favourable IMF%. This study supports the hypothesis that breed and gender influence meat quality traits.

Author Contributions: Conceptualization, R.M.H. and T.S.; methodology, R.M.H., T.S., P.A. and J.C.; lab analysis, J.C.; resources, R.M.H. and T.S.; data analysis, J.C. and J.V.O.; draft preparation, J.C.; writing—review and editing, R.M.H., P.A., A.C., J.V.O. and T.S.; project administration, R.M.H. and T.S.; funding acquisition, R.M.H. and T.S.

Funding: This work is funded by the BreedQuality project (11/SF/311) which is supported by The Irish Department of Food, Agriculture and the Marine (DAFM) under the National Development Plan 2007–2013.

Acknowledgments: We acknowledge the Irish Cattle Breeding Federation for supply of samples, Paula Reid for assistance with statistical analysis and Eugene Vesey for his assistance with sample collection.

Conflicts of Interest: The authors declare no conflict of interest.

References

1. Gagaoua, M.; Terlouw, E.M.C.; Micol, D.; Hocquette, J.F.; Moloney, A.P.; Nuernberg, K.; Picard, B. Sensory quality of meat from eight different types of cattle in relation with their biochemical characteristics. *J. Integr. Agric.* **2016**, *15*, 1550–1563. [CrossRef]

2. Cuvelier, C.; Clinquart, A.; Hocquette, J.F.; Cabaraux, J.F.; Dufrasne, I.; Istasse, L.; Hornick, J.L. Comparison of composition and quality traits of meat from young finishing bulls from Belgian Blue, Limousin and Aberdeen Angus breeds. *Meat Sci.* **2006**, *74*, 522–531. [CrossRef]

3. Mateescu, R.G. Genetics of meat quality. In *The Genetics of Cattle*, 2nd ed.; Garrick, D., Ruvinsky, A., Eds.; CABI: Oxfordshire, UK, 2015; pp. 544–562.

4. Dransfield, E.; Martin, J.F.; Bauchart, D.; Abouelkaram, S.; Lepetit, J.; Culioli, J.; Picard, B. Meat quality and composition of three muscles from French cull cows and young bulls. *Anim. Sci.* **2003**, *76*, 387–399. [CrossRef]

5. Hopkins, D.L.; Mortimer, S.I. Effect of genotype, gender and age on sheep meat quality and a case study illustrating integration of knowledge. *Meat Sci.* **2014**, *98*, 544–555. [CrossRef] [PubMed]

6. Thompson, J.M. The effects of marbling on flavour and juiciness scores of cooked beef, after adjusting to a constant tenderness. *Aust. J. Exp. Agric.* **2004**, *44*, 645–652. [CrossRef]

7. Melucci, L.M.; Panarace, M.; Feula, P.; Villarreal, E.L.; Grigioni, G.; Carduza, F.; Miquel, M.C. Genetic and management factors affecting beef quality in grazing Hereford steers. *Meat Sci.* **2012**, *92*, 768–774. [CrossRef] [PubMed]

8. O'Riordan, E.; Crosson, P.; McGee, M. Finishing male cattle from the beef suckler herd. *Irish Grassl. Assoc. J.* **2011**, *45*, 131–146.

9. Seideman, S.C.; Cross, H.R. Utilization of the Intact Male for Red Meat Production: A Review. *J. Anim. Sci.* **1982**, *55*, 826–840. [CrossRef]

10. Moran, L.; O'Sullivan, M.G.; Kerry, J.P.; Picard, B.; McGee, M.; O'Riordan, E.G.; Moloney, A.P. Effect of a grazing period prior to finishing on a high concentrate diet on meat quality from bulls and steers. *Meat Sci.* **2017**, *125*, 76–83. [CrossRef]

11. Guerrero, A.; Valero, M.V.; Campo, M.M.; Sañudo, C. Some factors that affect ruminant meat quality: From the farm to the fork. Review. *Acta Sci.* **2013**, 335–347. [CrossRef]

12. Park, S.J.; Beak, S.H.; Jung, D.J.S.; Kim, S.Y.; Jeong, I.H.; Piao, M.Y.; Baik, M. Genetic, management, and nutritional factors affecting intramuscular fat deposition in beef cattle—A review. *Asian-Australas. J. Anim. Sci.* **2018**, *31*, 1043–1061. [CrossRef] [PubMed]

13. Xie, X.; Meng, Q.; Cui, Z.; Ren, L. Effect of cattle breed on meat quality, muscle fiber characteristics, lipid oxidation and fatty acids in China. *Asian-Australas. J. Anim. Sci.* **2012**, *25*, 824–831. [CrossRef]

14. Chambaz, A.; Scheeder, M.R.L.; Kreuzer, M.; Dufey, P.A. Meat quality of Angus, Simmental, Charolais and Limousin steers compared at the same intramuscular fat content. *Meat Sci.* **2003**, *63*, 491–500. [CrossRef]

15. Coleman, L.W.; Hickson, R.E.; Schreurs, N.M.; Martin, N.P.; Kenyon, P.R.; Lopez-Villalobos, N.; Morris, S.T. Carcass characteristics and meat quality of Hereford sired steers born to beef-cross-dairy and Angus breeding cows. *Meat Sci.* **2016**, *121*, 403–408. [CrossRef]

16. Papaleo Mazzucco, J.; Goszczynski, D.E.; Ripoli, M.V.; Melucci, L.M.; Pardo, A.M.; Colatto, E.; Villarreal, E.L. Growth, carcass and meat quality traits in beef from Angus, Hereford and cross-breed grazing steers, and their association with SNPs in genes related to fat deposition metabolism. *Meat Sci.* **2016**, *114*, 121–129. [CrossRef] [PubMed]

17. Samootkwam, K.; Jaturasitha, S.; Tipnate, B.; Waritthitham, A.; Wicke, M.; Kreuzer, M. Effect of Improving Lamphun Cattle with Black Angus on Carcass and Meat Quality. *Agric. Agric. Sci. Procedia* **2015**, *5*, 145–150. [CrossRef]

18. Hanrahan, K. The Significance of Beef. In *Teagasc Beef Manual*; Available online: https://www.teagasc.ie/media/website/publications/2016/Beef-Manual-Section1.pdf (accessed on 10 December 2018).

19. Ilian, M.A.; Morton, J.D.; Bekhit, A.E.D.; Roberts, N.; Palmer, B.; Sorimachi, H.; Bickerstaffe, R. Effect of preslaughter feed withdrawal period on longissimus tenderness and the expression of calpains in the ovine. *J. Agric. Food Chem.* **2001**, *49*, 1990–1998. [CrossRef] [PubMed]

20. Warner, R.D.; Greenwood, P.L.; Pethick, D.W.; Ferguson, D.M. Genetic and environmental effects on meat quality. *Meat Sci.* **2010**, *86*, 171–183. [CrossRef]

21. Troy, D.J.; Kerry, J.P. Consumer perception and the role of science in the meat industry. *Meat Sci.* **2010**, *86*, 214–226. [CrossRef]

22. Troy, D.J.; Ojha, K.S.; Kerry, J.P.; Tiwari, B.K. Sustainable and consumer-friendly emerging technologies for application within the meat industry: An overview. *Meat Sci.* **2016**, *120*, 2–9. [CrossRef]

23. McCarthy, S.N.; Henchion, M.; White, A.; Brandon, K.; Allen, P. Evaluation of beef eating quality by Irish consumers. *Meat Sci.* **2017**, *132*, 118–124. [CrossRef]

24. Merlino, V.M.; Borra, D.; Girgenti, V.; Dal Vecchio, A.; Massaglia, S. Beef meat preferences of consumers from Northwest Italy: Analysis of choice attributes. *Meat Sci.* **2018**, *143*, 119–128. [CrossRef] [PubMed]

25. Ngapo, T.M.; Braña Varela, D.; Rubio Lozano, M.S. Mexican consumers at the point of meat purchase. Beef choice. *Meat Sci.* **2017**, *134*, 34–43. [CrossRef]

26. American Meat Science Association. Research Guidelines for Cookery, Sensory Evaluation, and Instrumental Tenderness Measurements of Meat. Available online: https://www.meatscience.org/docs/default-source/publications-resources/amsa-sensory-and-tenderness-evaluation-guidelines/research-guide/2015-amsa-sensory-guidelines-1-0.pdf?sfvrsn=6 (accessed on 14 January 2019).

27. AOAC 985.14. Moisture in meat and poultry products-rapid microwave drying method. In *Official Methods of Analysis of AOAC International*, 15th ed.; Cuniff, P., Ed.; AOAC: Arlington, VA, USA, 1991.

28. Honikel, K.; Hamm, R. Measurement of water holding capacity and juiciness. In *Advances in Meat Research*, 9th ed.; Pearson, A.M., Dutson, T.R., Eds.; Blackie Academic and Professional: London, UK, 1994; pp. 125–161.

29. Monteiro, A.C.G.; Gomes, E.; Barreto, A.S.; Silva, M.F.; Fontes, M.A.; Bessa, R.J.B.; Lemos, J.P.C. Eating quality of 'Vitela Tradicional do Montado'-PGI veal and Mertolenga-PDO veal and beef. *Meat Sci.* **2013**, *94*, 63–68. [CrossRef] [PubMed]

30. Ryan, M.T.; Hamill, R.M.; O'Halloran, A.M.; Davey, G.C.; McBryan, J.; Mullen, A.M.; McGee, C.; Gisbert, M.; Southwood, O.I.; Sweeney, T. SNP variation in the promoter of the *PRKAG3* gene and association with meat quality traits in pigs. *BMC Gen.* **2012**, *13*, 66. [CrossRef]

31. Huff-Lonergan, E.; Baas, T.J.; Malek, M.; Dekkers, J.C.M.; Prusa, K.; Rothschild, M.F. Correlations among selected pork quality traits. *J. Anim. Sci.* **2001**, *80*, 617–627. [CrossRef]

32. Nian, Y.; Allen, P.; Harrison, S.M.; Kerry, J.P. Effect of castration and carcass suspension method on the quality and fatty acid profile of beef from male dairy cattle. *J. Sci. Food Agric.* **2018**, *98*, 4339–4350. [CrossRef]

33. Eichhorn, J.; Bailey, C.; Blomquist, G. Fatty acid composition of muscle and adipose tissue from crossbred bulls and steers. *J. Anim. Sci.* **1985**, *61*, 892–904. [CrossRef]

34. Ronge, H.; Blum, J. Original article growth factor I during growth. *Reprod. Nutr. Dev.* **1989**, *29*, 105–111. [CrossRef] [PubMed]

35. Daix, M.; Pirotte, C.; Bister, J.L.; Wergifosse, F.; Cuvelier, C.; Cabaraux, J.F.; Paquay, R. Relationship between leptin content, metabolic hormones and fat deposition in three beef cattle breeds. *Vet. J.* **2008**, *177*, 273–278. [CrossRef]

36. Davis, M.; Boyles, S.; Moeller, S.; Simmen, R.C. Genetic Parameter Estimates for Serum Insulin-Like Growth Factor I Concentration and Performance Traits in Angus Beef Cattle. *Am. Soc. Anim. Sci.* **2003**, *81*, 2164–2170. [CrossRef] [PubMed]

37. Venkata Reddy, B.; Sivakumar, A.S.; Jeong, D.W.; Woo, Y.B.; Park, S.J.; Lee, S.Y.; Hwang, I. Beef quality traits of heifer in comparison with steer, bull and cow at various feeding environments. *Anim. Sci. J.* **2015**, *86*, 1–16. [CrossRef]

38. Lee, H.K.; Lee, J.K.; Cho, B. The Role of Androgen in the Adipose Tissue of Males. *World J. Mens. Health.* **2013**, *31*, 136. [CrossRef]

39. Xu, X.; De Pergola, G.; Björntorp, P. The Effects of Androgens on the Regulation of Lipolysis in Adipose Precursor Cells. *Endocrinology* **1990**, *126*, 1229–1234. [CrossRef]

40. Bong, J.J.; Jeong, J.Y.; Rajasekar, P.; Cho, Y.M.; Kwon, E.G.; Kim, H.C.; Baik, M. Differential expression of genes associated with lipid metabolism in longissimus dorsi of Korean bulls and steers. *Meat Sci.* **2012**, *91*, 284–293. [CrossRef]

41. Mberema, C.H.H.; Lietz, G.; Kyriazakis, I.; Sparagano, O.A.E. The effects of gender and muscle type on the mRNA levels of the calpain proteolytic system and beef tenderness during post-mortem aging. *Livest. Sci.* **2016**, *185*, 123–130. [CrossRef]

42. Gerrard, D.E.; Jones, S.J.; Aberle, E.D.; Lemenager, R.P.; Diekman, M.A.; Judge, M.D. Collagen Stability, Testosterone Secretion and Meat Tenderness in Growing Bulls and Steers. *J. Anim. Sci.* **1987**, *65*, 1236–1242. [CrossRef] [PubMed]

43. Avilés, C.; Martínez, A.L.; Domenech, V.; Peña, F. Effect of feeding system and breed on growth performance, and carcass and meat quality traits in two continental beef breeds. *Meat Sci.* **2015**, *107*, 94–103. [CrossRef]

44. Peña, F.; Avilés, C.; Domenech, V.; González, A.; Martínez, A.; Molina, A. Effects of stress by unfamiliar sounds on carcass and meat traits in bulls from three continental beef cattle breeds at different ageing times. *Meat Sci.* **2014**, *98*, 718–725. [CrossRef]

45. Marino, R.; Albenzio, M.; Della Malva, A.; Santillo, A.; Loizzo, P.; Sevi, A. Proteolytic pattern of myofibrillar protein and meat tenderness as affected by breed and aging time. *Meat Sci.* **2013**, *95*, 281–287. [CrossRef] [PubMed]

46. Ozawa, S.; Mitsuhashi, T.; Mitsumoto, M.; Matsumoto, S.; Itoh, N.; Itagaki, K.; Dohgo, T. The characteristics of muscle fiber types of longissimus thoracis muscle and their influences on the quantity and quality of meat from Japanese Black steers. *Meat Sci.* **2000**, *54*, 65–70. [CrossRef]

47. Mandell, I.B.; Gullett, E.A.; Wilton, J.W.; Kemp, R.A.; Allen, O.B. Effects of gender and breed on carcass traits, chemical composition, and palatability attributes in Hereford and Simmental bulls and steers. *Livest. Prod. Sci.* **1997**, *49*, 235–248. [CrossRef]

48. Waritthitham, A.; Lambertz, C.; Langholz, H.J.; Wicke, M.; Gauly, M. Muscle fiber characteristics and their relationship to water holding capacity of longissimus dorsi muscle in brahman and charolais crossbred bulls. *Asian-Australas. J. Anim. Sci.* **2010**, *23*, 665–671. [CrossRef]

49. Knight, T.W.; Cosgrove, G.P.; Death, A.F.; Anderson, C.B. Effect of interval from castration of bulls to slaughter on carcass characteristics and meat quality. *N. Z. J. Agric. Res.* **1999**, *42*, 269–277. [CrossRef]

50. Raes, K.; Balcaen, A.; Dirinck, P.; De Winne, A.; Claeys, E.; Demeyer, D.; De Smet, S. Meat quality, fatty acid composition and flavour analysis in Belgian retail beef. *Meat Sci.* **2003**, *65*, 1237–1246. [CrossRef]

 *foods*

Article

Is Meat of Breeder Turkeys so Different from That of Standard Turkeys?

Pascal Chartrin [1], **Thierry Bordeau** [1], **Estelle Godet** [1], **Karine Méteau** [2], **Jean-Christian Gicquel** [3], **Estelle Drosnet** [3], **Sylvain Brière** [4], **Marie Bourin** [5] **and Elisabeth Baéza** [1,*]

[1] Biologie des Oiseaux et Aviculture (BOA), Institut National de la Recherche Agronomique (INRA), Université de Tours, 37380 Nouzilly, France; pascal.chartrin@inra.fr (P.C.); thierry.bordeau@inra.fr (T.B.); estelle.godet@inra.fr (E.G.)

[2] Elevage, Alimentation et Santé des Monogastriques (EASM), INRA, Le Magneraud, Saint-Pierre d'Amilly, BP 52, 17700 Surgères, France; karine.meteau@inra.fr

[3] Société de Transformation des Volailles de l'Ouest (STVO), ZI les Riantières, BP 22, 44540 Saint-Mars la Jaille, France; Jean-Christian.GICQUEL@hendrix-genetics.com (J.-C.G.); estelle.DROSNET@hendrix-genetics.com (E.D.)

[4] Hendrix Genetics Turkeys France, La Bohardière, BP 1, St Laurent de la Plaine, 49290 Mauges Sur Loire, France; sylvain.briere@hendrix-genetics.com

[5] Institut Technique de l'Aviculture (ITAVI), BOA, INRA, Université de Tours, 37380 Nouzilly, France; bourin@itavi.asso.fr

* Correspondence: elisabeth.baeza-campone@inra.fr; Tel.: +33-2-47-42-78-41

Received: 28 November 2018; Accepted: 21 December 2018; Published: 24 December 2018

Abstract: The technological, nutritional, and sensorial quality of breasts and thighs with drumsticks of turkey male and female breeders was characterized by comparison with breasts and thighs with drumsticks of growing male and female turkeys from the Grademaker line (hybrid turkeys, $n = 20$ birds per sex and per physiological stage). The breeder turkeys were slaughtered at 397 and 410 days of age and 10.42 and 32.67 kg of body weight for the females and males, respectively. The standard turkeys were slaughtered at 75 and 103 days of age and 5.89 and 13.48 kg of body weight for the females and males, respectively. The differences observed between males and females on one hand and between standard and breeder turkeys on the other hand were mainly induced by differences in slaughter ages and sexual dimorphism on body weight. The meat of female breeders had characteristics close to those of female and male standard turkeys, whereas the meat of male breeders was clearly distinguishable, particularly by displaying lower tenderness and water holding capacity.

Keywords: male and female turkeys; breeders; broilers; carcass; meat; nutritional; sensorial and technological quality

---

1. Introduction

In France, the production of turkey meat reached 350,000 tec (tonnes equivalent carcasses) in 2015 [1]. The consumption of turkey meat was 4.6 kg per year and per capita in 2015 [2]. This production results essentially from standard turkeys. Females and males are slaughtered at 12 and 16 weeks of age, respectively, and 6–7 kg and 14–15 kg of body weight, respectively. The main part of the turkey production is cut, but there is a production of light turkeys sold under whole carcasses around Christmas and New Year holidays. Finally, when the period of reproduction of turkeys is finished, animals are slaughtered under industrial conditions and their meat is mainly valued under processed products. The French production of breeders was estimated to 8797 and 9174 tec in 2015 and 2016, respectively (Gicquel, personal communication). In the EU, production was estimated to 36,224 and 37,776 tec in 2015 and 2016, respectively (Gicquel, personal communication). There are many studies on the meat quality of standard turkeys. In other countries, standard turkeys are slaughtered at older ages (18 to 22 weeks for males, and 14 to 16 weeks for females) depending on the line (BUT Big 6, Hybrid Converter). The body weight at slaughter ranges between 16 and 22 kg for males, and between 9 and 11 kg for females [3–15]. The breast yield can vary between 22 and 30% and the yield of thighs with drumsticks can vary between 19 and 28% depending on the line, the slaughter age, and the sex. The range of ultimate pH comprises between 5.55 and 6.20 in breast muscle and between 5.75 and 6.30 in thigh muscle [3–5,9,16–28]. The juice loss can vary from 1 to 6% depending on the duration of cold storage and the cooking loss from 4% to 29% depending on the study and probably the cooking conditions and duration [7–10,12,14,15,19,23,25,28]. The shear force value of breast muscle ranges between 7 and 46 N depending on the study (cooked or raw meat). Some studies also reported the chemical composition of standard turkey meat. The water, protein, lipid, and ash contents of breast muscle vary between 72 and 75%, 21 and 28%, 0.4 and 4.0%, and 1.0 and 1.3%, respectively [3,5,7,9,10,12,14–16,20,24,26–28]. On the other hand, to our knowledge, no study has been published on the meat characteristics of male and female breeder turkeys. The aim of the present study was to evaluate the technological, nutritional, and sensorial quality of meat from breeder turkeys in comparison with that of standard turkeys.

2. Materials and Methods

2.1. Experimental Design

In order to realize this characterization, carcasses of male and female breeder turkeys from the Grademaker line (Hybrid Turkeys) were compared with carcasses of male and female standard turkeys from the same line ($n = 20$ per sex and per physiological stage). Hendrix Genetics Turkeys Company (Saint-Laurent de la Plaine, France) provided the animals reared, according the breeder recommendations. The turkeys were slaughtered according to standard procedures, which include immobilization by electrical stunning, followed by exsanguination, defeathering, and evisceration (=D0). STVO (Société de Transformation des Volailles de l'Ouest) Company (Saint-Mars la Jaille, France) cut the carcasses 24 h after slaughter and cold storage at 4 °C (=D1) in order to determine the meat yields. The thighs with drumsticks and breast muscles were then transported and stored under refrigerated conditions to the Research Unit BOA (Biologie des Oiseaux et Aviculture, INRA Nouzilly, France). From every left fillet, 4 cutlets of 200 g and 2 cm in thickness were individually cut and packed in a bag (sealed air cryovac, 60 μm) under vacuum (multivac P300, Cenpac, Chambray les Tours, France) and identified. The left thighs and drumsticks, with bones and skin, were individually packed in a bag, under vacuum, and identified. These samples were transported under refrigerated conditions to the experimental unit EASM (Elevage, Alimentation et Santé des Monogastriques, INRA Magneraud, Surgères, France) and stored at −20 °C for further sensorial analysis.

2.2. Analysis of Technological Quality of Meat

The right fillets and thighs with drumsticks were used to realize various measures and to take several samples on day 2. The ultimate pH (pHu) was determined by direct insertion of an electrode (pH meter Model 506, Crison Instruments, Barcelona, Spain) into the Pectoralis major (PM) and Iliotibialis superficialis (IT) muscles. The color was measured on the same muscles by using a Miniscan Spectrocolorimeter (Hunterlab, Reston, VA, USA) with the CIELAB thrichromatic system as lightness (L*), redness (a*), and yellowness (b*) values. The water holding capacity was estimated by measuring drip loss of a raw cutlet (around 150 g) placed in a plastic bag, hung from a hook, and stored at 4 °C for 6 days. The drip loss was expressed as the percentage of the initial cutlet weight. The cooking loss was measured on a thick cutlet (around 150–200 g) and packed in a bag under vacuum on day 3. The cutlets were cooked in a water bath at 85 °C for 16 min. They were then cooled during 15 min in crushed ice. The cooking loss was expressed as the percentage of the initial cutlet weight. The texture measurement was then realized on these cooked cutlets. The average Warner-Bratzler shear force value was determined on 3 strips (1 cm × 1 cm × 3 cm) for each cooked cutlet [29]. A piece of breast muscle (around 100 g) was processed into cured–cooked meat on day 4 in order to determine the technological yield [30].

2.3. Analysis of Nutritional Quality of Meat

In breast and thigh muscles, the content in haeminic pigments [31], protein content (Kjeldhal method), moisture content by differential weighing of 5 g of sample placed in steam room at 105 °C during 24 h, and lipid content were determined [32]. Lipids were then methylated [33] and the fatty acid composition was determined by gas chromatography (Perkin Elmer, Saint-Quentin en Yvelines, France) [34]. The classes of lipids were determined using Iatroscan (Iatron, Tokyo, Japan) based on thin-layer chromatography and a flame-ionisation-detector system (TLC–FID) [35]. The lipid peroxidation was evaluated [36] to determine the TBARS (Thio-Barbituric Acid Reactive Substances) value. The protein oxidation was evaluated by measuring the thiol and carbonyl content [37,38].

2.4. Analysis of Sensorial Quality of Meat

The sessions of sensory analysis were realized in the laboratory INRA of Magneraud (Surgères, France) in conditions corresponding to the standard [39]. The thighs with drumsticks and cutlets were defrosted before cooking for 24 to 48 h depending on the weight. The thighs with drumsticks were cooked in the oven (25 min at 250 °C, then maintained at 100 °C under wet heat in order to reach a core temperature of 80 °C). The cutlets placed between two aluminum foils were cooked in a steakhouse (5 min in 250 °C to reach a core temperature of 80 °C). Twelve panelists were first trained during three sessions and tasting one turkey per group and per session. Then, the panelists tasted the four groups in every session. Ten sessions were realized, the panelists testing one turkey per group and per session. For every criterion, the notation was made on a continuous scale limited from 0 to 10. Assessment criteria for the tasting of thighs were: Color, tenderness, juiciness, stringiness, compactness, oily sensation, global and rancid flavors, and global appreciation. Assessment criteria for the tasting of cutlets were: Color, tenderness, juiciness, stringiness, sticky, global and rancid flavors, and global appreciation.

2.5. Histological Analysis of PM Muscle

Samples of PM muscle were taken along a line parallel to the fiber axis on day 2 and frozen in isopentane cooled with liquid nitrogen and stored at −80 °C until histological analysis was performed [40]. Serial cross sections, 10 μm thick, were realized with a cryotome. The labeling of type VI collagen of chicken (Developmental Studies Hybridoma Bank, University of Iowa, Iowa City, IA, USA) was realized thanks to the kit Vectastain ABC elite (Laboratoires Eurobio/Abcys, Les Ulis, France) (Mouse IgG, Vector laboratories PK 6102 distributed by Eurobio Ingen (Les Ulis, France)) and revealed with DAB (Sigma, Saint-Quentin Fallavier, France). The cross-sectional area (CSA) of muscle fibers and the relative area occupied by collagen was determined using a computerized image analysis system (Visilog software, Noesis, Crolles, France) on 13 samples per group.

2.6. Statistical Analysis

Data were tested with a variance analysis using Statview software (SAS Institute Inc., Cary, NC, USA). The effects of sex, physiological stage, and their interaction were analyzed by comparing means with a t-test and a p value < 0.05. Pearson correlations were calculated between different measured parameters and considered significant with p < 0.05.

3. Results

3.1. Meat Yields (Table 1)

The breeders were slaughtered at older ages and they had a higher weight and higher yield of carcass and breast than standard birds. For the yield of thigh with drumstick, it was the opposite observation, particularly for the females. The most important observed differences concerned males compared to females and this whatever the physiological stage (standard or breeder). Indeed, the body weight of male breeders was 3.1 times higher than that of female breeders and the body weight of standard males was 2.3 times higher to that of standard females. Such differences were also reflected on the weight of carcass, breasts, and thighs with drumsticks. The carcass yield of male breeders was higher than that of female breeders, for which the ovaries removed during evisceration weighted approximately 270 g. The carcass yield of standard males was also higher than that of standard females. The yields of breasts and thighs with drumsticks of male breeders were higher than those of female breeders. It was the same for the standard turkeys.

Table 1. Meat yields of males and females of breeder and standard turkeys ($n = 20$).

	Female Breeders	Male Breeders	Standard Females	Standard Males	Physiological Stage Effect	Sex Effect	Interaction Effect
Slaughter age (days)	397	410	75	103			
Body weight at slaughter (kg)	10.42 ± 0.75 [c]	32.67 ± 1.51 [a]	5.89 ± 0.35 [d]	13.48 ± 0.56 [b]	0.001	0.001	0.001
Carcass weight (kg)	7.46 ± 0.56 [c]	25.08 ± 1.33 [a]	4.15 ± 0.31 [d]	9.74 ± 0.42 [b]	0.001	0.001	0.001
Breast weight (kg)	2.64 ± 0.23 [c]	9.20 ± 0.99 [a]	1.22 ± 0.13 [d]	2.97 ± 0.19 [b]	0.001	0.001	0.001
Thigh + drumstick weight (kg)	2.22 ± 0.18 [c]	7.83 ± 0.78 [a]	1.43 ± 0.10 [d]	3.40 ± 0.19 [b]	0.001	0.001	0.001
Carcass yield (%)	71.62 ± 1.09 [c]	76.75 ± 1.58 [a]	70.36 ± 1.59 [d]	72.25 ± 1.36 [b]	0.001	0.001	0.001
Breast yield (%)	25.31 ± 1.16 [b]	28.15 ± 2.51 [a]	20.69 ± 1.27 [d]	22.01 ± 1.02 [c]	0.001	0.001	0.001
Thigh + drumstick yield (%)	21.26 ± 0.70 [b]	23.98 ± 2.26 [a]	24.32 ± 0.73 [a]	25.19 ± 0.91 [a]	0.001	0.001	0.002

Yields are expressed as percentage of body weight at slaughter. The breasts include the two pectoral muscles (P. major and P. minor) without skin. The thighs and drumsticks include the bones and skin. [a, b, c, d] within a row, significant differences between groups with $p < 0.05$.

3.2. Technological Quality of Meat (Table 2)

The average pHu measured in the PM muscle was 5.70 whatever the sex or the physiological stage. The standard females had a higher pHu in PM muscle than that of standard males. The breeder turkeys had a lower pHu in IT muscle than that of standard turkeys. For the two physiological stages, the males had a lower pHu in IT than that of the females.

The standard males had darker PM muscle than the other groups and female breeders had darker IT muscle than the other groups. The standard turkeys had PM and IT muscles less red than those of the breeder turkeys, and the females had PM and IT muscles less red than those of the males. The male breeders had PM and IT muscles less yellow than those of the other groups. On the other hand, the females had IT muscles more yellow than the males.

The male breeders had PM muscles harder than those of the other groups. Globally, the drip loss of PM muscle after a storage at 4 °C was low. It was higher for the males compared to the females. The difference was mostly important for the breeder turkeys (×1.7). The cooking loss of PM muscle was higher for males compared to females and for breeder turkeys, particularly the males, compared to standard turkeys. The PM muscle of males had lower technological yield than that of females and the PM muscle of breeder turkeys, particularly the males, had lower technological yield than that of standard turkeys. Indeed, the female breeders had the highest technological yield. This confirmed the observations on the previous parameters concerning the water holding capacity of PM muscle. The coefficients of correlation between drip loss after a storage at 4 °C and cooking loss or technological yield were 0.46 and 0.63, respectively ($p < 0.05$). The coefficient of correlation between cooking loss and technological yield was 0.67 ($p < 0.05$).

The protein oxidation in meat was low. The carbonyl content determined in PM muscle of standard turkeys was higher than that measured in the PM muscle of breeder turkeys. However, the difference between the two physiological stages was low. The sex had no effect on the carbonyl content in PM muscle. In the Sartorius (SART) muscle, the carbonyl content was a bit higher than that measured in the PM muscle. In SART muscle, the physiological stage had no effect. The females had a higher carbonyl content in SART muscle than the males, but the difference was low. The physiological stage had no effect on the thiol content in PM and SART muscles. The sex had no effect on the thiol content in PM muscle. On the other hand, the females had lower thiol content in SART muscle than males, suggesting a higher level of protein oxidation, as the ability to release thiol was affected, and confirming the results obtained with carbonyl. The TBARS value was higher in SART muscle compared to PM muscle, and corroborating the contents in lipids and haemininc pigments. It was also higher in breeder turkeys compared to standard turkeys. The sex had no effect on the lipid peroxidation in PM and SART muscles.

Table 2. Technological quality of meat from males and females of breeder and standard turkeys ($n = 20$).

	Female Breeders	Male Breeders	Standard Females	Standard Males	Physiological Stage Effect	Sex Effect	Interaction Effect
pHu PM	5.68 ± 0.08 [b]	5.70 ± 0.13 [ab]	5.77 ± 0.07 [a]	5.68 ± 0.05 [b]	0.09	0.04	0.005
pHu IT	6.03 ± 0.11	5.79 ± 0.12	6.14 ± 0.14	5.92 ± 0.10	0.001	0.001	0.66
L* PM	49.14 ± 3.01 [a]	51.00 ± 3.24 [a]	49.30 ± 2.07 [a]	46.00 ± 2.21 [b]	0.001	0.23	0.001
a* PM	-0.38 ± 0.90	0.48 ± 1.02	-0.75 ± 0.73	-0.62 ± 0.86	0.001	0.01	0.07
b* PM	7.69 ± 1.44 [a]	6.27 ± 1.56 [b]	7.06 ± 1.23 [ab]	7.00 ± 1.31 [ab]	0.87	0.02	0.03
L* IT	36.90 ± 3.50 [b]	42.64 ± 4.48 [a]	42.85 ± 2.03 [a]	42.46 ± 2.02 [a]	0.001	0.001	0.001
a* IT	5.63 ± 1.04 [a]	5.97 ± 1.02 [a]	2.08 ± 0.99 [c]	4.03 ± 1.21 [b]	0.001	0.001	0.001
b* IT	4.00 ± 1.00	2.57 ± 1.74	4.66 ± 1.09	3.85 ± 0.80	0.001	0.001	0.26
PM texture (N/cm^2)	16.62 ± 1.65 [b]	27.00 ± 4.12 [a]	15.73 ± 1.63 [b]	16.68 ± 1.78 [b]	0.001	0.001	0.001
PM drip loss (%)	0.77 ± 0.34	1.29 ± 0.72	0.76 ± 0.31	0.97 ± 0.32	0.12	0.001	0.14
PM cooking loss (%)	11.08 ± 1.48 [b]	15.29 ± 2.43 [a]	8.59 ± 0.94 [c]	9.78 ± 1.08 [bc]	0.001	0.001	0.001
PM technological yield (%)	86.38 ± 2.16 [a]	72.38 ± 4.68 [c]	84.28 ± 2.72 [ab]	81.35 ± 2.03 [b]	0.001	0.001	0.001
PM carbonyl content	2.03 ± 0.38	2.14 ± 0.39	2.33 ± 0.30	2.26 ± 0.37	0.01	0.81	0.27
SART carbonyl content	3.43 ± 0.52	3.00 ± 0.42	3.21 ± 0.73	3.06 ± 0.56	0.52	0.03	0.27
PM thiol content	31.36 ± 6.96	35.76 ± 8.97	34.64 ± 6.84	35.99 ± 9.50	0.34	0.12	0.41
SART thiols content	34.84 ± 4.02	37.64 ± 4.81	37.03 ± 3.59	37.82 ± 4.40	0.21	0.06	0.29
PM TBARS value	0.80 ± 0.35	0.79 ± 0.25	0.44 ± 0.19	0.66 ± 0.34	0.001	0.13	0.09
SART TBARS value	1.34 ± 0.48	1.30 ± 0.45	0.89 ± 0.31	1.06 ± 0.38	0.001	0.48	0.26

PM = Pectoralis major; IT = Iliotibialis superficialis; SART = Sartorius; L* = lightness; a* = redness; b* = yellowness; PM texture was estimated by measuring the shear-force value (N/cm^2). The drip loss was measured after 6 days storage of breast cutlets at 4 °C. The technological yield was estimated on cured–cooked samples of breast muscle. The carbonyl content was expressed as nmol DNPH (2,4-dinitrophenylhydrazine) incorporated/mg protein. The thiol content was expressed as nmol/mg protein. The TBARS (Thio-Barbituric Acid Reactive Substances) value was expressed as mg equivalent MDA/g muscle. MDA = malondialdehyde; [a,b,c] within a row, significant differences between groups with $p < 0.05$.

3.3. Nutritional Quality of Meat (Tables 3–5)

The breeder turkeys had a higher lipid content in PM and IT muscle than standard animals (Table 3). The male breeders had a lower protein content in PM muscle than the other groups and the lowest protein content in IT muscle. The physiological stage had no effect on the iron and haeminic pigment (myoglobin and hemoglobin) contents in PM muscle whereas in IT muscle, the iron and haeminic pigment contents were higher in breeder turkeys compared to standard turkeys. The males had higher iron and haeminic pigment contents in PM and IT muscles than females. The iron and haeminic pigment contents in IT muscle were 2- to 3-fold higher to that measured in PM muscle. The coefficient of correlation between the haeminic pigment content in IT muscle and redness was 0.73 ($p < 0.05$).

Table 3. Nutritional quality of meat from males and females of breeder and standard turkeys ($n = 10$).

	Female Breeders	Male Breeders	Standard Females	Standard Males	Physiological Stage Effect	Sex Effect	Interaction Effect
Dry matter PM (%)	27.70 ± 0.87	24.11 ± 1.82	25.60 ± 0.70	26.16 ± 0.51	0.001	0.26	0.23
Proteins PM (%)	24.72 ± 0.87 [a]	21.68 ± 1.80 [b]	25.21 ± 0.89 [a]	25.41 ± 0.52 [a]	0.001	0.001	0.001
Lipides PM (%)	3.01 ± 0.86	2.51 ± 0.96	1.01 ± 0.33	1.02 ± 0.15	0.001	0.26	0.23
Ashes PM (%)	1.08 ± 0.04 [b]	1.05 ± 0.04 [b]	1.42 ± 0.05 [a]	1.10 ± 0.13 b	0.001	0.001	0.001
Iron (µg/g PM)	3.03 ± 0.56	3.42 ± 1.31	2.37 ± 1.15	3.77 ± 1.45	0.65	0.01	0.14
Myoglobin (µg/g PM)	922 ± 170	1042 ± 399	722 ± 349	1147 ± 441	0.65	0.01	0.14
Dry matter IT (%)	26.14 ± 0.46 [a]	24.92 ± 1.05 [ab]	23.69 ± 1.12 [b]	24.52 ± 0.75 [ab]	0.001	0.50	0.001
Proteins IT (%)	22.70 ± 0.31 [a]	20.25 ± 1.03 [b]	21.79 ± 0.67 [ab]	21.39 ± 0.95 [b]	0.65	0.001	0.001
Lipids IT (%)	3.50 ± 0.69	4.20 ± 1.13	2.56 ± 0.73	2.91 ± 0.77	0.001	0.06	0.52
Ash IT (%)	1.12 ± 0.04 [a]	1.08 ± 0.05 [ab]	1.02 ± 0.07 [b]	1.05 ± 0.04 [b]	0.001	0.57	0.02
Iron (µg/g IT)	10.33 ± 2.48	12.70 ± 2.51	5.58 ± 1.68	8.21 ± 2.35	0.001	0.001	0.84
Myoglobin (µg/g IT)	3143 ± 756	3865 ± 765	1699 ± 511	2500 ± 714	0.001	0.001	0.84
Triglycerides PM (%)	2.51 ± 0.81	2.02 ± 0.90	0.54 ± 0.21	0.64 ± 0.13	0.001	0.33	0.14
Cholesterol PM (%)	0.04 ± 0.02 [b]	0.07 ± 0.02 [a]	0.05 ± 0.02 [ab]	0.04 ± 0.01 [b]	0.31	0.15	0.001
Phospholipids PM (%)	0.46 ± 0.08	0.42 ± 0.07	0.40 ± 0.13	0.34 ± 0.07	0.01	0.09	0.77
Triglycerides IT (%)	2.76 ± 0.57	3.29 ± 0.95	1.77 ± 0.59	2.16 ± 0.70	0.001	0.05	0.77
Cholesterol IT (%)	0.07 ± 0.03 [b]	0.09 ± 0.04 [ab]	0.14 ± 0.04 [a]	0.08 ± 0.03 [b]	0.01	0.10	0.001
Phospholipids IT (%)	0.67 ± 0.12	0.82 ± 0.25	0.65 ± 0.11	0.67 ± 0.35	0.24	0.25	0.38

PM and IT = Pectoralis major and Iliotibialis superficialis muscles, respectively; [a,b] within a row, significant differences between groups with $p < 0.05$.

The sex had no effect on the triglyceride, cholesterol, and phospholipid contents of PM muscle (Table 3). The physiological stage had no effect on the cholesterol content of PM muscle. By contrast, the breeder turkeys had higher triglyceride and phospholipid contents in PM muscle than standard turkeys. The sex had no effect on the cholesterol and phospholipid contents of IT muscle. The males had a higher triglyceride content in IT muscle than the females. The physiological stage had no effect on the phospholipid content in IT muscle. The breeder turkeys had a higher triglyceride content and a lower cholesterol content than the standard turkeys.

The PM muscle had a high *n*-6 fatty acid (FA) content and the ratio *n*-6/*n*-3 FA was around 10–12 (Table 4). The sex had few effects on the FA composition of PM muscle. However, the breeder females had a higher saturated FA (SFA) content than breeder males. The breeder turkeys had a higher mon-unsaturated FA (MUFA) content and lower SFA and poly-unsaturated FA (PUFA) contents.

Table 4. Fatty acid (FA) composition of P. major muscle from males and females of breeder and standard turkeys (% total FA; $n = 10$).

	Female Breeders	Male Breeders	Standard Females	Standard Males	Physiological Stage Effect	Sex Effect	Interaction Effect
C14:0	0.90 ± 0.54 [a]	0.54 ± 0.06 [ab]	0.41 ± 0.06 [b]	0.52 ± 0.08 [ab]	0.006	0.18	0.01
C14:1	0.10 ± 0.04 [ab]	0.14 ± 0.06 [a]	0.09 ± 0.06 [ab]	0.06 ± 0.02 [b]	0.004	0.74	0.05
C16:0	26.11 ± 0.83 [a]	22.05 ± 1.15 [c]	24.03 ± 1.03 [bc]	25.60 ± 0.53 [ab]	0.001	0.18	0.03
C16:1	3.93 ± 0.79 [a]	4.12 ± 0.76 [a]	2.07 ± 0.68 [b]	1.28 ± 0.46 [b]	0.001	0.18	0.03
C18:0	7.09 ± 0.54	7.11 ± 1.26	11.21 ± 1.49	9.81 ± 1.46	0.001	0.09	0.08
C18:1	35.63 ± 1.65	34.94 ± 1.62	24.88 ± 1.77	26.06 ± 0.97	0.001	0.61	0.06
C18:2 *n*-6	22.02 ± 1.23 [c]	25.89 ± 2.20 [bc]	29.38 ± 1.77 [ab]	29.68 ± 1.43 [a]	0.001	0.001	0.01
C18:3 *n*-3	1.17 ± 0.15	1.95 ± 0.40	1.77 ± 0.36	2.15 ± 0.40	0.001	0.001	0.07
C20:0	0.05 ± 0.03	0.07 ± 0.02	0.09 ± 0.02	0.11 ± 0.02	0.001	0.10	0.14
C20:1	0.20 ± 0.06	0.25 ± 0.06	0.17 ± 0.02	0.17 ± 0.02	0.001	0.10	0.14
C20:4 *n*-6	1.53 ± 0.46 [c]	1.92 ± 1.20 [bc]	4.53 ± 1.05 [a]	3.42 ± 0.91 [ab]	0.001	0.24	0.02
C20:5 *n*-3	0.62 ± 0.59	0.48 ± 0.74	0.10 ± 0.07	0.09 ± 0.04	0.01	0.60	0.66
C22:4 *n*-6	0.11 + 0.03	0.10 + 0.07	0.22 + 0.05	0.20 + 0.04	0.001	0.37	0.52
C22:5 *n*-3	0.12 ± 0.03 [c]	0.28 ± 0.15 [bc]	0.69 ± 0.19 [a]	0.57 ± 0.13 [ab]	0.001	0.62	0.01
C22:6 *n*-3	0.41 ± 0.17	0.16 ± 0.15	0.36 ± 0.11	0.27 ± 0.08	0.49	0.001	0.07
SFA	34.16 ± 0.82 [a]	29.77 ± 1.79 [b]	35.74 ± 2.16 [a]	36.04 ± 1.55 [a]	0.001	0.001	0.001
MUFA	39.86 ± 1.43	39.45 ± 2.10	27.20 ± 2.21	27.57 ± 1.38	0.001	0.96	0.50
PUFA	25.99 ± 1.05 [c]	30.78 ± 2.55 [b]	37.05 ± 2.63 [a]	36.38 ± 1.28 [a]	0.001	0.01	0.001
n-6 FA	23.66 ± 0.98 [c]	27.91 ± 2.18 [b]	34.13 ± 2.31 [a]	33.31 ± 1.11 [a]	0.001	0.01	0.001
n-3 FA	2.32 ± 0.57	2.87 ± 0.89	2.92 ± 0.36	3.08 ± 0.24	0.03	0.06	0.29
n-6 FA/*n*-3 FA	10.70 ± 2.47	10.27 ± 2.14	11.79 ± 0.91	10.87 ± 0.68	0.13	0.23	0.65

SFA, MUFA, PUFA = Saturated, mono-unsaturated, and poly-unsaturated fatty acids, respectively; [a,b,c] within a row, significant differences between groups with $p < 0.05$.

The IT muscle had a fatty acid composition close to that described for PM muscle (Table 5). The ratio *n*-6/*n*-3 FA varied between 11 and 15. The sex had more effect on the FA composition of IT muscle, but the differences observed between males and females for one given physiological stage were low. The males had a content in C18:3 *n*-3 higher to that of females. The effect of physiological stage was significant for all FA except C20:5 *n*-3 and C22:4 *n*-6. The IT muscle of breeder turkeys had a higher MUFA content and lower SFA and PUFA contents to that of standard turkeys.

Table 5. Fatty acid (FA) composition of I. superficialis muscle from males and females of breeder and standard turkeys (% total FA, *n* = 10).

	Female Breeders	Male Breeders	Standard Females	Standard Males	Physiological Stage Effect	Sex Effect	Interaction Effect
C14:0	0.68 ± 0.04 [a]	0.54 ± 0.08 [b]	0.52 ± 0.04 [b]	0.60 ± 0.04 [a]	0.01	0.07	0.001
C14:1	0.12 ± 0.03	0.13 ± 0.02	0.07 ± 0.02	0.05 ± 0.02	0.001	0.97	0.14
C16:0	24.60 ± 0.49 [ab]	21.32 ± 1.05 [c]	23.32 ± 1.15 [bc]	25.13 ± 0.62 [a]	0.001	0.01	0.001
C16:1	3.44 ± 0.57 [ab]	3.89 ± 0.77 [a]	2.28 ± 0.78 [bc]	1.41 ± 0.48 [c]	0.001	0.33	0.01
C18:0	8.39 ± 0.51	7.95 ± 0.84	9.64 ± 0.93	8.89 ± 0.93	0.001	0.03	0.57
C18:1	32.52 ± 0.73	33.26 ± 1.24	24.69 ± 2.03	25.87 ± 0.84	0.001	0.03	0.60
C18:2 *n*-6	25.51 ± 1.06 [b]	27.91 ± 1.52 [b]	32.55 ± 2.17 [a]	32.10 ± 0.72 [a]	0.001	0.001	0.01
C18:3 *n*-3	1.17 ± 0.13 [c]	1.85 ± 0.23 [b]	2.43 ± 0.27 [a]	2.67 ± 0.19 [a]	0.001	0.001	0.01
C20:0	0.08 ± 0.01	0.08 ± 0.01	0.10 ± 0.02	0.11 ± 0.02	0.001	0.21	0.10
C20:1	0.22 ± 0.03 [ab]	0.27 ± 0.04 [a]	0.18 ± 0.02 [b]	0.19 ± 0.02 [b]	0.001	0.001	0.01
C20:4 *n*-6	2.37 ± 0.29 [ab]	2.33 ± 0.70 [b]	3.35 ± 0.95 [a]	2.32 ± 0.56 [b]	0.03	0.02	0.03
C20:5 *n*-3	0.07 ± 0.06	0.06 ± 0.04	0.08 ± 0.09	0.10 ± 0.04	0.15	0.98	0.49
C22:4 *n*-6	0.19 + 0.04 [a]	0.11 + 0.03 [b]	0.15 + 0.03 [ab]	0.12 + 0.04 [b]	0.35	0.001	0.04
C22:5 *n*-3	0.12 ± 0.02 [c]	0.21 ± 0.06 [b]	0.45 ± 0.13 [a]	0.32 ± 0.08 [ab]	0.001	0.52	0.001
C22:6 *n*-3	0.54 ± 0.14 [a]	0.12 ± 0.04 [c]	0.21 ± 0.08 [b]	0.12 ± 0.05 [c]	0.001	0.001	0.001
SFA	33.75 ± 0.94 [a]	29.88 ± 1.16 [b]	33.58 ± 1.19 [a]	34.74 ± 0.82 [a]	0.001	0.001	0.001
MUFA	36.29 ± 0.97 [a]	37.55 ± 1.85 [a]	27.21 ± 2.75 [b]	27.52 ± 1.29 [b]	0.001	0.38	0.01
PUFA	29.89 ± 0.97 [b]	32.51 ± 1.93 [b]	39.13 ± 3.20 [a]	37.65 ± 1.08 [a]	0.001	0.38	0.01
n-6 FA	28.07 ± 0.94 [b]	30.34 ± 1.93 [b]	36.05 ± 2.85 [a]	34.54 ± 1.03 [a]	0.001	0.02	0.03
n-3 FA	1.83 ± 0.12 [c]	2.18 ± 0.19 [b]	3.09 ± 0.38 [a]	3.10 ± 0.11 [a]	0.001	0.02	0.03
n-6 FA/*n*-3 FA	15.41 ± 0.98	14.05 ± 1.65	11.76 ± 0.84	11.14 ± 0.42	0.001	0.01	0.28

SFA, MUFA, PUFA = saturated, mono-unsaturated, and poly-unsaturated fatty acids, respectively; [a,b,c] within a row, significant differences between groups with $p < 0.05$.

3.4. Sensorial Quality of Meat (Table 6)

The thighs of breeder turkeys were judged more colored, less soft, less juicy, stringier, and more compact than those of standard animals. They were also less appreciated. There was no effect of the physiological stage on the oily sensation of cooked thighs nor on the global and rancid flavors, whose scores were very low. The thighs of males were judged more colored, juicier, and stringier than those of females.

The breasts of breeder turkeys were judged less tender, less juicy, stringier, and less sticky to those of standard turkeys. They also had a flavor less acid and they were less appreciated, particularly those of male breeders. There was no effect of physiological stage on the color of cooked breast and global and rancid flavors, whose scores were very low. The breasts of males, particularly those of breeders, were less tender, stringier, and less sticky than the breasts of females. Their global flavor was lower, and they were less appreciated than those of females.

The coefficients of correlation between the tenderness and stringiness scores and the shear-force value were −0.84 and 0.73, respectively ($p < 0.05$).

Table 6. Sensorial analysis of cooked thigh and breast meat from males and females of breeder and standard turkeys ($n = 10$).

	Female Breeders	Male Breeders	Standard Females	Standard Males	Physiological Stage Effect	Sex Effect	Interaction Effect
Thighs							
Colour	4.78 ± 1.41 [b]	6.29 ± 1.49 [a]	3.04 ± 1.32 [c]	3.43 ± 1.43 [c]	0.001	0.001	0.001
Tenderness	2.45 ± 1.03 [c]	2.99 ± 1.45 [b]	4.38 ± 1.34 [a]	4.01 ± 1.21 [a]	0.001	0.46	0.001
Juiciness	1.54 ± 0.85	1.81 ± 1.21	2.22 ± 0.97	2.36 ± 1.32	0.001	0.05	0.56
Stinginess	2.18 ± 1.15	2.86 ± 1.40	1.47 ± 1.18	1.99 ± 1.21	0.001	0.001	0.47
Compactness	3.03 ± 1.75	2.60 ± 1.97	1.64 ± 1.11	1.76 ± 1.25	0.001	0.29	0.07
Oily sensation	1.35 ± 1.19	1.50 ± 1.42	1.54 ± 1.29	1.47 ± 1.14	0.51	0.73	0.37
Global flavour	3.92 ± 1.05	4.08 ± 1.01	3.83 ± 1.00	3.88 ± 0.95	0.13	0.29	0.55
Rancid flavour	0.37 ± 0.41	0.37 ± 0.37	0.36 ± 0.39	0.32 ± 0.29	0.36	0.45	0.63
Global appreciation	2.52 ± 1.26	2.51 ± 1.21	3.38 ± 1.54	3.58 ± 1.33	0.001	0.46	0.39
Breast							
Colour	2.25 ± 1.43	2.49 ± 1.35	2.23 ± 1.37	2.45 ± 1.36	0.83	0.07	0.93
Tenderness	3.79 ± 1.16 [b]	2.07 ± 1.14 [c]	4.71 ± 1.39 [a]	4.41 ± 1.33 [a]	0.001	0.001	0.001
Juiciness	3.05 ± 1.46	2.78 ± 1.61	3.33 ± 1.50	3.15 ± 1.31	0.02	0.11	0.74
Stringiness	2.28 ± 1.26 [b]	3.15 ± 1.61 [a]	1.69 ± 1.24 [c]	2.03 ± 1.21 [bc]	0.001	0.001	0.04
Sticky	1.63 ± 1.13	1.11 ± 1.00	1.83 ± 1.33	1.71 ± 1.19	0.001	0.004	0.07
Global flavour	3.41 ± 0.94 [a]	3.00 ± 1.04 [b]	3.37 ± 1.03 [a]	3.38 ± 0.99 [a]	0.06	0.03	0.03
Acid flavour	1.11 ± 0.80	1.03 ± 1.00	1.36 ± 1.13	1.26 ± 1.02	0.01	0.31	0.93
Rancid flavour	0.33 ± 0.29	0.37 ± 0.42	0.33 ± 0.34	0.29 ± 0.28	0.23	0.82	0.23
Global appréciation	2.27 ± 1.16 [a]	1.60 ± 1.03 [b]	2.28 ± 1.23 [a]	2.42 ± 1.25 [a]	0.001	0.02	0.001

[a,b,c] within a row, significant differences between groups with $p < 0.05$.

3.5. Histological Characteristics of Pectoralis Major Muscle (Table 7, Figures 1 and 2)

The females, particularly the standard ones, had lower CSA of muscle fibers and higher relative area occupied by collagen than males. The breeder turkeys, particularly the males, had a higher CSA of muscle fibers and lower relative area occupied by collagen than standard turkeys. The breeder females and the standard males had a comparable average CSA. The CSA of muscle fibers was correlated ($p < 0.05$) with the weight (0.87), the shear-force value (0.74), and the tenderness (-0.83) and stringiness (0.73) scores of PM muscle. The relative area occupied by collagen was correlated ($p < 0.05$) with the weight (-0.40), the shear-force value (-0.34), and CSA of muscle fibers (-0.42) of PM muscle.

Table 7. Relative area occupied by collagen and cross-sectional area (CSA) of muscle fibers of Pectoralis major muscle from males and females of breeder and standard turkeys ($n = 13$).

	Female Breeders	Male Breeders	Standard Females	Standard Males	Physiological Stage Effect	Sex Effect	Interaction Effect
Collagen, %	19.09 ± 3.20	17.06 ± 2.37	20.75 ± 1.88	18.85 ± 3.50	0.03	0.02	0.94
AST, μm²	1826 ± 419 [b]	3695 ± 593 [a]	1181 ± 238 [c]	2117 ± 439 [b]	0.001	0.001	0.001

[a, b, c] within a row, significant differences between groups with $p < 0.05$.

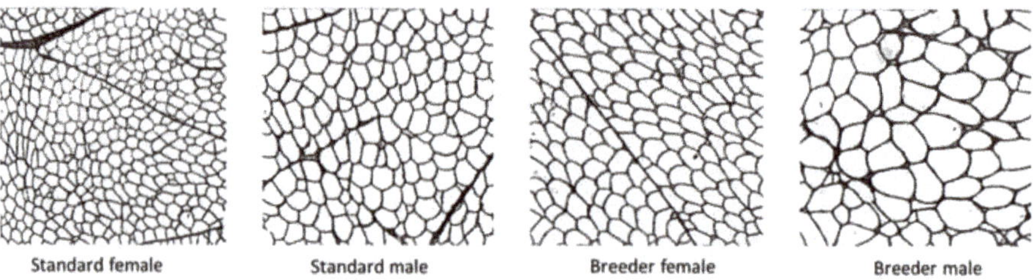

Figure 1. Cross-sectional area of Pectoralis major muscle from males and females of breeder and standard turkeys ($n = 13$, magnification ×5).

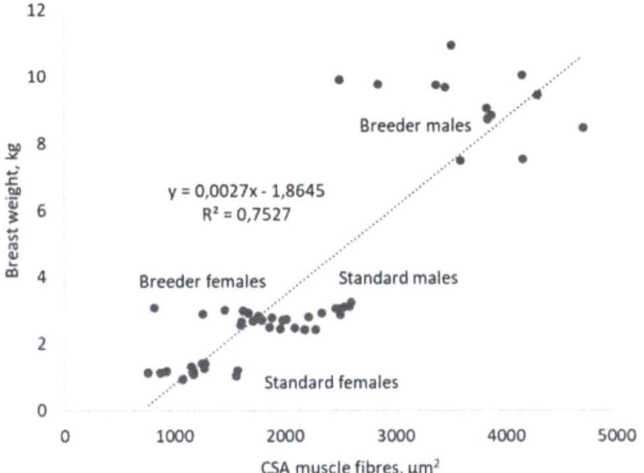

Figure 2. Relation between CSA of muscle fibers and breast weight from males and females of breeder and standard turkeys ($n = 13$).

4. Discussion

4.1. Sex Effect

The turkey is characterized by a strong sexual dimorphism on body weight resulting in a higher weight of carcass and cut pieces for males compared to females. It was the same for the yields expressed relative to body weight. The coefficient multiplier between the breast weight of male breeders and that of female breeders was 3.48. For the CSA of muscle fibers of PM muscle, this coefficient was 2.02. For the standard turkeys, these coefficients were 2.43 and 1.79, respectively. This means that the breast weight difference between males and females was partially due to the hypertrophy of the muscle fibers but also due to a more important number and/or length of muscle fibers.

Concerning the technological quality, males slaughtered at older ages presented a lower pHu in PM and IT muscles than that measured for females, suggesting higher glycogen reserves [41]. The PM and IT muscles of males had higher haeminic pigment content and they were redder than those of females. The males had PM muscles displaying higher drip loss during a storage at 4 °C and higher cooking loss, resulting in a lower technological yield than that of females [41]. On the other hand, the protein oxidation in SART muscle during a storage at 4 °C was lower for males compared to females.

4.2. Effect of Physiological Stage

The breeder turkeys slaughtered at older ages had higher body weight and weight of cut pieces than those of standard turkeys. Their carcass and breast yields were also higher, whereas the yield of thigh with drumstick was lower than those of standard turkeys. A study concerning the meat valuation of hens at the end of their laying cycle had shown rather the opposite, with a carcass yield relative to body weight of 60% and a breast yield of 11% [42]. The PM and IT muscles of breeder turkeys were redder and the content in haeminic pigment of IT muscle was higher than those of standard turkeys. During the sensory analysis, the thighs of breeder turkeys were also judged more colored. The thigh and breast muscles of hens and cocks slaughtered at 64 weeks of age were also redder compared to those of male and female chickens slaughtered at 6 weeks of age [43]. The same observation was reported for a comparison between breeder ducks slaughtered at 500 days of age and growing ducks slaughtered at 38 days of age [44]. The breast muscles of breeder turkeys were less tender than those of standard turkeys. They were also judged stringier. These differences were mainly explained by the difference in CSA of muscle fibers. The breast and thigh muscles of breeder ducks were also judged less tender than those of growing ducks [44]. The PM muscle of breeder males also had a lower protein content and a lower technological yield after the cured and cooked process. During the sensorial analysis, the thighs of breeder turkeys were judged less juicy than those of standard turkeys. The PM and IT muscles of breeder turkeys had higher lipid, triglyceride, and MUFA contents than those of standard turkeys. However, this had no effect on the flavor of cooked thighs evaluated during the sensorial analysis. On the reverse, the breeder males had the lowest flavor score. A lower meat tenderness and a higher carcass and intramuscular fatness were recorded in laying hens slaughtered at 56 weeks of age compared to broiler chickens [45]. The breast and thigh muscles of breeder ducks had higher protein, lipid, and MUFA contents than those of growing ducks [44]. They also had a higher flavor score. The breast muscles of breeders 64 weeks old were less juicy, less tender, and less tasty than those of chickens 6 weeks old [43]. The flavor score was also lower.

5. Conclusions

The differences observed between males and females on one hand and between standard and breeder turkeys on the other hand mainly resulted from the difference in slaughter ages and sexual dimorphism on body weight. The meat of female breeders had characteristics close to those of standard turkeys, whereas the meat of male breeders was clearly distinguishable, particularly by displaying lower tenderness and water holding capacity. For the latter, the transformation and/or cooking processes must be clearly adapted by suggesting, for example. a use in minced meat to make loaf, or in cut pieces associated with a long cooking. The storage at -20 °C of breast and thigh muscles vacuum-packed for 5 to 7 months did not induce oxidation phenomena, which was confirmed by the oxidation measures of proteins and lipids.

Author Contributions: All the authors prepared the experimental design and the project. S.B. organized the rearing and slaughtering of turkeys and made the choice of the animals used in the present study. J.-C.G. and E.D. took in charge the determination of carcass composition. P.C., T.B., E.G., E.B., and M.B. participated to sample collection and the realization of the different laboratory measures. K.M. realized the sensorial analysis of breast and thigh meat. E.B. realized the statistical analysis, interpreted the data, and wrote the manuscript. P.C. and M.B. participated to the writing of the manuscript.

Funding: Hendrix Genetics Turkeys Company funded this study.

Conflicts of Interest: S.B., E.D., and J.-C.G. participated to the design of the study and the decision to publish the results.

References

1.   FranceAgriMer. *Données et Bilans: Les Filières Animales Terrestres et Aquatiques*; Bilan 2015 et Perspectives 2016; FranceAgriMer: Paris, France, 2016; p. 137.

2.   Institut Technique de l'Aviculture (ITAVI). Situation de la Production et des Marchés des Volailles de Chair. *Bilan* **2015**. Available online: https://www.google.com/search?q=2.+ITAVI.+Situation+de+la+Production+ et+des+March%C3%A9s+des+Volailles+de+Chair.+Bilan+2015&rlz=1C1GCEU_frFR821FR821&oq=2. +ITAVI.+Situation+de+la+Production+et+des+March%C3%A9s+des+Volailles+de+Chair.+Bilan+2015& aqs=chrome..69i57j69i59.4742j1j9&sourceid=chrome&ie=UTF-8 (accessed on 23 December 2018).

3.   Werner, C.; Riegel, J.; Wicke, M. Slaughter performance of four different turkey strains with special focus on the muscle fiber structure and the meat quality of the breast muscle. *Poult. Sci.* **2008**, *87*, 1849–1859. [CrossRef] [PubMed]

4.   Werner, C.; Janisch, S.; Kuembet, U.; Wicke, M. Comparative study of the quality of broiler and turkey meat. *Br. Poult. Sci.* **2009**, *50*, 318–324. [CrossRef] [PubMed]

5.   Konca, Y.; Kirkipinar, F.; Mert, S. Effect of mannan-oligosaccharides and live yeast in diets on the carcass, cut yields, meat composition and colour of finishing turkeys. *Asian Australas J. Anim. Sci.* **2009**, *22*, 550–556. [CrossRef]

6.   Krischek, C.; Janisch, S.; Günther, R.; Wicke, M. Nutrient composition of broiler and turkey breast meat in relation to age, gender and genetic line of the animals. *Arch. Lebensmittelhyg.* **2011**, *62*, 76–81.

7.   Jankowski, J.; Lecewicz, A.; Chwastowska-Siwiecka, I.; Juskiewicz, J.; Zdunczyk, Z. Performance, slaughter value and meat quality of turkeys fed diets with different content of sunflower. *Arch. Geflügelk.* **2011**, *75*, 104–112.

8.   Jankowski, J.; Juskiewicz, J.; Lichtorowicz, K.; Zdunczyk, Z. Effects of the dietary level and source of sodium on growth performance, gastrointestinal digestion and meat characteristics in turkeys. *Anim. Feed Sci. Technol.* **2012**, *178*, 74–83. [CrossRef]

9.   Mikulski, D.; Jankowski, J.; Zdunczyk, Z.; Juskiewicz, J.; Slominski, B.A. The effect of different dietary levels of rapeseed meal on growth performance, carcass traits and meat quality in turkeys. *Poult. Sci.* **2012**, *91*, 215–223. [CrossRef]

10.   Damaziak, K.; Pietrzak, D.; Michalczuk, M.; Mroczek, J.; Niemiec, J. Effect of genotype and sex on selected quality attributes of turkey meat. *Arch. Geflügelk.* **2013**, *77*, 206–214.

11.   Damaziak, K.; Michalczuk, M.; Zdanowska-Sasiadek, Z.; Niemiec, J.; Gozdowski, D. Variation in growth performance and carcass yield of pure and reciprocal crossbred turkeys. *Ann. Anim. Sci.* **2015**, *15*, 51–66. [CrossRef]

12.   Damaziak, K.; Pietrzak, D.; Michalczuk, M.; Adamczak, L.; Chmiel, M.; Florowski, T.; Gozdowski, D.; Niemiec, J. Early and 24 h post-mortem changes in breast muscle quality traits of two turkey genotypes and their reciprocal crosses raised under semi-confined conditions. *Br. Poult. Sci.* **2016**, *57*, 51–62. [CrossRef]

13.   Murawska, D.; Kozlowski, K.; Tomaszewska, K.; Brzozowski, W.; Zawacka, M.; Michalik, D. Age-related chages in the tissue composition of carcass parts and in the distribution of lean meat, fat with skin and bones in turkey carcasses. *Eur. Poult. Sci.* **2015**, *79*. [CrossRef]

14.   Krawczyk, M.; Mikulski, D.; Przywitowski, M.; Jankowski, J. The effect of dietary yellow lupine (*L. luteus* cv. Baryt) on growth performance, carcass characteristics, meat quality and selected serum parameters of turkeys. *J. Anim. Feed Sci.* **2015**, *24*, 61–70.

15.   Przywitowski, M.; Mikulski, D.; Zdunczyk, Z.; Rogiewicz, A.; Jankowski, J. The effect of dietary high-tannin and low-tannin faba bean (*Vicia faba* L.) on the growth performance, carcass traits and breast meat characteristics of finisher turkeys. *Anim. Feed Sci. Technol.* **2016**, *221*, 124–136. [CrossRef]

16.   Fernandez, X.; Santé, V.; Baéza, E.; Le Bihan-Duval, E.; Berri, C.; Rémignon, H.; Babilé, R.; Le Pottier, G.; Millet, N.; Berge, P.; et al. Post mortem muscle metabolism and meat quality in three genetic types of turkey. *Br. Poult. Sci.* **2001**, *42*, 462–469. [CrossRef] [PubMed]

17.   Fernandez, X.; Santé, V.; Baéza, E.; Le Bihan-Duval, E.; Berri, C.; Rémignon, H.; Babilé, R.; Le Pottier, G.; Astruc, T. Effects of the rate of muscle post mortem pH fall on the technological yield of turkey meat. *Br. Poult. Sci.* **2002**, *43*, 245–252. [CrossRef] [PubMed]

18.   Le Bihan-Duval, E.; Berri, C.; Baéza, E.; Santé, V.; Astruc, T.; Rémignon, H.; Le Pottier, G.; Bentley, J.; Beaumont, C.; Fernandez, X. Genetic parameters of meat technological quality traits in a grand-parental commercial line of turkey. *Gen. Select. Evol.* **2003**, *35*, 623–635. [CrossRef]

19.   El Rammouz, R.; Babilé, R.; Fernandez, X. Effect of ultimate pH on the physico-chemical and biochemical characteristics of turkey breast muscle showing normal rate of postmortem pH fall. *Poult. Sci.* **2004**, *83*, 1750–1757. [CrossRef] [PubMed]

20.   Karwowska, M.; Stadnik, J.; Dolatowski, Z.J.; Grela, E.R. Effect of protein-xanthophylls (PX) concentrate of alfalfa supplementation on physico-chemical properties of turkey breast and thigh muscles during ageing. *Meat Sci.* **2010**, *86*, 486–490. [CrossRef]

21.   Sarica, M.; Ocak, N.; Turhan, S.; Kop, C.; Yamak, U.S. Evaluation of meat quality from 3 turkey genotypes reared with or without outdoor access. *Poult. Sci.* **2011**, *90*, 1313–1323. [CrossRef]

22.   Aslam, M.L.; Muhammad, L.; Bastiaansen, J.W.M.; Crooijmans, R.P.M.A.; Ducro, B.J.; Vereijken, A.; Groenen, M.A.M. Genetic variances, heritabilities and maternal effects on body weight, breast meat yield, meat quality traits and the shape of the growth curve in turkey birds. *BMC Genet.* **2011**, *12*, 14. [CrossRef]

23.   Jankowski, J.; Zdunczyk, Z.; Mikulski, D.; Juskiewicz, J.; Naczmanski, J.; Pomianowski, J.F.; Zdunczyk, P. Fatty acid profile, oxidative stability and sensory properties of breast meat from turkeys fed diets with a different n-6/n-3 PUFA ratio. *Eur. J. Lipid Sci. Technol.* **2012**, *114*, 1025–1035. [CrossRef]

24.   Carvalho, R.H.; Soares, A.L.; Honorato, D.C.B.; Guarnieri, P.D.; Pedrao, M.R.; Paiao, F.G.; Oba, A.; Ida, E.I.; Shimokomaki, M. The incidence of pale, soft and exudative (PSE) turkey meat at a Brazilian commercial plant and the functional properties in its meat product. *LWT-Food Sci. Technol.* **2014**, *59*, 883–888. [CrossRef]

25. Ozturk, B.; Serdaroglu, M. Quality characteristics of PSE-like turkey Pectoralis major muscles generated by high post-mortem temperature in a local Turkish slaughterhouse. *Korean J. Food Sci. Anim. Resour.* **2015**, *35*, 524–532. [CrossRef] [PubMed]
26. Celen, M.F.; Sögüt, B.; Zorba, O.; Demirulus, H.; Tekell, A. Comparison of normal and PSE turkey breast meat for chemical composition, pH, color, myoglobin and drip loss. *Rev. Bras. Zoot.* **2016**, *45*, 441–444. [CrossRef]
27. Patterson, B.A.; Matarneh, S.K.; Stufft, K.M.; England, E.M.; Scheffler, T.L.; Preisser, R.H.; Shi, H.; Stewart, E.C.; Eilert, S.; Gerrard, D.E. Pectoralis major muscle of turkey displays divergent function as correlated with meat quality. *Poult. Sci.* **2017**, *96*, 1492–1503. [PubMed]
28. Soglia, F.; Baldi, G.; Laghi, L.; Mudalal, S.; Cavani, C.; Petracci, M. Effect of white striping on turkey breast meat quality. *Animal* **2018**, *12*, 2198–2204. [CrossRef] [PubMed]
29. Honikel, K.O. Reference methods for the assessment of physical characteristics of meat. *Meat Sci.* **1998**, *49*, 447–457. [CrossRef]
30. Berri, C.; Le Bihan-Duval, E.; Baéza, E.; Chartrin, P.; Picgirard, L.; Jehl, N.; Quentin, M.; Picard, M.; Duclos, M.J. Further processing characteristics of breast and leg meat from fast-, medium- and slow-growing commercial chickens. *Anim. Res.* **2005**, *54*, 123–134. [CrossRef]
31. Hornsey, H.C. The colour of cooked cured pork. Estimation of the nitric oxide-haem pigments. *J. Sci. Food Agric.* **1956**, *7*, 534–540. [CrossRef]
32. Folch, J.; Lees, M.; Sloane Stanley, G.H. A simple method for the isolation and purification of total lipids from animal tissues. *J. Biol. Chem.* **1957**, *226*, 497–509.
33. Morrisson, W.R.; Smith, M.L. Preparation of fatty acid methyl esters and dimethylacetates from lipid with boron trifluoride methanol. *J. Lipid Res.* **1964**, *5*, 600–608.
34. Chartrin, P.; Berri, C.; Le Bihan-Duval, E.; Quentin, M.; Baéza, E. Lipid and fatty acid composition of fresh and cured-cooked breast meat of standard, certified and label chickens. *Arch. Geflügelk.* **2005**, *69*, 219–225.
35. Mares, P.; Ranny, M.; Sedlacek, J.; Skorepa, J. Chromatography analysis of blood lipids. Comparison between gas chromatography and thin layer chromatography with flame ionisation detector. *J. Chromatogr.* **1983**, *277*, 295–305.
36. Lynch, S.M.; Frei, B. Mechanisms of copper- and iron-dependent oxidative modification of human low-density lipoprotein. *J. Lipid Res.* **1993**, *34*, 1745–1753. [PubMed]
37. Mercier, Y.; Gatellier, P.; Viau, M.; Rémignon, H.; Renerre, M. Effect of dietary fat and vitamin E on colour stability and on lipid and protein oxidation in turkey meat during storage. *Meat Sci.* **1998**, *48*, 301–318. [CrossRef]
38. Morzel, M.; Gatellier, P.; Sayd, T.; Renerre, M.; Laville, E. Chemical oxidation decreases proteolytic susceptibility of skeletal muscle myofibrillar proteins. *Meat Sci.* **2006**, *73*, 536–543. [CrossRef] [PubMed]
39. AFNOR. *Analyse Sensorielle*; AFNOR: La Plaine Saint-Denis, France, 2007; p. 648.
40. Baéza, E.; Marché, G.; Wacrenier, N. Effect of sex on muscular development of Muscovy ducks. *Reprod. Nutr. Dev.* **1999**, *39*, 675–682. [CrossRef] [PubMed]
41. Petracci, M.; Soglia, F.; Berri, C. Muscle metabolism and meat quality abnormalities. In *Poultry Quality Evaluation: Quality Attributes and Consumer Values*; Woodhead Publishing: Duxford, UK, 2017; pp. 51–75.
42. Guerder, F.; Parafita, E.; Debut, M.; Vialter, S. Première approche de la caractérisation de la qualité technologique de la viande de poule. In *8èmes Journées de la Recherche Avicole*; ITAVI: Saint-Malo, France, 2009; pp. 507–511.
43. Kokoszynski, D.; Bernacki, Z.; Steczny, K.; Saleh, M.; Wasilewski, P.D.; Kotowicz, M.; Wasilewski, R.; Biegniewska, M.; Grzonkowska, K. Comparison of carcass composition, physicochemical and sensory traits of meat from spent broiler breeders with broilers. *Eur. Poult. Sci.* **2016**, *80*. [CrossRef]
44. Qiao, Y.; Huang, J.; Chen, Y.; Chen, H.; Zhao, L.; Huang, M.; Zhou, G. Meat quality, fatty acid composition and sensory evaluation of Cherry Valley, spent layer and crossbred ducks. *Anim. Sci. J.* **2017**, *88*, 156–165. [CrossRef]
45. Puchala, M.; Krawczyk, J.; Calik, J. Influence of origin of laying hens on the quality of their carcasses and meat after the first laying period. *Ann. Anim. Sci.* **2014**, *14*, 685–696. [CrossRef]

 *foods*

Review

Predicting the Quality of Meat: Myth or Reality?

Cécile Berri [1,*], Brigitte Picard [2], Bénédicte Lebret [3], Donato Andueza [2], Florence Lefèvre [4], Elisabeth Le Bihan-Duval [1], Stéphane Beauclercq [1], Pascal Chartrin [1], Antoine Vautier [5], Isabelle Legrand [6] and Jean-François Hocquette [2]

1 UMR Biologie des Oiseaux et Aviculture, INRA, Université de Tours, 37380 Nouzilly, France; elisabeth.duval@inra.fr (E.L.B.-D.); s.beauclercq@gmail.com (S.B.); pascal.chartrin@inra.fr (P.C.)
2 UMR Herbivores, INRA, VetAgro Sup, Theix, 63122 Saint-Genès Champanelle, France; brigitte.picard@inra.fr (B.P.); donato.andueza@inra.fr (D.A.); jean-francois.hocquette@inra.fr (J.-F.H.)
3 UMR Physiologie, Environnement et Génétique pour l'Animal et les Systèmes d'Élevage, INRA, AgroCampus Ouest, 35590 Saint-Gilles, France; benedicte.lebret@inra.fr
4 Laboratoire de Physiologie et Génomique des poissons, INRA, 35000 Rennes, France; florence.lefevre@inra.fr
5 Institut du porc, La motte au Vicomte, 35651 Le Rheu, CEDEX, France; antoine.vautier@ifip.asso.fr
6 Institut de l'Elevage, Maison Régionale de l'Agriculture—Nouvelle Aquitaine, 87000 Limoges, France; isabelle.legrand@idele.fr
* Correspondence: cecile.berri@inra.fr; Tel.: +33-2-47-42-78-43

Received: 6 September 2019; Accepted: 20 September 2019; Published: 24 September 2019

Abstract: This review is aimed at providing an overview of recent advances made in the field of meat quality prediction, particularly in Europe. The different methods used in research labs or by the production sectors for the development of equations and tools based on different types of biological (genomic or phenotypic) or physical (spectroscopy) markers are discussed. Through the various examples, it appears that although biological markers have been identified, quality parameters go through a complex determinism process. This makes the development of generic molecular tests even more difficult. However, in recent years, progress in the development of predictive tools has benefited from technological breakthroughs in genomics, proteomics, and metabolomics. Concerning spectroscopy, the most significant progress was achieved using near-infrared spectroscopy (NIRS) to predict the composition and nutritional value of meats. However, predicting the functional properties of meats using this method—mainly, the sensorial quality—is more difficult. Finally, the example of the MSA (Meat Standards Australia) phenotypic model, which predicts the eating quality of beef based on a combination of upstream and downstream data, is described. Its benefit for the beef industry has been extensively demonstrated in Australia, and its generic performance has already been proven in several countries.

Keywords: meat; quality; prediction; biological marker; spectroscopy; phenotypic model

---

1. Introduction

The control of meat quality, especially sensory traits, remains an important issue for any farm animal production. This is the case for ruminant meat, in particular beef, but also for poultry and pork, although, for them, controlling the technological quality (processability) is at least as important [1]. Research efforts over many years, particularly in Europe, have led to a better understanding of the impact of the various production factors on the muscle characteristics and the quality of the meat obtained. At the same time, it highlights the complex determinism of the biological characteristics of muscles and meat, which is most often driven by many factors related to genetics as well as animal husbandry and slaughter systems [2]. Under these conditions, having reliable indicators to predict meat quality is a major challenge for the meat industry. These indicators would facilitate the selection of animals that are capable of producing good quality meat. In addition, they would improve breeding and slaughtering practices with the aim of optimizing the intrinsic qualities of muscles. With this in mind, many studies have explored the possibilities offered by different methodologies in the fields of biology, physical chemistry, and modeling to predict meat quality. This review aims to present, through examples, the different approaches developed in Europe (with the exception of genetic tests) to better predict the sensory and technological quality of meat and meat products. It also provides a critical analysis, in the light of the results obtained or the obstacles identified, to the deployment of these new tools at the industrial level.

2. Seeking the Genes that Control the Quality of Pork and Chicken Meat

Pork and poultry are the two most widely consumed meats in the world, in the form of a wide variety of products, both fresh and processed. Therefore, the notion of meat quality in these species is complex, and the technological and sensory traits, which have often common determinants, are generally considered simultaneously to evaluate product quality [3,4]. As with all species, the quality of poultry and pork is the result of interactions between genetic, breeding, slaughtering, and processing factors. Although many genetic or breeding factors influencing meat quality have been identified, this complex phenotype remains variable and difficult to predict. Therefore, the search for quality ante- or post-mortem biomarkers has started in these species in order to predict—in vivo or quickly after slaughter—the technological or sensory quality of meat or carcasses to optimize their use in different processing sectors. There are many potential applications: the selection of breeding animals but also, in the case of pork, the sorting of carcasses or parts of carcasses at the slaughterhouse (reviewed in [5]). In these two species, several studies have sought to find muscle transcripts for meat quality prediction.

2.1. Biomarkers of Meat Quality in Pork

In pork, the identification of meat quality biomarkers through transcriptomic approaches was performed on the loin and ham muscles. Studies focusing on the main sensory or technological traits of pork were carried out on different animal models, including experimental groups with extreme intramuscular fat (IMF) content, shear force value of cooked loin, pale, soft, exudative (PSE) defect, or affected or not by the presence of 'destructured' (also called "PSE-like") zones found especially in the *Semimembranosus* muscle. This quality defect is a major issue that seriously impairs the processing yield and the sensory quality of cooked ham.

In 2006, Damon et al. [6] identified the first molecular markers of 'destructured' ham muscle. They included several genes encoding the myofibrillar proteins involved in actin–myosin interactions and sarcomere integrity (tropomodulin, ankyrin, myomesin) and the enzymes involved in the glycolytic pathway. In pork, IMF is an important phenotypic trait that contributes to the tenderness and juiciness of meat. Liu et al. and Hamill et al. [7,8] showed that the IMF content of the *Longissimus* muscle (loin) at the slaughter stage is associated with the expression of genes involved in different functions: carbohydrate and lipid metabolism, cell communication, binding, response to stimulus, cell assembly, and organization. These results strengthen the fact that IMF content depends on the regulation of various metabolic and cellular pathways. These studies also highlighted the significant role of genes involved in adipogenesis regulation during animal growth (70 kg live weight) determining the IMF content at slaughter (110 kg). This reinforces the hypothesis that the inter-individual variability in IMF content depends on the early expression of genes regulating the development of intramuscular adipocytes [7]. Overall, this explains the difficulty in finding robust biomarkers of IMF content that can be measured from muscle samples taken at slaughter.

Several studies aimed to find predictive biomarkers of loin tenderness, assessed either by a trained sensory panel or by measuring the Warner–Bratzler shear force (WBSF) of cooked samples. The microarray transcriptomic profiles of loin muscle varying in WBSF were compared, and classification or regression analyses were used to identify 63 genes strongly associated with WBSF [9,10]. Integration of these transcriptomic results with proteomic data obtained from the same samples [11] showed that low WBSF values (i.e., tender meat) are associated with genes involved in the regulation of lipid metabolism, while high WBSF (i.e., tough meat) is related to genes controlling the morphology of muscle fibers (number, size, sarcomere, etc.). Furthermore, analysis of the transcriptomic data using random forest methodology identified 12 genes that are most important in determining the tenderness of cooked loin [9,10].

Since the sensory and technological qualities of pork are complex phenotypes that are determined by different traits (color, pH, drip loss, marbling or IMF, tenderness, juiciness, etc.), further studies aimed at identifying predictive biomarkers of pork quality simultaneously considered the variability of several meat quality traits within or between breeds or between production systems. Furthermore, and contrary to what has been done before, these studies included further validation steps to test the generic biomarker prediction on additional meat samples taken from the same population used for their identification as well as others. For instance, using an experimental design involving two breeds produced in different farming systems, it was possible to generate gradual and high variability in the technological and sensory qualities of pork [12]. This set of animal samples was used to identify and further validate biomarkers of pork quality. The muscle transcriptome profiles of 50 loins sampled 30 minutes after slaughter were associated with several meat quality traits: ultimate pH, drip loss, lightness (L*), redness (a*), hue angle (h°), IMF, as well as the WBSF, tenderness, and juiciness of the cooked meat. The wide range of meat traits and gene expression patterns made it possible to establish thousands of correlations between gene expression and meat quality traits (140 for a* up to 2892 for tenderness). Then, 40 genes selected for their high correlation coefficient value or relevant biological process terms regarding muscle development and meat quality were considered for a first "technical" validation step from the same 50 loin samples, assessing gene expression by real-time PCR, which is a sensitive and fast method that can be used when developing molecular diagnosis tools. At this step, 113 transcript-trait associations were confirmed, of which 60 were further validated ($R^2 \leq 0.46$, $p < 0.05$) for eight meat quality traits using 50 additional animals from the same experimental design. This means that among the genes identified, the level of expression of one gene could explain up to 46% of the variability of one meat quality trait [13]. Finally, an external validation was carried out on 100 commercial pigs (Duroc × Landrace × Yorkshire), validating 19 of these biomarkers ($R^2 \leq 0.24$, $p < 0.05$) that were correlated to the ultimate pH of meat (6), drip loss (4), L * (5), h° (2), IMF content (1), and tenderness (1) [14]. In addition to single correlations, multiple regression models were calculated from the quantification of gene expression by real-time PCR. Models with three to five genes explained up to 59% of the phenotypic variability of meat quality traits. The best accuracy (highest R^2) was found for meat color (h°), ultimate pH, drip loss, and IMF [13], but their predictive value when tested on commercial pigs (Duroc × Landrace × Yorkshire) was quite moderate ($R^2 \leq 0.23$, $p \leq 0.01$ between the predicted and measured value).

Until now, several biomarkers of individual pork quality traits have been identified, and some have been validated in different pig populations. However, their predictive capacity still needs to be improved before considering using for diagnosis purposes in the pork industry. Therefore, another strategy was considered. It aims to identify and validate biomarkers of a meat quality level combining both sensory and technological dimensions (instead of predictors of single meat quality traits), with the ultimate objective of proposing molecular tools to classify carcasses or primary cuts soon after slaughter in the meat industries, according to their predicted quality level. Using the meat quality data for pork loin samples from the experimental design presented above (two breeds and production systems, $n = 100$) and combining scientific expertise and statistical approaches, three classes differing in both sensory and technological qualities were specified: low (defective), acceptable, and extra. The expression levels (qRT-PCR) of 40 genes previously obtained on these 100 loin samples were used as predictive variables in a generalized linear model (stepwise selection) to discriminate quality classes. The best predictive model included 12 genes corresponding to different biological functions associated with meat quality development: mitochondrial energy metabolism, lipid and carbohydrate metabolism, gene expression control, cell regulation and apoptotic processes, calcium transport, protein transport, muscle structure and contraction, and muscle hypertrophy. After cross-validation using the leave-one-out method, this model exhibits an overall correct classification rate of 76%, with 88% for defective samples and 82% for extra samples [15].

2.2. Biomarkers of Meat Quality in Chicken

In chicken, several studies have sought to find muscle transcripts that predict the quality of the breast meat. They took advantages of several experimental animal designs. These are groups of individuals with extreme muscle characteristics (acid meats or DFD for "dark, firm, dry") from a single population of divergent lines selected for body composition or meat quality, or cross-breeds between lines to access individuals who are extreme in term of meat quality, but have a homogeneous genetic background for other traits such as production. The main traits studied were related to the post-mortem (p.m.) pH drop, in particular the pH measured at 15 min p.m. (pH15) and the breast meat pH measured at 24 h p.m. (ultimate pH, or pHu). Both influence the physicochemical and functional properties of proteins and affect a large number of quality parameters: color, water-retention capacity, hardness after cooking, technological yield, and susceptibility to oxidation [16–18].

The first network of genes identified as being related to the breast meat pHu was obtained by studying lean or fat chicken lines. In chicken, selection for low carcass fattening has led to changes in muscle properties: lean animals have lower glycogen reserves than fat animals, resulting in a higher meat pHu and better technological yield [19]. The results highlighted the involvement of several important pathways for glycogen control in muscle, such as the AMP dependent pathway involving the AMP-activated protein kinase (AMPK) complex as well as the cyclic AMP-dependent signaling pathways, and pathways involved in the control of carbohydrate availability in muscle [20]. To overcome the demonstrated links between peripheral fattening and glycogen storage ability in chicken muscle [19], a divergent selection based on a modern commercial broiler line was made, allowing the creation of two divergent lines specifically for breast pHu [17]. Transcriptome analysis of the breast muscle revealed very different metabolic statuses and energy production modes between these two lines (pHu+ and pHu−). The pHu of the breast muscles mainly use their high reserve of carbohydrate, while those of the pHu+ line use alternative catabolic pathways leading to significant remodeling of the muscle tissue [21,22]. From the transcriptome, including 1436 genes identified as differential between pHu+ and pHu− individuals, sPLS (sparse partial least squares) models were adjusted to predict pHu. The fitted models have good explanatory and predictive ability of the pHu (R^2Y = 0.77–0.87, Q^2 = 0.68–0.79). Twenty-one genes from this model supplemented by 27 other biomarker candidate genes were selected for high-throughput qRT-PCR validation (Fluidigm technology) in a population of 280 animals from both lines (pHu range 5.41–6.50). After a step of elimination of the genes with low explanatory abilities, a final PLS model including 20 confident genes was adjusted, which could be used to predict the pHu of the breast meat with an explanatory power (R^2) of 0.65, a predictive power (Q^2) of 0.62, and an error rate of 16% [22].

Recently, studies carried out on animals with pedigree information have made it possible to combine positional (quantitative trait loci, QTL) and expressional (transcribed) data with two objectives: identifying the genetic markers or mutations responsible for the variation of meat quality traits and facilitating the identification of fine molecular phenotypes for diagnosis and selection purposes. This strategy has already demonstrated its effectiveness in a study on the color of chicken meat. The detection of an expression QTL (or eQTL) confirmed that the gene BCMO1 (which encodes β-carotene 15,15′-monooxygenase 1) was responsible for variations in the yellow color of chicken meat and accelerated the identification of causal mutations within its promoter region [23]. These results led to the development of a patented genetic test [24] currently available to breeders who wish to control the yellow color of chicken breast meat in response to variations in the composition of feedstuffs. Studies on the interactions with feed have demonstrated the possibility of modulating the deposition of xanthophyll pigments and therefore the coloration of meat through this test [25,26]. Recently, a similar approach took advantage of two broiler lines divergently selected for the ultimate pH of the pectoralis major muscle to decipher the genetic control of this trait. By combining the detection of selection signatures and QTL with whole transcriptome analysis, it has identified genomic regions and a major candidate gene for chicken breast meat ultimate pH: PPP1R3A. It codes for a muscle-specific regulatory subunit of protein phosphatase 1 (PP1) that promotes the dephosphorylation of glycogen synthase (GS) and glycogen phosphorylase (GP), and thus glycogen synthesis. It was differentially expressed between the pHu+ and pHu− lines [22] and was located close to the most significant single-nucleotide polymorphism (SNP) for pHu [27].

These promising results (listed in Table 1) make it possible to foresee the development of tools to sort animals or carcasses based on their quality level. Indeed, while it is unlikely that diagnostic tests based on gene expression measurement will be used commercially, the identification of these intermediate molecular phenotypes will facilitate the future development of more accessible and/or less invasive techniques that are useful for breeding or industrial purposes.

Table 1. Studies dedicated to the search for gene biomarkers of meat quality in pork and chicken.

Species	Meat	Animal Model	Parameters	Reference
Pork	Ham	Normal and defected (destructured) groups within genotype	Destructured ham	[6]
Pork	Loin	Low and high-IMF groups within genotype	IMF	[7,8]
Pork	Loin	Low and high-WBSF groups within genotype	WBSF	[9–11]
Pork	Loin	Gradual variability of meat quality using two breeds produced in different farming systems	pHu, color, drip loss, IMF, WBSF, tenderness, and juiciness	[12,13]
Pork	Loin	Gradual variability in meat quality using commercial pigs (Duroc × Landrace × Yorkshire)	pHu, color, drip loss, IMF, WBSF, tenderness, and juiciness	[14]
Pork	Loin	Gradual variability of meat quality using two breeds produced in different farming systems	Meat quality index combining several technological and sensory parameters	[15]
Chicken	Breast	Lean and fat experimental lines	pHu	[19]
Chicken	Breast	F2 cross between the lean and fat experimental lines	pHu	[20]
Chicken	Breast	Low and high-pHu experimental lines	pHu	[22]
Chicken	Breast	Experimental slow-growing line	Color	[23]
Chicken	Breast	Low and high-pHu experimental lines	pHu	[27]

IMF: intramuscular fat content. pHu: ultimate pH. WBSF: Warner–Bratzler shear force.

3. Quantification of Proteins to Predict the Tenderness of Beef

One of the main objectives of beef research has long been to control and predict meat tenderness. Although many studies have clarified the role of muscle components such as muscle fibers, connective tissue, lipids, proteases [28], and the well-documented effects of muscle type, animal type or breed, and meat aging [2], meat tenderness remains poorly controlled by the beef industry. Currently, the techniques available to evaluate this trait are mechanical measurements, sensory analysis by trained panels, or consumer tests that are generally performed after meat aging and on cooked meat. These methods are quite complex, time-consuming, and expensive to apply, and the meat industry is still waiting for objective criteria and tools to evaluate and predict meat tenderness in live animals or quickly after slaughter to improve carcass valuation and limit consumer dissatisfaction.

One strategy implemented over the past 10 years has consisted of identifying protein biomarkers of tenderness with the objective of proposing a molecular test to evaluate beef tenderness [5,29,30]. Protein analysis provides additional information that gene or transcript analysis cannot give, because the expression of a gene does not always mean that the corresponding protein is proportionally expressed in the tissue of interest. Moreover, a protein can exist in several forms (isoforms) and can undergo post-translational modifications. This can only be observed at the protein level after separation by two-dimensional electrophoresis, as already performed by Bouley et al. [31].

As for the transcriptome, the approach used to identify protein biomarkers of tenderness is largely based on approaches developed by the medical community to find protein biomarkers of pathologies. It consists of comparing different samples with extreme tenderness scores (measured by mechanical measurements and/or sensory analysis) to highlight proteins whose abundance varies according to the trait studied [32,33]. This strategy has been applied in several experiments and has led to the proposal of a list of candidate biomarkers, subsequently completed by proteins revealed by bioinformatic analysis that have functional interactions with them [34,35]. Then, the relationships between tenderness and protein abundance were tested on a large number of cattle of various types (age, sex, breed) using a specifically developed immunological technique [36]. Among the 20 quantified proteins, the most confident biomarkers (four to five) were used to build equations, whose explanatory ability and error vary according to the muscle and animal type considered. For example, the variability of the shear force of the *Semitendinosus* muscle was better explained than that of the *Longissimus thoracis* muscle, which was itself better explained than the variability of its tenderness score determined by trained panelists [37]. Interestingly, the relationship between the abundance of some biomarkers and tenderness appeared to be specific to a muscle or an animal type. For instance, proteins related to fast glycolytic contractile activity were positively related to tenderness in the *Semitendinosus* muscle, but negatively in the *Longissimus thoracis* muscle of the same animal. In contrast, the Hsp70-1B was negatively related to the tenderness of these two muscles in two types of animals: young bulls from French beef breeds and Aberdeen Angus, which is known to have a more oxidative muscle metabolism than the French breeds [37]. Finally, structural proteins such as alpha-actin, F-actin-capping protein, or desmin have been identified as positive biomarkers of tenderness in different breeds and muscles by several authors [32,33].

The identification of protein biomarkers was first applied to tenderness, but is also applicable to other beef quality criteria. For example, protein biomarkers of the fat content of meat have been recently identified using the same methodology. Some protein biomarkers of tenderness also appear to be good biomarkers of other meat quality traits such as pH, color, juiciness, or flavor [38,39]. In particular, the abundance of the PRDX6 protein (peroxyredoxin), which had been identified as a biomarker of tenderness by several authors [37,40], was found to be positively related to the initial rate of p.m. pH drop and negatively to pHu in the *Longisssimus thoracis* of young Blonde d'Aquitaine bulls [38]. It was also positively related to meat redness in this same muscle. These authors also showed that the Hsp70-1B protein and the calcium-dependent protease µ-calpain were related to the color parameters: lightness, redness, and yellowness. Therefore, it is possible to develop an explanatory model based on protein abundance for most of the meat sensory qualities. Some of these proteins are also associated with muscular hypertrophy [41], suggesting joint control of the amount of meat and its sensory quality. The relationship between sensory and nutritional value has also been studied, in particular by quantification of the proteins involved in oxidative stress [42], thus opening up new opportunities for predicting the nutritional value of beef.

The first result of these studies is the considerable progress in the knowledge of the biological processes involved in determining meat tenderness [29,30]. The next step is to develop reliable tools to measure the abundance of proteins associated with tenderness to better predict this trait. Ongoing research offers hope for the rapid development of such a tool, mainly because protein quantification techniques are progressing rapidly, leading to considerable methodological simplifications [43].

4. Blood Biomarkers: First Encouraging Results

Having confident predictive blood markers would greatly facilitate the development of phenotyping methods in live animals. The pHu+ and pHu− lines model mentioned above was used to identify blood and muscle metabolite predictors of the pHu of chicken breast meat. A first step was to analyze by high-resolution nuclear magnetic resonance or NMR (proton and phosphorus NMR for muscle, and proton NMR for serum) the muscle (breast) and serum extracts from extreme animals belonging to both lines. These analyses revealed very specific metabolomic signatures of the two groups in blood and muscle that enabled an almost perfect discrimination between them [21]. A total of 20 and 26 discriminant metabolites between the two lines were identified by multivariate OPLS-DA (orthogonal projections to latent structures discriminant analysis) in serum and muscle, respectively. Three independent models were fitted with good explanatory (R^2Y) and predictive (Q^2) abilities for pHu (R^2Y = 0.63–0.82, Q^2 = 0.45–0.76). A multiblock model, including muscle and blood metabolites, was subsequently developed with even better explanatory (R^2Y = 0.91) and predictive (Q^2 = 0.86) power. To develop a test that could be routinely used on live animals, the study focused specifically on the metabolites identified in the blood. Thus, a model including seven metabolites (acetylglutamine, arginine, formate, glucose, hypoxanthine, phenylalanine, and xanthine) always provides good discrimination (R^2Y = 0.73, Q^2 = 0.64, Figure 1) while limiting the number of biological tests for diagnosis as much as possible. However, the predictive potential of this set of serum biomarkers must be validated on other chicken populations that are representative of the pHu variability observed in slaughterhouses and have different genetic backgrounds from the pHu+ and pHu− lines. If this validation step is successful, these biomarkers could be used in selection to exclude from parental stocks the individuals predisposed to produce high-pH or low-pH meat, or in research to evaluate innovations related to animal husbandry practices.

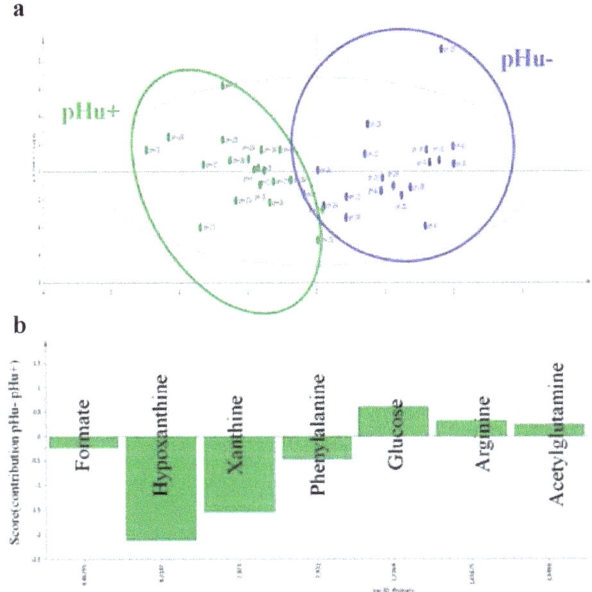

Figure 1. (a) Projection of individuals according to major principal components based on an OPLS-DA (orthogonal projections to latent structures discriminant analysis) model with an explanatory ability (R^2Y) of 0.73 and a predictive value (Q^2) of 0.64 (the pHu− and pHu + individuals are shown in green and blue, respectively); (b) Contribution of the seven metabolites identified by the OPLS-DA model (pHu−/pHu +). Illustration based on results published in [21].

In cattle, the search for plasma biomarkers of the sensory qualities of meat was initiated by proteomic analysis. In addition, bioinformatics tools have identified secreted proteins that could be good potential candidates for quantification in plasma [44]. These blood-based approaches are of considerable interest for the analysis of meat quality biomarkers in live animals. For instance, significant negative correlations were observed between blood retinol content and marbling score ($R = -0.47$, $p \leq 0.01$), and between blood aspartic acid transaminase content and *longissimus* muscle area at the 13th rib ($R = -0.67$, $p \leq 0.01$) in the finishing phase of Hanwoo steers [45]. Metabolomics approaches are also powerful, not only to differentiate meat from different species (cattle, pigs, and chickens), but also from different cattle breeds [46]. Furthermore, the evolution of key metabolites and associated pathways has been studied during the post-mortem aging of beef [47]. Some specific muscle metabolites have also been described to well differentiate muscles that differ in the composition of their muscle fiber types. The content of metabolites involved in the Krebs cycle change differently in different muscles according to the aging time [48].

5. Spectroscopic Methods: Physical Chemistry to Decipher Biology

Spectroscopy can be defined as the study of the interaction between light and matter. Among spectroscopic techniques, near-infrared (NIR) and Raman spectroscopy are currently the most widely used techniques for predicting meat quality. The background of near-infrared spectroscopy (NIRS) was described in the review by Bertrand [49], while Raman spectroscopy was reviewed by Yang and Ying [50]. Briefly, these technologies are physical methods of analysis based on the property of light absorbed by organic molecules at specific frequencies. The relationships between the chemical composition and the absorbance values or their derivatives are then established (Figure 2).

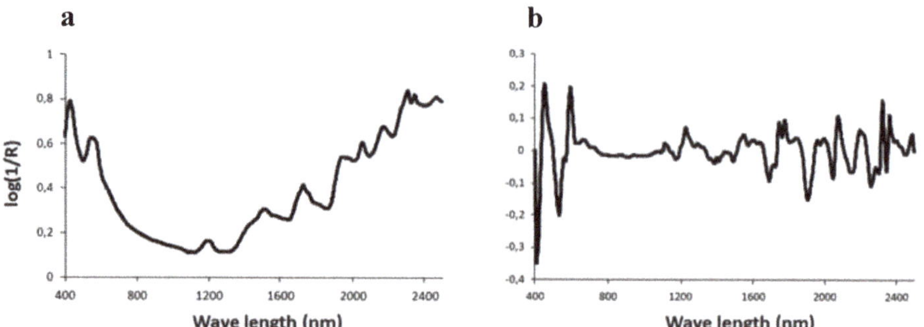

Figure 2. Visible/near infrared spectrum and first derivative between 400–2500 nm of a sample of bovine muscle *(Rectus abdominis)* after grinding (**a**) and lyophilization (**b**).

5.1. Difficulties in Predicting Sensory Quality

The number of studies published in recent years and the number of companies recently equipped with NIR instruments show the importance of spectroscopic technologies and in particular that of NIRS for estimating meat sensorial quality [51,52].

The sensory characteristics of meat derive from the amount of different chemical compounds or from biological parameters (lipid content, collagen, muscle fiber typology, pH, etc. [28]). Consequently, it may be possible to predict sensorial determinations using spectroscopic methods [53,54]. However, the results reported in the literature do not allow validating this hypothesis due to different reasons: According to Liu et al. [55], in a sensory analysis, the use of a narrow scale for the intensity of sensory characteristics could reduce the precision and accuracy of their predictions. Furthermore, the samples scanned by NIRS are not exactly the same as those tested by the tasting panels. The high heterogeneity of meat traits within a muscle can contribute to generating some significant bias between the NIR-predicted values and the measured values. These observations may also explain some of the difficulties in predicting meat-eating quality using the biomarkers described above. Contrary to the studies cited above, Ripoll et al. [56] reported R^2 values of 0.98 for models built to predict the tenderness of beef. These models were developed using meat obtained from different breeds and maturity levels, thus increasing the variability of the data sets used to build the models.

The prediction of shear force by NIRS or by Raman spectroscopy gives very variable results (R^2 values from 0.01 to 0.74 [57–62] (Table 2). Elsewhere, Liao et al. [63] reported R^2 values in calibration of 0.72, whereas the R^2 value in validation was only 0.27. Other authors have shown that NIR is able to predict the tenderness of pork after cooking with similar variability to that shown above (R^2 = 0.20 to 0.72 [63,64]).

The on-line prediction of *Longissimus lumborum* tenderness in slaughterhouses by NIRS was tested by Rust et al. [65]. They observed that the proportion of muscles correctly classified as tender was 70%. According to Leroy et al., Liu et al., and De Marchi et al. [55,59,66], these results could be explained not only by the variability of samples and data sets, but also by the variability within replications for the Warner–Braztler shear force method [67,68]. De Marchi et al. [68] also reported some changes in the spectral information of the infrared segment due to the modification of the muscle structure by the impact of grinding. Consequently, the R^2 values of the models were not improved by the effect of grinding, and the values of this statistic may even deteriorate at times.

Table 2. Statistical parameters of Warner–Braztler shear force (WBSF) and tenderness prediction in meat by visible/near-infrared (VIS/NIR) and Raman spectroscopy.

Method	Meat	Parameter	R^2c	SEc	R^2cv	SEcv	R^2p	SEp	Reference
VIS/NIR (R)	beef	WBSF	0.72	0.84					[55]
VIS/NIR (R)	beef	Tenderness	0.98	0.37			0.98	0.35	[56]
		WBSF	0.74	0.66			0.74	1.06	
NIR (R)	beef	WBSF	0.65	2.30	0.53	2.67			[57]
NIR (R)	beef	WBSF	0.21	0.48					[58]
NIR (T) (intact)	beef	WBSF			0.31	3.07			
NIR (T) (ground)	beef	WBSF			0.12	3.48			[68]
VIS/NIR (R) (intact)	beef	WBSF			0.34	9.39			
VIS/NIR (R) (ground)	beef	WBSF			0.13	10.74			
Raman	beef	WBSF	0.94	2.00	0.79	3.90	0.23	8.80	[60]
Raman		WBSF			0.75	0.63			[61]
		Tenderness			0.65	0.97			
Raman	lamb	WBSF			0.06	13.60			[62]
VIS/NIR (R) (intact) on line	pork	WBSF	0.72	0.23			0.27	0.36	[63]
VIS/NIR (R) (ground)	pork	WBSF	0.48	4.22	0.30	4.98	0.25	5.51	[64]
NIR (R)	beef	WBSF	0.45	9.32		10.00			[67]
NIR (R)	beef	WBSF	0.17	15.69		15.89			
NIR (R)	beef	WBSF			0.25	11.19			[66]
NIR (T)	beef				0.41	9.59			
NIR (R) freeze dried	beef	WBSF	0.20	4.65	0.12	4.99			[59]
NIR (R) fresh minced	beef	WBSF	0.08	5.09	0.03	5.21			

VIS/NIR (R): Visible/near infrared in reflectance. VIS/NIR (T): Visible/near infrared in transmission. R^2c: Coefficient of determination of calibration. SEc: Standard error of calibration. R^2cv: Coefficient of determination of cross-validation. SEcv: Standard error of cross-validation. R^2p: Coefficient of determination of prediction. SEp: Standard error of prediction.

In conclusion, the use of spectroscopic techniques is not fully effective in predicting the sensory quality (tenderness or shear force) of meat. The high variability of the results obtained can be partially explained by the low repeatability of the reference method measurements, as with any predictive method (including the search for biomarkers above). However, some promising results indicate that further research is needed to obtain suitable models for predicting the sensory quality of meat.

5.2. Routine Uses for Nutritional Quality

Although it is still difficult to predict the sensory quality of meat using NIRS, this technology is nonetheless recognized and used to determine the chemical composition and therefore the nutritional value of meat.

For instance, the lipid content of chicken or duck breast (from lean or fat ducks) can be routinely measured by NIRS (replacing chemical methods) thanks to the development of very robust prediction equations. In chickens, the coefficient of determination (R^2) is between 0.8–1.0, and calibration standard deviations are close to 0.2 (R^2 = 0.98, [69,70], R^2 = 0.83, [71]); in ducks, an R^2 of 0.94 and a standard deviation of 0.31 were published [72]. The fatty acid composition is an important trait that can influence the nutritional, sensory, and technological quality of meat. It is possible to estimate by NIRS technology the content of the main fatty acids and their different types (polyunsaturated, monounsaturated, and saturated) with excellent coefficients of determination (>0.9 [70] for freeze-dried samples, although those obtained on thawed samples are lower (approximately 0.6 [73]). The protein and dry matter content of chicken meat can also be estimated by NIRS on freeze-dried samples using models with high R^2 values (approximate values of R^2 = 0.98 and standard deviation = 0.2 [70]). In chicken, NIRS models constructed from thigh muscles (fatter and more variable than breast muscles) to predict chemical composition have even higher R^2 values than those developed from breast meat [74].

In the case of pork, many studies have focused on predicting its chemical composition. Calibration for predicting intramuscular fat content was of variable quality, with R^2 values ranging from 0.28 to 0.88. The error was close to 1% regardless of the type of sample, i.e., intact meat, crushed or salted/dried meat [64,75–78]. The composition of polyunsaturated, monounsaturated, and saturated fatty acids of backfat can also be predicted by NIRS with good precision (R^2 = 0.61 to 0.99) [79–82]. The accuracy is roughly comparable regardless of the type of fatty acid, with an error of about/approximately 1%. However, the quality of the prediction by NIRS in backfat depends on the content of the various fatty acids in the product: the relative error is less than 5% when the saturated and monounsaturated fatty acid content is high, while it is close to 10% when the polyunsaturated fatty acid content is high. It is also possible to predict the fatty acid profile of the intramuscular fat in Iberian pork loin with very good precision (R^2 > 0.99 and error of about 1%) for the prediction of polyunsaturated, monounsaturated, and saturated fatty acids [83]. Promising results in predicting the IMF content and fatty acid composition of pork from various local breeds were also obtained by Bozzi et al. [84], using the Fourier transformation NIRS technology (FT-NIRS). This technology improves the signal-to-noise ratio, spectral resolution, and wavelength accuracy, and reduces scan time [85]. Partial least square regression models were established and validated on external data using FT-NIRS. At the validation step, high determination coefficients and low errors were obtained for IMF content ($R^2 \geq 0.96$, root mean square error (RMSE) ≤ 0.66) and polyunsaturated fatty acid proportion ($R^2 \geq 0.87$, RMSE ≤ 0.70). Performance was slightly lower, but still valuable for monounsaturated fatty acid ($R^2 \geq 0.77$, RMSE ≤ 1.13), and saturated fatty acid proportions ($R^2 \geq 0.77$, RMSE ≤ 0.97) [84]. Thus, FT-NIRS seems promising to estimate the principal parameters of fatty acid groups on pig muscle samples and therefore the nutritional composition of pork.

5.3. High Expectations for the Technological Quality of Meat

The value of NIRS technology to evaluate the technological properties of meat has been mainly studied in pork and chicken. In chicken, the best predicted criteria are color indexes (redness and yellowness) with correlation coefficients greater than 0.8 [86,87]. However, the interest of using NIRS to predict these parameters is limited, because it is very easy to measure them by spectrometry. The prediction performance for pHu, drip loss during storage, and cooking losses is generally lower, with correlation coefficients between 0.6–0.8 [70,86]. However, these levels of correlation suggest prospects for improving these characteristics. This is not the case for other traits such as pH15 min p.m. or shear force [86,87], for which correlation coefficients <0.5 have been obtained.

The selection of raw meat according to its technological potential is now widespread in the pork industry. However, sorting as it is practiced today (mainly based on the pHu value) only covers certain aspects of meat quality, and lacks precision in industrial conditions. Rapid acquisition alternatives such as NIRS or vision/hyperspectral imaging show strong potential for predicting quality. The majority of the studies focusing on predicting the pHu have reported calibrations with R^2 values between

0.65–0.87 [64,77,88,89]. However, the pHu prediction error remains high (0.05 to 0.18) compared to the repeatability of the reference method (0.03). Numerous studies show that it is also possible to establish calibrations that are sufficiently accurate to predict the water retention capacity of meat. This is the case for drip loss, for which R^2 values between 0.31–0.76 (often around 0.60) have been reported [75,77,78,88,90–93]. However, the prediction error of drip loss remains high, between 1% and 2%. NIRS prediction of the processing yield has also been studied in pork. The few publications available show satisfactory accuracy for ham and loin (R^2 = 0.57 to 0.78) [94,95].

Vision and hyperspectral imagery are currently considered as high-potential technologies for the slaughter/processing sector. Vision systems are automatic alternatives to measure meat color, using a camera (contactless) rather than a colorimeter, which requires a contact probe and an operator. The camera has to be previously calibrated to obtain a reliable color measurement, and the RGB camera signal is converted to L*a*b* color space using colorchecker tiles. Before obtaining the L*a*b* value of the meat, the images must be processed to extract the color of a specific region of interest (ROI) from the meat. Hyperspectral systems use a very similar approach, but a NIR spectrum is obtained for each pixel, so after obtaining a mean spectrum on a specific ROI, classical chemometrics are applied to perform prediction models, such as NIRS. Compared to NIRS, vision and hyperspectral imagery have the advantage of a contactless measurement, and can be easily integrated into a production line because they do not require operator intervention. The fields of application are the same as for NIRS, although there are currently not enough publications to draw conclusions regarding the accuracy of these techniques. Properly adjusted vision calibrations have been established to predict the ultimate pH of pork (R^2 = 0.49 to 0.72) [96,97], while the prediction of the processing yield of ham by this same technique lacks sufficient accuracy to be operational (R^2 = 0.31 to 0.43). Hyperspectral analysis, which can be considered as the combination of vision and NIRS, may be useful for predicting drip loss (R^2 = 0.60) [97] and for classifying meat according to its technological quality. For example, 84% of pork was correctly classified as PSE, RFN (red, firm and non-exudative), and RSE (red, soft, and exudative) meat [98]. This still undeveloped prediction technique will require further work to assess its relevance in predicting the quality of meat from different species.

6. Development of Phenotypic Models for Beef Evaluation

6.1. Principles

The Australian beef industry and researchers have jointly built, as part of a common and collective strategy, the MSA (Meat Standards Australia) grading scheme, which is a mathematical model for predicting the eating quality of beef for each "muscle × cooking method" combination. This model was constructed from a large database of consumer tests using a standard protocol [99,100]. A dozen parameters with a statistically significant effect on eating quality, such as traits characterizing animals (physiological maturity, weight, genetic type, sex, etc.), pre-slaughter and slaughter conditions (carcass hanging method, etc.), meat (pH, color, marbling, etc.), and post-mortem events (aging time, cooking method, etc.) are considered in the model as well as the interactions between them. Some stakeholders of the beef industry are already using this model, and it appears that its use has contributed to reducing the decline in beef consumption in Australia.

In practice, the slaughterhouse is the backbone of the system. A specific grader who is accredited after training, and receives recurrent trainings, grades the carcasses. Then, the MSA model predicts an overall quality score called MQ4 (for "Meat Quality 4") on a scale of 0 to 100, for each piece of meat associated with a specific cooking method and aging time. This score is a linear combination of consumer scores for tenderness, flavor liking, juiciness, and overall liking. In sensory testing, to make the best link with the quality ranking of meat also given by consumers, four quality classes are used: unsatisfactory, good every day (3*), better than every day (4*), and premium (5*) (Figure 3). The values of the overall MQ4 score defining the limits between each quality class are precisely calculated for each data set and regularly refined: they are about 40 (between unsatisfactory and 3*), 60 (between 3* and 4*), and 80 (between 4* and 5*) on a scale of 0 to 100 [99,100].

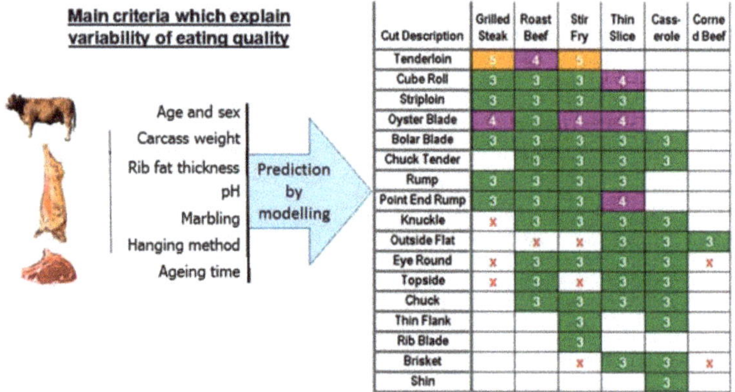

Cut Description	Grilled Steak	Roast Beef	Stir Fry	Thin Slice	Casserole	Corned Beef
Tenderloin	5	4	5			
Cube Roll	3	3	3	4		
Striploin	3	3	3	3		
Oyster Blade	4	3	4	4		
Bolar Blade	3	3	3	3	3	
Chuck Tender		3	3	3	3	
Rump	3	3	3	3		
Point End Rump	3	3	3	4		
Knuckle	x	3	3	3	3	
Outside Flat		x	x	3	3	3
Eye Round	x	3	3	3	3	x
Topside	x	3	x	3	3	
Chuck		3	3	3	3	
Thin Flank			3		3	
Rib Blade			3			
Brisket			x	3	3	x
Shin					3	

The model allows the prediction of eating quality for each cut x cooking method combination according to different quality classes: unsatisfactory (x), good every day (3 stars), better than every day (4 stars) and premium (5 stars).

Figure 3. Prediction of the overall beef eating quality score (combining tenderness, flavor liking, juiciness, and overall liking) from different traits related to animals, carcasses, and cuts using the "Meat Standards Australia" (MSA) grading scheme.

6.2. Applications

The principle of the MSA system has been evaluated in various countries such as South Korea, the USA, Japan, South Africa, New Zealand, Northern Ireland, Poland, France, and the Republic of Ireland [101–106]. The general conclusion is that the MSA methodology is relevant in all these countries. However, the relative weighting coefficients for tenderness, flavor liking, juiciness, and overall liking in the optimal calculation of the MQ4 score vary slightly between countries, and the optimal limits between quality classes can be refined for each country or each group of consumers.

In the French context, experts have judged it to be rigorous, relevant, and credible. Experts have recognized that the MSA system has helped to federate a large number of Australian professionals and scientific stakeholders. This approach, based primarily on real consumer satisfaction, is likely to upset the traditional attitudes and political positions of stakeholders in the beef sector. We should not consider the MSA system as a new official quality label, but rather as a rigorous tool to better use the existing official quality signs [107]. Two studies [106,108] experimentally tested the MSA system in France. They concluded that the MSA system provides a fairly good prediction of the eating quality of French beef, despite differences in animal type (cows and young bulls in France versus steers and heifers in Australia) and the degree of cooking (55 °C for "rare" cooking in France versus 74 °C for "well-done" meat in Australia). Using the MSA system, about 70% of the French meat was correctly ranked according to the different classes (unsatisfactory, 3*, 4*, or 5*). This rate is as good as if not better than that observed in other countries where the MSA system has been studied, including Australia. Prediction using the MSA system is even better than prediction methods based on muscle biochemistry [109] or genomic data [110].

In the case of a binary use of the MSA system in Europe (i.e., unsatisfactory quality versus acceptable quality with no distinction between classes 3*, 4*, or 5*), the probability of really disappointing the consumer—that is to say, erroneously classifying a poor quality sample as 3* or greater, is only 7% based on data from different European countries (Northern Ireland, Ireland, France, and Poland). Considering that currently, according to the same data set, the sensory quality of about 25% of meat is deemed unsatisfactory, a binary use of the MSA system would therefore constitute an important step forward, as it could theoretically reduce customer dissatisfaction from 25% to 7% [111]. This is particularly important for an expensive product such as beef. A European beef quality assurance system similar to MSA would need to be simple, effective, and sufficiently flexible to allow companies to develop their own brands [112]. Furthermore, it was proposed to consider in the European model both the gender (female, castrated male, or entire male) and breed type (dairy or meat breed), because recent work has shown that their effects on the sensory quality of beef are not fully explained by animal age or carcass characteristics (weight, fatness, etc.). It was also suggested to assess the physiological maturity of animals that are included in the MSA system using the degree of carcass ossification (as in Australia) for young animals and the age of animals (as in Europe) for older ones [113]. All these improvements would certainly help to increase the accuracy of the prediction of beef eating quality in Europe. Further research will be conducted in these directions as part of the activities of the International Meat Research 3G Foundation (https://imr3gfoundation.org/) recently launched under the auspices of the United Nations Economic Commission for Europe (UNECE).

6.3. Perspectives

Carcass quality criteria (i.e., conformation and fat scores based on the EUROP grid) according to which farmers are paid do not have a clear and systematic relationship with the eating quality of beef [114]. This partly explains why a consumer can buy very expensive beef without necessarily being satisfied, and vice versa [115]. In general, consumers are willing to pay more if they are sure that the product will be of higher quality. Therefore, one can expect that premium products are likely to generate significant profits compared to "good for everyday" products, regardless of the market. However, this difference may vary from country to country, French and Japanese consumers being the most likely to pay more for premium quality products [111,116]. In Australia, implementation of the MSA system has generated significant profits that were distributed to various stakeholders in the sector: producers, slaughterers, and others. It was calculated that $12.50 of additional revenue was generated for every dollar invested over five years (2010/2011 to 2014/2015). Therefore, there is a real financial incentive in Australia to produce premium-eating quality meat, which is not yet the case in many countries that continue to pay producers according to carcass characteristics, i.e., conformation and fatness [117].

The MSA prediction model also opens up new possibilities for animal breeding. The potential of a muscle to produce beef of a predicted quality, weighted by the relative weight of this muscle in the carcass, makes it possible to calculate a global MSA index of sensory quality for the carcass [118]. This MSA index can be considered as a new phenotype to evaluate the animal's potential to produce meat of a given eating quality. This index could potentially be introduced into genetic selection schemes to include the sensory quality potential of animals, which has never been achieved so far.

7. Advances and Barriers to the Development of Predicting Tools

The meat sector is significant in Europe. However, it is facing a difficult economic context, particularly due to a steady decline in per-capita meat consumption, especially red meat. The reasons for this decline in consumption are numerous, and include the high variability of sensory quality as mentioned in this article as well as criticism of the nutritional qualities. These qualities, particularly tenderness in cattle, are sometimes or often considered insufficient by professionals or consumers. All these observations fully justify the analytical or technological research carried out on the prediction of these quality components. However, the quality of the predictions of the different quality criteria is often modest. In addition, operators in the meat sector, from livestock to processing, have small profit margins, and are therefore greatly concerned about the economic profitability of their activities. For cost/benefit reasons, they are reluctant to incorporate certain advances in research and development into their practices, which is noted and regretted by researchers [119]. Unfortunately, the economic crisis mentioned above is likely to reinforce this phenomenon.

In addition, it is clear that marketing channels are increasingly complex, disconnected from the animal and carcass, and include many intermediaries between the producers and consumers. Consumer behavior is also changing in terms of the place of purchase (less and less in the butcher's shop and more and more in supermarkets) and consumption (more away from home), the nature of the products consumed (more elaborate products or individual portions), and in terms of expectations, which have diversified over the years, now including social concerns related in particular to animal welfare and protection of the environment [120]. Thus, consumer preferences, behavior, and their perception of meat and meat products are heterogeneous and depend not only on the appearance and eating qualities of the meat, but also on psychological and marketing aspects [121]. Therefore, research and development (R & D) must broaden its scope [122]. While research activities have largely focused on the intrinsic characteristics of meat, which are the sensory, technological, nutritional (subject of this article), and health qualities, the extrinsic qualities associated with the product, which meet broad societal expectations, must now be taken into account in interaction with the former [2].

In general, research directly related to consumer expectations is increasingly necessary [123], particularly to objectively predict the intrinsic qualities of meat, as well as its extrinsic qualities [124]. However, research and development are generally carried out over a long time scale, which may not be compatible with the short-term concerns of professionals [119]. At the same time, scientists must be able to take into account the expectations of professionals and consumers in order to guide their work as effectively as possible and to encourage the appropriation of technological innovations by the stakeholders in the sector.

There is also a pressing need for innovation with regard to trading, especially for export. In this perspective, the Australian methodology (the MSA system) is promising, and could become an international standard under the United Nations Economic Commission for Europe (UNECE) [125]. The search for biological predictors of quality, combined with recent technological evolutions, makes it possible to envisage practical applications in the medium term. It is undeniable that all of this research has led to considerable advances in the knowledge of the genetic and biological mechanisms governing the establishment of the different components of meat quality [126]. This improved knowledge may contribute to the development of selection tools (such as the genetic test of meat coloring developed for chicken) as well as decision support tools to propose innovative production strategies adapted to the biological potential of animals and the production objectives of the various sectors. For example, the development of spectral methods to authenticate the animal feed of animals or qualify meat in terms of nutrition is also promising. In order to be routinely applicable in a professional context, the diagnosis tool must be easy to use, give a result quickly after sampling, and have a limited cost. Some methods developed so far do not yet meet all these criteria [126].

Finally, this review also opens up new perspectives on the possibility of combining different types of the above-mentioned technologies. So, could we consider looking for biomarkers of phenotypes based on spectral methods? Similarly, would there be any interest in including in the MSA model some biological predictors resulting from high-throughput molecular approaches or spectral-based methods? This integration work remains to be done, but should undoubtedly contribute to the improvement of meat quality prediction tools and the subsequent application thereof.

8. Conclusions

Numerous technological innovations based on genomics or modeling approaches, or spectral or physical methods, have been described in this article to predict the specific intrinsic qualities of meat such as sensory (tenderness, flavor), technological (defects related to tissue integrity, ultimate pH, processing yield), or nutritional (lipid content, fatty acid composition) traits. Most of these approaches require additional work and methodological developments to be routinely applicable. However, their ability to analyze a large number of samples at a reduced cost, as well as their ability to predict the desired quality criteria, are fundamental elements for attracting the interest of the stakeholders of the sector. The appropriation of these methodologies by researchers is a key issue, and some laboratories already value these tools as part of their work on the impact of production factors on the quality of meat. Appropriation by the professionals will occur in a second phase and require more dialogue between professionals and researchers to properly define the objectives to be achieved as well as the conditions required for developing these tools (in terms of cost/benefit ratio in particular). In addition, research must also focus on integrative approaches to comprehensively predict all the desired quality criteria, which implies the combination of innovations described in this article on the one hand, but also insertion thereof into a more global thought process that includes a sociological dimension of research questions. The ultimate goal is to respond better to the expectations of industrials and consumers through better appropriation of innovations by the industry.

Funding: This research received no external funding.

Conflicts of Interest: The authors declare no conflict of interest.

References

1. Lebret, B.; Picard, B. Les principales composantes de la qualité des carcasses et des viandes dans les différentes espèces animales [The main components of carcasses and meat quality in various animal species]. *INRA Prod. Anim.* **2015**, *28*, 93–98.
2. Lebret, B.; Prache, S.; Berri, C.; Lefèvre, F.; Bauchart, D.; Picard, B.; Corraze, G.; Médale, F.; Faure, J.; Alami Durante, H. Qualités des viandes: Influences des caractéristiques des animaux et de leurs conditions d'élevage [Meat quality: Influence of animals' characteristics and rearing conditions]. *INRA Prod. Anim.* **2015**, *28*, 151–168.
3. Berri, C. La viande de volaille: Des attentes pour la qualité qui se diversifient et des défauts spécifiques à corriger [Poultry: Diversified expectations for quality and specific defects to be corrected]. *INRA Prod. Anim.* **2015**, *28*, 115–118.
4. Lebret, B.; Faure, J. La viande et les produits du porc: Comment satisfaire des attentes qualitatives variées? [Pork and pork products: How to fulfill a variety of quality demands?]. *INRA Prod. Anim.* **2015**, *28*, 111–114.
5. Picard, B.; Lebret, B.; Cassar Malek, I.; Liaubet, L.; Berri, C.; Le Bihan-Duval, E.; Hocquette, J.F.; Renand, G. Recent advances in omic technologies for meat quality management. *Meat Sci.* **2015**, *109*, 18–26. [CrossRef] [PubMed]
6. Damon, M.; Vincent, A.; Cherel, P.; Frank, M.; Le Roy, P. Transcriptomic analysis of destructured ham. In Proceedings of the 1st Conference on Pig Genomics, Lodi, Italy, 20–21 February 2006.
7. Liu, J.; Damon, M.; Guitton, N.; Guisle, I.; Ecolan, P.; Vincent, A. Differentially expressed genes in pig Longissimus muscles with contrasting levels of fat, as identified by combined transcriptomic, reverse transcription PCR, and proteomic analyses. *J. Agric. Food Chem.* **2009**, *57*, 3808–3817. [CrossRef] [PubMed]
8. Hamill, R.M.; McBryan, J.; McGee, C.; Mullen, A.M.; Sweeney, T.; Talbot, A. Functional analysis of muscle gene expression profiles associated with tenderness and intramuscular fat content in pork. *Meat Sci.* **2012**, *92*, 440–450. [CrossRef] [PubMed]
9. Lobjois, V.; Liaubet, L.; SanCristobal, M.; Le Roy, P.; Cherel, P.; Hatey, F. Etude d'un critère de qualité de la viande, la tendreté, par l'analyse du transcriptome du muscle porcin (*Longissimus dorsi*) [A study of a meat quality trait, tenderness, by the transcriptome analysis in the pig *Longissimus dorsi* muscle]. *J. Rech. Porc.* **2006**, *38*, 97–104.
10. Lobjois, V.; Liaubet, L.; San Cristobal, M.; Glenisson, J.; Feve, K.; Rallieres, J.; Le Roy, P.; Milan, D.; Cherel, P.; Hatey, F. A muscle transcriptome analysis identifies positional candidate genes for a complex trait in pig. *Anim. Genet.* **2008**, *39*, 147–162. [CrossRef] [PubMed]
11. Laville, E.; Sayd, T.; Terlouw, C.; Chambon, C.; Damon, M.; Larzul, C.; Leroy, P.; Glénisson, J.; Chérel, P. Comparison of sarcoplasmic proteomes between two groups of pig muscles selected for shear force of cooked meat. *J. Agric. Food Chem.* **2007**, *55*, 5834–5841. [CrossRef]
12. Lebret, B.; Ecolan, P.; Bonhomme, N.; Méteau, K.; Prunier, A. Influence of production system in local and conventional pig breeds on stress indicators at slaughter, muscle and meat traits and pork eating quality. *Animal* **2015**, *9*, 1404–1413. [CrossRef] [PubMed]

13. Damon, M.; Denieul, K.; Vincent, A.; Bonhomme, N.; Wyszynska-Koko, J.; Lebret, B. Associations between muscle gene expression pattern and technological and sensory meat traits highlight new biomarkers for pork quality assessment. *Meat Sci.* **2013**, *95*, 744–754. [CrossRef] [PubMed]

14. Lebret, B.; Denieul, K.; Vincent, A.; Bonhomme, N.; Wyszynska-Koko, J.; Kristensen, L.; Young, J.F.; Damon, M. Identification par transcriptomique de biomarqueurs de la qualité de la viande de porc [Identification by transcriptomics of biomarkers of pork quality]. *J. Rech. Porc.* **2013**, *45*, 97–102.

15. Lebret, B.; Castellano-Perez, R.; Vincent, A.; Faure, J.; Kloareg, M. Molecular biomarkers to discriminate pork quality classes based on sensory and technological attributes. In Proceedings of the 61st International Congress of Meat Science and Technology (ICoMST), Clermont-Ferrand, France, 23–28 August 2015; Volume 61, p. 99.

16. Le Bihan-Duval, E.; Debut, M.; Berri, C.M.; Sellier, N.; Sante-Lhoutellier, V.; Jégo, Y.; Beaumont, C. Chicken meat quality: Genetic variability and relationship with growth and muscle characteristics. *BMC Genet.* **2008**, *9*, 53. [CrossRef] [PubMed]

17. Alnahhas, N.; Berri, C.; Boulay, M.; Baéza, E.; Jégo, Y.; Baumard, Y.; Chabault, M.; Le Bihan-Duval, E. Selecting broiler chickens for ultimate pH of breast muscle: Analysis of divergent selection experiment and phenotypic consequences on meat quality, growth, and body composition traits. *J. Anim. Sci.* **2014**, *92*, 3816–3824. [CrossRef] [PubMed]

18. Alnahhas, N.; Le Bihan-Duval, E.; Baéza, E.; Chabault, M.; Chartrin, P.; Bordeau, T.; Cailleau-Audouin, E.; Méteau, K.; Berri, C. Impact of divergent selection for ultimate pH of pectoralis major muscle on biochemical, histological, and sensorial attributes of broiler meat. *J. Anim. Sci.* **2015**, *93*, 4524–4531. [CrossRef]

19. Sibut, V.; Le Bihan-Duval, E.; Tesseraud, S.; Godet, E.; Bordeau, T.; Cailleau-Audouin, E.; Chartrin, P.; Duclos, M.J.; Berri, C. Adenosine monophosphate-activated protein kinase involved in variations of muscle glycogen and breast meat quality between lean and fat chickens. *J. Anim. Sci.* **2008**, *86*, 2888–2896. [CrossRef]

20. Sibut, V.; Hennequet Antier, C.; Le Bihan-Duval, E.; Marthey, S.; Duclos, M.J.; Berri, C. Identification of differentially expressed genes in chickens differing in muscle glycogen content and meat quality. *BMC Genom.* **2011**, *12*, 112. [CrossRef]

21. Beauclercq, S.; Nadal-Desbarats, L.; Hennequet Antier, C.; Collin, A.; Tesseraud, S.; Bourin, M.; Le Bihan-Duval, E.; Berri, C. Serum and Muscle Metabolomics for the Prediction of Ultimate pH, a Key Factor for Chicken-Meat Quality. *J. Proteome Res.* **2016**, *15*, 1168–1178. [CrossRef]

22. Beauclercq, S.; Hennequet-Antier, C.; Praud, C.; Godet, E.; Collin, A.; Tesseraud, S.; Metayer-Coustard, S.; Bourin, M.; Moroldo, M.; Martins, F.; et al. Muscle transcriptome analysis reveals molecular pathways and biomarkers involved in extreme ultimate pH and meat defect occurrence in chicken. *Sci. Rep.* **2017**, *7*, 6447. [CrossRef]

23. Le Bihan-Duval, E.; Nadaf, J.; Berri, C.; Pitel, F.; Graulet, B.; Godet, E.; Leroux, S.; Demeure, O.; Lagarrigue, S.; Duby, C.; et al. Detection of a Cis eQTL controlling *BCMO1* gene expression leads to the identification of a QTG for chicken breast meat color. *PLoS ONE* **2011**, *6*, e14825. [CrossRef]

24. Le Bihan-Duval, E.; Nadaf, J.; Berri, C.; Duclos, M.; Pitel, F. Marqueurs génétiques pour la coloration de la viande, 2010. International Patent EP2161345A1; first deposit, 25 August 2008.

25. Jlali, M.; Graulet, B.; Chauveau-Duriot, B.; Chabault, M.; Godet, E.; Leroux, S.; Praud, C.; Le Bihan-Duval, E.; Duclos, M.J.; Berri, C. A mutation in the promoter of the chicken β, β-carotene 15,15′-monooxygenase 1 gene alters xanthophyll metabolism through a selective effect on its mRNA abundance in the breast muscle. *J. Anim. Sci.* **2012**, *90*, 4280–4288. [CrossRef] [PubMed]

26. Jlali, M.; Graulet, B.; Chauveau-Duriot, B.; Godet, E.; Praud, C.; Simoes Nunes, C.; Le Bihan-Duval, E.; Berri, C.; Duclos, M.J. Nutrigenetics of carotenoid metabolism in the chicken: A polymorphism at the β, β-carotene 15,15′-mono-oxygenase 1 (*BCMO1*) locus affects the response to dietary β-carotene. *Br. J. Nutr.* **2014**, *111*, 2079–2088. [CrossRef] [PubMed]

27. Le Bihan-Duval, E.; Hennequet-Antier, C.; Berri, C.; Beauclercq, S.A.; Bourin, M.C.; Boulay, M.; Demeure, O.; Boitard, S. Identification of genomic regions and candidate genes for chicken meat ultimate pH by combined detection of selection signatures and QTL. *BMC Genom.* **2018**, *19*, 294. [CrossRef] [PubMed]

28. Listrat, A.; Lebret, B.; Louveau, I.; Astruc, T.; Bonnet, M.; Lefaucheur, L.; Bugeon, J. Comment la structure et la composition du muscle déterminent la qualité des viandes ou chairs [How muscle structure and composition determine meat quality]. *INRA Prod. Anim.* **2015**, *28*, 125–136.

29. Ouali, A.; Gagaoua, M.; Boudida, Y.; Becila, S.; Boudjellal, A.; Herrera-Mendez, C.H.; Sentandreu, M.A. Biomarkers of meat tenderness: Present knowledge and perspectives in regards to our current understanding of the mechanisms involved. *Meat Sci.* **2013**, *95*, 854–870. [CrossRef] [PubMed]

30. Picard, B.; Gagaoua, M. Proteomic investigations of beef tenderness. In *Proteomics in Food Science*; Elsevier: Amsterdam, The Netherlands, 2017; pp. 177–197.

31. Bouley, J.; Chambon, C.; Picard, B. Mapping of bovine skeletal muscle proteins using two-dimensional gel electrophoresis and mass spectrometry. *Proteomics* **2004**, *4*, 1811–1824. [CrossRef]

32. Morzel, M.; Terlouw, C.; Chambon, C.; Micol, D.; Picard, B. Muscle proteome and meat eating qualities of *Longissimus thoracis* of "Blonde d'Aquitaine" young bulls: Central role of HSP27 isoforms. *Meat Sci.* **2008**, *78*, 297–304. [CrossRef]

33. Chaze, T.; Hocquette, J.F.; Meunier, B.; Renand, G.; Jurie, C.; Chambon, C.; Journaux, L.; Rousset, S.; Denoyelle, C.; Lepetit, J.; et al. Biological Markers for Meat Tenderness of the Three Main French Beef Breeds Using 2-DE and MS Approach. In *Proteomics in Foods*; Toldrá, F., Nollet, L.M.L., Eds.; Springer: Boston, MA, USA, 2013; pp. 127–146.

34. Guillemin, N.; Cassar-Malek, I.; Hocquette, J.F.; Jurie, C.; Micol, D.; Listrat, A.; Levéziel, H.; Renand, G.; Picard, B. La maîtrise de la tendreté de la viande bovine: Un futur proche. I. Approche biologique et identification de marqueurs. *INRA Prod. Anim.* **2009**, *22*, 331–344.

35. Guillemin, N.; Bonnet, M.; Jurie, C.; Picard, B. Functional analysis of beef tenderness. *J. Proteom.* **2011**, *75*, 352–365. [CrossRef]

36. Guillemin, N.; Meunier, B.; Jurie, C.; Cassar-Malek, I.; Hocquette, J.F.; Leveziel, H.; Picard, B. Validation of a Dot-Blot quantitative technique for large scale analysis of beef tenderness biomarkers. *J. Physiol. Pharmacol.* **2009**, *60*, 91–97.

37. Picard, B.; Gagaoua, M.; Micol, D.; Cassar-Malek, I.; Hocquette, J.F.; Terlouw, C.E. Inverse relationships between biomarkers and beef tenderness according to contractile and metabolic properties of the muscle. *J. Agric. Food Chem.* **2014**, *62*, 9808–9818. [CrossRef] [PubMed]

38. Gagaoua, M.; Terlouw, E.M.; Micol, D.; Boudjellal, A.; Hocquette, J.F.; Picard, B. Understanding early *post-mortem* biochemical processes underlying meat color and pH decline in the *Longissimus thoracis* muscle of young Blond d'Aquitaine bulls using protein biomarkers. *J. Agric. Food Chem.* **2015**, *63*, 6799–6809. [CrossRef] [PubMed]

39. Gagaoua, M.; Couvreur, S.; Le Bec, G.; Aminot, G.; Picard, B. Associations among protein biomarkers and pH and color traits in longissimus thoracis and rectus abdominis muscles in protected designation of origin Maine-Anjou cull cows. *J. Agric. Food Chem.* **2017**, *65*, 3569–3580. [CrossRef]

40. Jia, X.; Veiseth-Kent, E.; Grove, H.; Kuziora, P.; Aass, L.; Hildrum, K.I.; Hollung, K. Peroxiredoxin-6—a potential protein marker for meat tenderness in bovine longissimus thoracis muscle. *J. Anim. Sci.* **2009**, *87*, 2391–2399. [CrossRef] [PubMed]

41. Bouley, J.; Meunier, B.; Chambon, C.; De Smet, S.; Hocquette, J.F.; Picard, B. Proteomic analysis of bovine skeletal muscle hypertrophy. *Proteomics* **2015**, *5*, 490–500. [CrossRef]

42. Gobert, M.; Sayd, T.; Gatellier, P.; Santé-Lhoutellier, V. Application to proteomics to understand and modify meat quality. *Meat Sci.* **2014**, *98*, 539–543. [CrossRef]

43. Gagaoua, M.; Bonnet, M.; De Koning, L.; Picard, B. Reverse phase protein array for the quantification and validation of protein biomarkers of beef qualities: The case of meat color from Charolais breed. *Meat Sci.* **2018**, *145*, 308–319. [CrossRef]

44. Bonnet, M.; Tournayre, J.; Cassar-Malek, I. Integrated data mining of transcriptomic and proteomic datasets to predict the secretome of adipose tissue and muscle in ruminants. *Mol. Biosyst.* **2016**, *12*, 2722–2734. [CrossRef]

45. Moon, Y.H.; Cho, W.K.; Lee, S.S. Investigation of blood biomarkers related to meat quality and quantity in Hanwoo steers. *Asian-Australas. J. Anim. Sci.* **2018**, *31*, 1923–1929. [CrossRef]

46. Ueda, S.; Iwamoto, E.; Kato, Y.; Shinohara, M.; Shirai, Y.; Yamanoue, M. Comparative metabolomics of Japanese Black cattle beef and other meats using gas chromatography-mass spectrometry. *Biosci. Biotechnol. Biochem.* **2019**, *83*, 137–147. [CrossRef]

47. Muroya, S.; Oe, M.; Ojima, K.; Watanabe, A. Metabolomic approach to key metabolites characterizing *postmortem* aged loin muscle of Japanese Black (Wagyu) cattle. *Asian-Australas. J. Anim. Sci.* **2019**, *32*, 1172–1185. [CrossRef]

48. Yu, Q.Q.; Tian, X.J.; Shao, L.L.; Li, X.M.; Dai, R.T. Targeted metabolomics to reveal muscle-specific energy metabolism between bovine *longissimus lumborum* and *psoas major* during early postmortem periods. *Meat Sci.* **2019**, *156*, 166–173. [CrossRef]

49. Bertrand, D. La spectroscopie proche infrarouge et ses applications dans les industries de l'alimentation animale [Near infrared spectroscopy: Principles and applications in the animal feed industry]. *INRA Prod. Anim.* **2002**, *15*, 209–219.

50. Yang, D.; Ying, Y. Applications of Raman spectroscopy in agricultural products and food analysis: A review. *Appl. Spectrosc. Rev.* **2011**, *46*, 539–560. [CrossRef]

51. Andueza, D.; Mourot, B.P.; Aït-Kaddour, A.; Prache, S.; Mourot, J. Utilisation de la spectroscopie dans le proche infrarouge et de la spectroscopie de fluorescence pour estimer la qualité et la traçabilité de la viande. The use of non-invasive methods for the estimation of meat quality: Near infrared spectroscopy and fluorescence spectroscopy. *INRA Prod. Anim.* **2015**, *28*, 197–208.

52. Andueza, D.; Mourot, B.P.; Hocquette, J.F.; Mourot, J. Phenotyping of animals and their meat: Applications of low-power ultrasounds, near-infrared spectroscopy, raman spectroscopy and hyperspectral imaging. In *Lawrie's Meat Science*, 8th ed.; Fidel, T., Ed.; Elsevier Ltd.: Amsterdam, The Netherlands, 2017; pp. 501–519.

53. Venel, C.; Mullen, M.; Downey, G.; Troy, D.J. Prediction of tenderness and other quality attributes of beef by near infrared reflectance spectroscopy between 750 and 1100 nm. *J. Near Infrared Spectrosc.* **2001**, *9*, 185–198. [CrossRef]

54. Prieto, N.; Roehe, R.; Lavín, P.; Batten, G.; Andrés, S. Application of near infrared reflectance spectroscopy to predict meat and meat products quality: A review. *Meat Sci.* **2009**, *83*, 175–183. [CrossRef]

55. Liu, Y.; Lyon, B.G.; Windham, W.R.; Realini, C.E.; Pringle, T.D.D.; Duckett, S. Prediction of color, texture, and sensory characteristics of beef steaks by visible and near infrared reflectance spectroscopy. A feasibility study. *Meat Sci.* **2003**, *65*, 1107–1115. [CrossRef]

56. Ripoll, G.; Albertí, P.; Panea, B.; Olleta, J.L.; Sañudo, C. Near infrared reflectance spectroscopy for predicting chemical, instrumental and sensory quality of beef. *Meat Sci.* **2008**, *80*, 697–702. [CrossRef]

57. Andrés, S.; Silva, A.; Soares-Pereira, A.L.; Martins, C.; Bruno-Soares, A.M.; Murray, I. The use of visible and near infrared reflectance spectroscopy to predict beef *M. longissimus thoracis et lumborum* quality attributes. *Meat Sci.* **2008**, *78*, 217–224. [CrossRef]

58. Cecchinato, A.; De Marchi, M.; Penasa, M.; Albera, A.; Bittante, G. Near-infrared reflectance spectroscopy predictions as indicator traits in breeding programs for enhanced beef quality. *J. Anim. Sci.* **2011**, *89*, 2687–2695. [CrossRef]

59. De Marchi, M.; Berzaghi, P.; Boukha, A.; Mirisola, M.; Gallo, L. Use of near infrared spectroscopy for assessment of beef quality traits. *Ital. J. Anim. Sci.* **2007**, *6*, 421–423. [CrossRef]

60. Bauer, A.; Scheier, R.; Eberle, T.; Schmidt, H. Assessment of tenderness of aged bovine gluteus medius muscles using Raman spectroscopy. *Meat Sci.* **2016**, *115*, 27–33. [CrossRef]

61. Beattie, R.J.; Bell, S.J.; Farmer, L.J.; Moss, B.W.; Patterson, D. Preliminary investigation of the application of Raman spectroscopy to the prediction of the sensory quality of beef silverside. *Meat Sci.* **2004**, *66*, 903–913. [CrossRef]

62. Fowler, S.M.; Schmidt, H.; van de Ven, R.; Wynn, P.; Hopkins, D.L. Raman spectroscopy compared against traditional predictors of shear force in lamb *m. longissimus lumborum. Meat Sci.* **2014**, *98*, 652–656. [CrossRef]

63. Liao, Y.T.; Fan, Y.X.; Cheng, F. On-line prediction of fresh pork quality using visible/near-infrared reflectance spectroscopy. *Meat Sci.* **2010**, *86*, 901–907. [CrossRef]

64. Balage, J.; Silva, S.L.; Abdalla Gomide, C.; Bonin, M.N.; Figueira, A.C. Predicting pork quality using Vis/NIR spectroscopy. *Meat Sci.* **2015**, *108*, 37–43. [CrossRef]

65. Rust, S.R.; Price, D.M.; Subbiah, J.; Kranzler, G.; Hilton, G.G.; Vanoverbeke, D.L.; Morgan, J.B. Predicting beef tenderness using near-infrared spectroscopy. *J. Anim. Sci.* **2008**, *86*, 211–219. [CrossRef]

66. Leroy, B.; Lambotte, S.; Dotreppe, O.; Lecocq, H.; Istasse, L.; Clinquart, A. Prediction of technological and organoleptic properties of beef *longissimus thoracis* from near infrared reflectance and transmission spectra. *Meat Sci.* **2004**, *66*, 45–54. [CrossRef]

67. Prieto, N.; Andrés, S.; Giráldez, F.J.; Mantecón, A.R.; Lavín, P. Ability of near infrared reflectance spectroscopy (NIRS) to estimate physical parameters of adult steers (oxen) and young cattle meat samples. *Meat Sci.* **2008**, *79*, 692–699. [CrossRef] [PubMed]

68. De Marchi, M.; Penasa, M.; Cecchinato, A.; Bittante, G. The relevance of different near infrared technologies and sample treatments for predicting meat quality traits in commercial beef cuts. *Meat Sci.* **2013**, *93*, 329–335. [CrossRef] [PubMed]

69. Abeni, F.; Bergoglio, G. Characterization of different strains of broiler chicken by carcass measurements, chemical and physical parameters and NIRS on breast muscle. *Meat Sci.* **2001**, *57*, 133–137. [CrossRef]

70. Berzaghi, P.; Dalle Zotte, A.; Jansson, L.M.; Andrighetto, I. Near-infrared reflectance spectroscopy as a method to predict chemical composition of breast meat and discriminate between different n-3 feeding sources. *Poult. Sci.* **2005**, *84*, 128–136. [CrossRef] [PubMed]

71. Chartrin, P.; Rousseau, X.; Gigaud, V.; Bastianelli, D.; Baéza, E. Near-infrared reflectance spectroscopy for predicting lipid content in chicken breast meat. In Proceedings of the 13th WPSA European Poultry Conference, Tours, France, 23–27 August 2010; p. 4.

72. Bastianelli, D.; Bonnal, L.; Chartrin, P.; Bernadet, M.D.; Marie-Etancelin, C.; Baéza, E. Near-infrared reflectance spectroscopy for predicting lipid content in duck breast meat. In Proceedings of the XIXth WPSA European Symposium on the Quality of Poultry Meat, Turku, Finland, 21–25 September 2009; pp. 1–9.

73. De Marchi, M.; Riovanto, R.; Penasa, M.; Cassandro, M. Feasibility of the direct application of near-infrared reflectance spectroscopy on intact chicken breasts to predict meat color and physical traits. *Meat Sci.* **2011**, *90*, 653–657. [CrossRef] [PubMed]

74. Cozzolino, D.; Murray, I.; Paterson, R. Visible and near infrared reflectance spectroscopy for the determination of moisture, fat and protein in chicken breast and thigh muscle. *J. Near Infrared Spectrosc.* **1996**, *4*, 216–223. [CrossRef]

75. Brondum, J.; Munck, L.; Henckel, P.; Karlsson, A.; Tornberg, E.; Engelsen, S.B. Prediction of water-holding capacity and composition of porcine meat by comparative spectroscopy. *Meat Sci.* **2000**, *55*, 177–185. [CrossRef]

76. Hoving-Bolink, A.H.; Vedder, H.W.; Merks, J.W.M.; de Klein, W.J.H.; Reimert, H.G.M.; Frankhuizen, R.; van den Broek, W.H.A.M.; Lambooij, E. Perspective of NIRS measurements early post mortem for prediction of pork quality. *Meat Sci.* **2005**, *69*, 417–423. [CrossRef]

77. Savenije, B.; Geesink, G.H.; van der Palen, J.G.P.; Hemke, G. Prediction of pork quality using visible/near-infrared reflectance spectroscopy. *Meat Sci.* **2006**, *73*, 181–184. [CrossRef]

78. Prevolnik, M.; Škrlep, M.; Janeš, L.; Velikonja-Bolta, Š.; Škorjanc, D.; Čandek-Potokar, M. Accuracy of near infrared spectroscopy for prediction of chemical composition, salt content and free amino acids in dry-cured ham. *Meat Sci.* **2011**, *88*, 299–304. [CrossRef]

79. Ripoche, A.; Guillard, A.S. Determination of fatty acid composition of pork fat by Fourier transform infrared spectroscopy. *Meat Sci.* **2001**, *58*, 299–304. [CrossRef]

80. Pérez-Marín, D.; De Pedro Sanz, E.; Guerrero-Ginel, J.E.; Garrido-Varo, A. A feasibility study on the use of near-infrared spectroscopy for prediction of the fatty acid profile in live Iberian pigs and carcasses. *Meat Sci.* **2009**, *83*, 627–633. [CrossRef]

81. Mairesse, G.; Douzenel, P.; Mourot, J.; Vautier, A.; Le Page, R.; Goujon, J.M.; Poffo, L.; Sire, O.; Chesneau, G. La spectroscopie proche infrarouge: Outil d'analyse rapide sur carcasse de la teneur en acides gras polyinsaturés n-3 des gras de bardière du porc charcutier [Near-infrared spectrometry: A rapid analytical tool for n-3 polyunsaturated fatty acid measurement on backfat of pig carcass]. *J. Rech. Porc.* **2012**, *44*, 211–212.

82. Zamora-Rojas, E.; Garrido-Varo, A.; De Pedro-Sanz, E.; Guerrero-Ginel, J.E.; Pérez-Marín, D. Prediction of fatty acids content in pig adipose tissue by near infrared spectroscopy: At-line versus in-situ analysis. *Meat Sci.* **2013**, *95*, 503–511. [CrossRef]

83. Gonzalez-Martın, I.; Gonzalez-Perez, C.; Alvarez-Garcıa, N.; Gonzalez-Cabrera, J.M. On-line determination of fatty acid composition in intramuscular fat of Iberian pork loin by NIRs with a remote reflectance fiber optic probe. *Meat Sci.* **2005**, *69*, 243–248. [CrossRef] [PubMed]

84. Bozzi, R.; Parrini, S.; Crovetti, A.; Pugliese, C.; Bonelli, A.; Gasparini, S.; Karolyi, D.; Martins, J.M.; Garcia-Gasco, J.M.; Panella-Riera, N.; et al. Determination of fatty acid groups in intramuscular fat of various local pig breeds by FT-NIRS. In *Book of Abstracts of the 69th Annual Meeting of the European Association for Animal Production, Dubrovnik, Croatia; 26–31 August 2018*; Wageningen Academic Publishers: Wageningen, The Netherlands, 2018; p. 492.

85. Dvořáček, V.; Prohasková, A.; Chrpová, J.; Štočková, L. Near infrared spectroscopy for deoxynivalenol content estimation in intact wheat grain. *Plant Soil Environ.* **2012**, *58*, 196–203. [CrossRef]

86. De Marchi, M.; Pulici, C.; Battagin, C.; Penasa, M.; Rizzi, C.; Cassandro, M. Prediction of physical and colour characteristics of breast meat by near infrared spectroscopy. In Proceedings of the XIIIth European Poultry Conference, Tours, France, 23–27 August 2010.

87. Liu, Y.; Lyon, B.G.; Windham, W.R.; Lyon, C.E.; Savage, E.M. Principal component analysis of physical, color, and sensory characteristics of chicken breasts deboned at two, four, six, and twenty-four hours *postmortem*. *Poult. Sci.* **2004**, *83*, 1467–1474. [CrossRef]

88. Candek-Potokar, M.; Prevolnik, M.; Skrlep, M. Ability of near infrared spectroscopy to predict pork technological traits. *J. Near Infrared Spectrosc.* **2006**, *14*, 269–277. [CrossRef]

89. Kapper, C.; Klont, R.E.; Verdonk, J.M.A.J.; Urlings, H.A.P. Prediction of pork quality with near infrared spectroscopy (NIRS).1. Feasibility and robustness of NIRS measurements at laboratory scale. *Meat Sci.* **2012**, *91*, 294–299. [CrossRef]

90. Forrest, J.C.; Morgan, M.T.; Borggaard, C.; Rasmussen, A.J.; Jespersen, B.L.; Andersen, J.R. Development of technology for the early post mortem prediction of water holding capacity and drip loss in fresh pork. *Meat Sci.* **2000**, *55*, 115–122. [CrossRef]

91. Pedersen, D.K.; Morel, S.; Andersen, H.J.; Engelsen, S.B. Early prediction of water-holding capacity in meat by multivariate vibrational spectroscopy. *Meat Sci.* **2003**, *65*, 581–592. [CrossRef]

92. Boschetti, L.; Ottavian, M.; Facco, P.; Barolo, M.; Serva, L.; Balzan, S.; Novelli, E. A correlative study on data from pork carcass and processed meat (Bauernspeck) for automatic estimation of chemical parameters by means of near-infrared spectroscopy. *Meat Sci.* **2013**, *95*, 621–628. [CrossRef]

93. Prevolnik, M.; Candek-Potokar, M.; Novic, M.; Škorjanc, D. An attempt to predict pork drip loss from pH and colour measurements or near infrared spectra using artificial neural networks. *Meat Sci.* **2009**, *83*, 405–411. [CrossRef]

94. Vautier, A.; Gault, E.; Lhommeau, T.; Bozec, A. Meat quality mapping of the loin: pH vs. In NIR spectroscopy to predict the cooking yield. In Proceedings of the 60th International Congress of Meat Science and Technology, Punta del Este, Uruguay, 17–22 August 2014.

95. Vautier, A.; Lhommeau, T.; Daumas, G. A feasability study for the prediction of the technological quality of ham with NIR spectroscopy. In *Book of Abstracts of the 64th European Federation of Animal Science (EAAP), Nantes, France; 26–20 August 2013*; Wageningen Academic Publishers: Wageningen, The Netherlands, 2013; Session 2, Theatre 16; p. 108.

96. Chmiel, M.; Slowinski, M. The use of computer vision system to detect pork defect. *Food Sci. Technol.* **2016**, *73*, 473–480. [CrossRef]

97. Qiao, J.; Wang, N.; Ngadi, M.O.; Gunenc, A.; Monroy, M.; Gariepy, C.; Prasher, S.O. Prediction of drip-loss, pH, and color for pork using a hyperspectral imaging technique. *Meat Sci.* **2007**, *76*, 1–8. [CrossRef]

98. Liu, L.; Ngadi, M.O.; Prasher, S.O.; Gariépy, C. Categorization of pork quality using Gabor filter-based hyperspectral imaging technology. *J. Food Eng.* **2010**, *99*, 284–293. [CrossRef]

99. Watson, R.; Gee, A.; Polkinghorne, R.; Porter, M. Consumer assessment of eating quality—Development of protocols for Meat Standards Australia (MSA) testing. *Aust. J. Exp. Agric.* **2008**, *48*, 1360–1367. [CrossRef]

100. Watson, R.; Polkinghorne, R.; Thompson, J.M. Development of the Meat Standards Australia (MSA) prediction model for beef palatability. *Aust. J. Exp. Agric.* **2008**, *48*, 1368–1379. [CrossRef]

101. Hocquette, J.F.; Van Wezemael, L.; Chriki, S.; Legrand, I.; Verbeke, W.; Farmer, L.; Scollan, N.D.; Polkinghorne, R.J.; Rødbotten, R.; Allen, P.; et al. Modelling of beef sensory quality for a better prediction of palatability. *Meat Sci.* **2014**, *97*, 316–322. [CrossRef]

102. Guzek, D.; Glabska, D.; Gutkowska, K.; Wierzbicki, J.; Wozniak, A.; Wierzbicka, A. Influence of cut and thermal treatment on consumer perception of beef in polish trials. *Pak. J. Agric. Sci.* **2015**, *52*, 521–526.

103. McCarthy, S.N.; Henchion, M.; White, A.; Brandon, K.; Allen, P. Evaluation of beef eating quality by Irish consumers. *Meat Sci.* **2017**, *132*, 118–124. [CrossRef]

104. Bonny, S.; O'Reilly, A.; Pethick, D.W.; Gardner, G.E.; Hocquette, J.F.; Pannier, L. Update of Meat Standards Australia and the cuts based grading scheme for beef and sheepmeat. *J. Integr. Agric.* **2018**, *17*, 1641–1654. [CrossRef]
105. Polkinghorne, R.; Nishimura, T.; Neath, K.E.; Watson, R. Japanese consumer categorisation of beef into quality grades, based on Meat Standards Australia methodology. *Anim. Sci. J.* **2011**, *82*, 325–333. [CrossRef]
106. Legrand, I.; Hocquette, J.F.; Polkinghorne, R.J.; Pethick, D.W. Prediction of beef eating quality in France using the Meat Standards Australia system. *Animal* **2013**, *7*, 524–529. [CrossRef]
107. Hocquette, J.F.; Legrand, I.; Jurie, C.; Pethick, D.W.; Micol, D. Perception in France of the Australian system for the prediction of beef quality (MSA) with perspectives for the European beef sector. *Anim. Prod. Sci.* **2011**, *51*, 30–36. [CrossRef]
108. Legrand, I.; Hocquette, J.F.; Polkinghorne, R.J.; Wierzbicki, J. Comment prédire la qualité de la viande bovine en Europe en s'inspirant du système australien MSA? [How to predict beef eating quality in Europe through the adaptation of the Australian system MSA?]. *Innov. Agron.* **2017**, *55*, 171–182.
109. Bonny, S.P.F.; Gardner, G.E.; Pethick, D.W.; Legrand, I.; Polkinghorne, R.J.; Hocquette, J.F. Biochemical measurements of beef are a good predictor of untrained consumer sensory scores across muscles. *Animal* **2015**, *9*, 179–190. [CrossRef]
110. Hocquette, J.F.; Bernard-Capel, C.; Vidal, V.; Jesson, B.; Levéziel, H.; Cassar-Malek, I. The GENOTEND chip: A new tool to analyse gene expression in muscles of beef cattle for beef quality prediction. *BMC Vet. Res.* **2012**, *8*, 135. [CrossRef] [PubMed]
111. Bonny, S.P.F.; Hocquette, J.F.; Pethick, D.W.; Legrand, I.; Wierzbicki, J.; Allen, P.; Farmer, L.J.; Polkinghorne, R.J.; Gardner, G.E. The variability of the eating quality of beef can be reduced by predicting consumer satisfaction. *Animal* **2018**, *12*, 2434–2442. [CrossRef]
112. Farmer, L.; Farrell, D. Review: Beef-eating quality: A European journey. *Animal* **2018**, *12*, 2424–2433. [CrossRef]
113. Bonny, S.P.F.; Pethick, D.W.; Legrand, I.; Wierzbicki, J.; Allen, P.; Farmer, L.J.; Polkinghorne, R.J.; Hocquette, J.F.; Gardner, G.E. The maturity estimate most accurate in predicting eating quality depends on the age range of the cattle examined. *Animal* **2016**, *10*, 718–728. [CrossRef]
114. Bonny, S.P.F.; Pethick, D.W.; Legrand, I.; Wierzbicki, J.; Allen, P.; Farmer, L.J.; Polkinghorne, R.J.; Hocquette, J.F.; Gardner, G.E. European conformation and fat scores have no relationship with eating quality. *Animal* **2016**, *10*, 996–1006. [CrossRef]
115. Normand, J.; Rubat, E.; Evrat-Georgel, C.; Turin, F.; Denoyelle, C. A national survey of beef tenderness in France. *Viandes Prod. Carnés* **2014**, *30*, 5.
116. Lyford, C.; Thompson, J.; Polkinghorne, R.; Miller, M.; Nishimura, T.; Neath, K.; Allen, P.; Belasco, E. Is willingness to pay (WTP) for beef quality grades affected by consumer demographics and meat consumption preferences? *Aust. Agribus. Rev.* **2010**, *18*, 1–17.
117. Polkinghorne, R.J.; Thompson, J.M. Meat standards and grading. *Meat Sci.* **2010**, *86*, 227–235. [CrossRef]
118. McGilchrist, P.; Polkinghorne, R.; Ball, A.; Thompson, J. The Meat Standards Australia Index indicates beef carcass quality. *Animal* **2019**, in press. [CrossRef]
119. Troy, D.J.; Kerry, J.P. Consumer perception and the role of science in the meat industry. *Meat Sci.* **2010**, *86*, 214–226. [CrossRef]
120. Scollan, N.D.; Greenwood, P.L.; Newbold, C.J.; Yáñez Ruiz, D.R.; Shingfield, K.J.; Wallace, R.J.; Hocquette, J.F. Future research priorities for animal production in a changing world. *Anim. Prod. Sci.* **2011**, *51*, 1–5. [CrossRef]
121. Font-i-Furnols, M.; Guererro, L. Consumer preference, behavior and perception about meat and meat products: An overview. *Meat Sci.* **2014**, *98*, 361–371. [CrossRef]
122. Legrand, I.; Hocquette, J.F.; Denoyelle, C.; Bièche-Terrier, C. La gestion des nombreux critères de qualité de la viande bovine: Une approche nécessairement complexe [Management of the many quality criteria for beef: A complex approach]. *INRA Prod. Anim.* **2016**, *29*, 185–200.
123. Verbeke, W.; Perez-Cueto, F.J.A.; de Barcellos, M.D.; Krystallis, A.; Grunert, K.G. European citizen and consumer attitudes and preferences regarding beef and pork. *Meat Sci.* **2010**, *84*, 284–292. [CrossRef]
124. Greenheck, J.; Johnson, B.; Graves, A.; Oak, A. Giving meat meaning: Creating value-based connections with consumers. *Anim. Front.* **2018**, *8*, 11–15. [CrossRef]

125. Polkinghorne, R.J. From commodity, to customer, to consumer: The Australian beef industry evolution. *Anim. Front.* **2018**, *8*, 47–52. [CrossRef]
126. Moloto, K.W.; Frylinck, L.; Modika, K.Y.; Pitse, T.; Strydom, P.E.; Koorsen, G. Is there a Possibility of Meat Tenderness Protein-Biomarkers on the Horizon? *Int. J. Agric. Innov. Res.* **2017**, *6*, 467–472.

Article

Online Prediction of Physico-Chemical Quality Attributes of Beef Using Visible—Near-Infrared Spectroscopy and Chemometrics

Amna Sahar [1], Paul Allen [1], Torres Sweeney [2], Jamie Cafferky [1], Gerard Downey [1], Andrew Cromie [3] and Ruth M. Hamill [1,*]

1   Teagasc Food Research Centre Ashtown, Dublin D15 KN3K, Ireland; Amna.sahar@teagasc.ie (A.S.); paul.allen@teagasc.ie (P.A.); jamie.cafferky@teagasc.ie (J.C.); Gerard.Downey@teagasc.ie (G.D.)
2   UCD School of Veterinary Medicine, University College Dublin, Belfield, Dublin D04 W6F6, Ireland; torres.sweeney@ucd.ie
3   Irish Cattle Breeders Federation, Highfield House, Shinagh, Bandon, Co. Cork P72 X050, Ireland; acromie@icbf.com
*   Correspondence: ruth.hamill@teagasc.ie

Received: 6 September 2019; Accepted: 16 October 2019; Published: 23 October 2019

Abstract: The potential of visible–near-infrared (Vis–NIR) spectroscopy to predict physico-chemical quality traits in 368 samples of bovine musculus longissimus thoracis et lumborum (LTL) was evaluated. A fibre-optic probe was applied on the exposed surface of the bovine carcass for the collection of spectra, including the neck and rump (1 h and 2 h post-mortem and after quartering, i.e., 24 h and 25 h post-mortem) and the boned-out LTL muscle (48 h and 49 h post-mortem). In parallel, reference analysis for physico-chemical parameters of beef quality including ultimate pH, colour (L, a*, b*), cook loss and drip loss was conducted using standard laboratory methods. Partial least-squares (PLS) regression models were used to correlate the spectral information with reference quality parameters of beef muscle. Different mathematical pre-treatments and their combinations were applied to improve the model accuracy, which was evaluated on the basis of the coefficient of determination of calibration (R^2C) and cross-validation (R^2CV) and root-mean-square error of calibration (RMSEC) and cross-validation (RMSECV). Reliable cross-validation models were achieved for ultimate pH (R^2CV: 0.91 (quartering, 24 h) and R^2CV: 0.96 (LTL muscle, 48 h)) and drip loss (R^2CV: 0.82 (quartering, 24 h) and R^2CV: 0.99 (LTL muscle, 48 h)) with lower RMSECV values. The results show the potential of Vis–NIR spectroscopy for online prediction of certain quality parameters of beef over different time periods.

Keywords: meat; quality; near-infrared spectroscopy; on-line; monitoring

1. Introduction

A wide range of factors interactively affect the quality of meat, including sex, genotype, rearing conditions, feeding practices, transport, slaughtering and post-mortem handling of the carcass. Meat quality is a complex set of parameters including physico-chemical, chemical and sensory quality. Quality parameters like aroma, flavour, mouth-feel and tenderness can be evaluated by sensory analysis. In addition to these traits, other quality attributes such as colour, water-holding capacity, texture and pH can be studied using instrumental techniques. These technological traits are extremely important as they provide data on the development of ultimate meat quality and also convey information on appreciation of the product and its value, providing specific and important evidence on overall meat quality as it varies among individuals of a population [1]. Meat colour is important to the consumer as a key cue in perception at the point of sale and, therefore, has a major bearing on the decision to purchase [2]. Drip loss is exudate lost from meat through cutting, heating and pressing [3], and losses of ~5% are common in beef [2]. Water-holding-capacity traits such as drip loss and cook loss are important, as they represent variability in economic losses for the processor and furthermore, nutritional losses for the consumers [4]. Variation in pH fall and ultimate pH during the conversion of muscle to meat influence water-holding capacity, colour, and, through influencing the ultimate contractile state and proteolytic enzyme activity post-mortem, tenderness.

In order to improve the overall population for meat quality, through incorporation of meat quality into breeding programmes, a means of providing information on quality, routinely and for as many animals as possible, is required. The assessment of technological meat quality traits using conventional approaches is disadvantageous in terms of time consumption, sample destruction, sample preparation, requirement of expert analysts, chemical utilization and lack of on-site facilities in processing plants for detailed quality evaluation, which adds to operational costs. Subsequently, these traditional methods lack the potential to be applied online for the prediction of meat quality attributes in the industry [5]. It is therefore required to introduce some rapid, non-invasive, non-destructive, chemical-free and more reliable approaches for accurate and online determination of meat quality, which enables quality assessment based on a multivariate approach.

Non-destructive methods include spectroscopic techniques, use of biosensors, electronic noses, ultrasound methods, microscopy, microwave characterization, nuclear magnetic resonance and dielectric methods [6]. Among these approaches, spectroscopic methods have gained significant popularity regarding the prediction of numerous quality attributes of meat in the last decade. In the case of spectroscopy, electromagnetic radiations in the ultraviolet, visible, near-, mid- and far-infrared regions interact with matter, providing fingerprints of the samples under consideration, which can further be processed to extract useful qualitative and quantitative information [7]. Globally, spectroscopic methods have been extensively employed to assess the quality of muscle foods. For instance, Fourier-transform infrared spectroscopy (FTIR) has been successfully employed for meat quality and fraud detection [8–10]. Similarly, Raman spectroscopy has also shown its potential to provide structural information about muscle proteins [11,12]. Likewise, visible spectroscopy can be used as a non-invasive method for tissue characterization [13]. Furthermore, fluorescence spectroscopy can be employed for quality evaluation of meat and meat-based products [7,14–17]. Recently, near-infrared spectroscopy and hyperspectral imaging have also been investigated for non-destructive prediction of quality and compositional analysis of meat [18–22] and for the determination of adulterants in meat [23].

The application of visible–near-infrared (Vis–NIR) spectroscopy directly on the meat carcass is advantageous because it does not require the preparation of the sample before analysis and it is applicable to the prediction of quality online using a fibre-optic probe. Vis–NIR spectra would allow a timely prediction of meat quality traits which could potentially assist the processors to sort out and classify carcasses accordingly. Consequently, the objective of this study was to model the physico-chemical parameters of beef quality including ultimate pH, colour, cook loss and drip loss, using visible–near infrared spectroscopic profiles collected at various time points post-mortem. Spectra were recorded by directly applying a fibre-optic probe on the exposed surface of the carcass (neck and rump) in the abattoir (1 h and 2 h post-mortem), on the cut surface of carcass in the abattoir immediately and one hour after quartering (24 h and 25 h post-mortem), on the cut surface of musculus longissimus thoracis et lumborum (LTL) in the laboratory (48 h and 49 h post-mortem).

2. Materials and Methods

2.1. Animals and Meat Samples Preparation

A total of three 368 cross-bred beef animals, reared under the same environmental and feeding conditions in the Irish Cattle Breeders Federation Tully Progeny Test Centre were used for the study over a time span of 18 months. The animals were slaughtered in 10 batches from February 2014 until September 2015 in a commercial EU-licensed abattoir in Ireland. All carcasses were quartered at the 8th rib on pistol hind 24 h post-mortem, and the loins muscles were deboned 48 h post-mortem. A total of 12 steaks (2.54 cm thickness) were sliced from the right side of the LTL muscle and vacuum-packed at 4 °C for further analysis. Then, 48 h post-mortem, the loins were transported from the factory to the Teagasc Food Research Centre, Ashtown for further analysis.

2.2. Spectra Collection

Vis–NIR spectra were collected using a portable Vis–NIR spectrophotometer (ASD Inc., Boulder Colorado, CO, USA) with detection waveband range from 350 to 2500 nm, using the Indico Pro program. A high-intensity contact probe was used to transmit the light reflected from the surface of the carcass to the internal detector. Prior to spectral acquisition, the instrument was calibrated using a Spectralon tile as the white reference. Spectra were collected on the day of slaughtering from neck and rump, specifically, one hour after slaughtering (1 h post-mortem) and two hours after slaughtering (2 h post-mortem). Spectra were also collected from the quartered surface of the carcass (5th rib) at the time of quartering (24 h and 25 h post-mortem, after 1 h blooming in the chill room). Spectra were also collected from the LTL muscle in the laboratory (48 h and 49 h post-mortem, after 1 h blooming in the chill room). The spectra were collected in triplicate from three representative sites of the transverse surface of the LTL muscle (the method was described in detail [24]). For each of these three scans, 20 spectra were automatically collected by the instrument consecutively and averaged to reduce noise. Spectral data were exported as a JCAMP file to The Unscrambler X version 10.3 (CAMO ASA, Oslo, Norway) for further chemometric analysis.

2.3. Chemical and Physical Analyses

2.3.1. Ultimate pH (pHu)

The ultimate pH was determined in 366 beef carcasses between the 12th and the 13th rib (48 h post-mortem). A portable pH meter was employed (Hanna Instrument HI 9126, Woonsocket, RI, USA) to record both the pH and the temperature. Each sampling day, the pH meter was calibrated with standardized buffers at pH 7.0 and pH 4.0.

2.3.2. Colour

The equipment employed was the UltraScan® PRO with a dual-beam xenon flash spectrophotometer (λ: 350–1050 nm, $\Delta\lambda$: 5 nm, D65, 8°). The steaks (48 h post-mortem) obtained from LTL muscles were wrapped in oxygen-permeable transparent film and keptfor 1 h of blooming before being measured with the uppermost side placed to the light. The spectrophotometer was calibrated with a black and white baseline. L* (brightness) (varies from 100 for perfect white to 0 for black), a* (redness) (a negative value indicates green, while a positive value indicates red) and b* (yellowness) (a negative value indicates blue, and a positive value indicates yellow) are the coordinates which describe the colour of meat. EasyMach QC software (Hunter Associates Laboratory, Inc., Reston, VA, USA) was used, and the colour coordinates values were obtained as the average of three measurements performed on different locations of each LTL muscle slice.

2.3.3. Cooking Loss Percentage

The samples (14-day-aged steaks) were cooked in a circulating water bath to an internal temperature of 70 °C. The temperature was monitored continuously during the cooking process until a plateau was achieved at 70 °C, using a temperature probe (Eirelec Ltd, Dublin, Ireland) inserted into the geometric centre of the steak. Samples were weighted before and after cooking in order to determine the cooking loss percentage.

$$\text{Cook-loss (\%)} = (\text{raw weight} - \text{cooked weight})/\text{raw weight} \times 100 \tag{1}$$

2.3.4. Drip Loss

Drip loss was determined on LTL muscles, 48 h post-mortem. A meat slice of approximately 100 g weight (2.5 cm thickness, 7.5 cm length, 5.0 cm width) was cut from the LTL muscle and hung in the chill room at 4 °C. The samples were weighed after 96 hours, and the drip loss percentage was calculated [25].

$$\text{Drip-loss (\%)} = (\text{Initial weight} - \text{Final weight})/\text{initial weight} \times 100 \qquad (2)$$

2.4. Data Analysis

Partial least-squares regression was performed using The UNSCRAMBLER program (version 8.5.0, Camo, Trondheim, Norway). After visual inspection, the detection of anomalous spectra was accomplished using the H-statistic, which indicates how different a sample spectrum is from the average spectrum of the set [26]. A sample with an H statistic of standardized units from the mean spectrum was defined as a global H outlier and was eliminated from the population. Baseline correction was applied to the spectra because the NIR spectra are affected by light scatter and path-length variation, and pre-treatments of the spectral data improve the accuracy of calibration. In these cases, spectral data pre-treatments such as standard normal variate (SNV) were applied to the spectra to reduce the noise and light scattering effects. Partial least-squares (PLS) regression was used for predicting the chemico-physical properties using Vis–NIR spectra as independent variables. Internal full cross-validation was performed to avoid overfitting the PLS equations; thus, the optimal number of factors in each equation was determined as the number of factors after which the standard error of cross-validation no longer decreased substantially. The accuracy of prediction was evaluated in terms of coefficient of determination (R^2C and R^2CV) and root-mean-square error of calibration (RMSEC) and cross-validation (RMSECV).

3. Results and Discussion

3.1. Spectral Profiles

Figure 1 illustrates the average Vis–NIR spectra measured for neck and rump (1 h post-mortem), quartered LTL muscle (24 h post-mortem) and loin muscle (48 h post-mortem). A remarkable difference was seen in the absorbance between the mean spectra of the sites. The variation in the spectra was potentially due to differences in the time of spectral collection (1 h to 48 h post-mortem) and type of muscle. Neck and rump showed lower absorbance as compared to the quartered surface of the LTL muscle and loin muscle in the 1300–2500 nm range. The spectra that were collected during quartering (24 h post-mortem) showed resemblance with the spectra that were collected from the LTL muscle (48 h post-mortem). This is due the fact that, although the spectra were collected in 24 h intervals, they were from the same muscle. Various spectral absorbance bands were identified from the average Vis–NIR range of the analysed samples. These bands could be easily identified by visual inspection of the wavebands at which the highest absorbance values were found. Spectral collection from the neck and quartered surface showed intense peaks at 415 nm and in the 540–580 nm range. Less intense peaks at 1445 nm and 1940 nm were observed for the neck and rump. However, spectral data collected at the time of quartering and in loin muscles showed an inverse pattern of peaks in this region. These wavelengths correspond to different specific functional bonds. Prominent peaks could be seen at 415 nm, 540–580 nm, 1449 nm and 1933 nm [27]. The visual spectral features were analogous to the findings of Andrés et al. [28].

The average spectrum of 10 samples with high ultimate pH values and that of 10 spectrum with low ultimate pH values, compared to the average spectrum of all samples, determined using Vis–NIR spectroscopy, are shown in Figure 2a. The graph depicts higher absorbance value for the samples having low ultimate pH in comparison with those having high ultimate pH. Low-pH-sample spectra showed peak intensity closer to that of the samples with average values of pH, in agreement with the fact that most of the samples in our study had relatively lower ultimate pH and samples with high ultimate pH were more unusual. The absorbance bands at 415 nm and 546 nm were quite prominent and correspond to myoglobin [29]. These results show the efficiency of Vis–NIR spectroscopy in analysing the variation of the ultimate pH of beef.

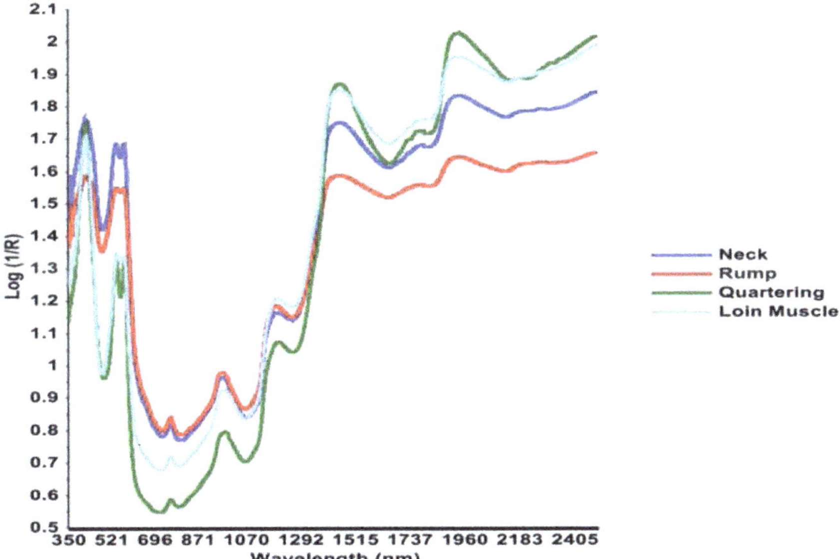

Figure 1. Average visible–near-infrared (Vis–NIR) spectra measured from neck and rump (1 h post-mortem), on quartering from the musculus longissimus thoracis et lumborum (LTL) muscle (24 h post-mortem) and from the loin muscle (48 h post-mortem).

Figure 2b shows the spectral results concerning the percentage of drip loss for 10 high-drip-loss and 10 low-drip-loss samples. Differences in the absorbance values of the mean samples can be seen. High-drip-loss samples showed less abundant peaks in the spectral range of 1200–2400 nm that corresponds to moisture content [18]. Low-drip-loss samples showed peak intensities closer to those of samples with average drip loss values, especially in the 1200–2400 nm range, showing that most of the samples in our study had relatively low drip loss.

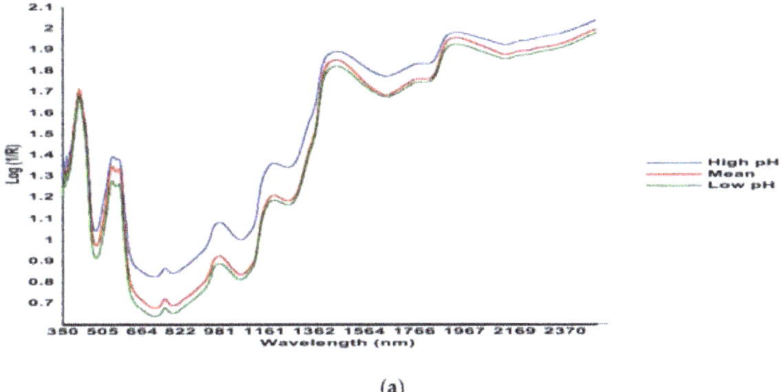

(a)

Figure 2. *Cont.*

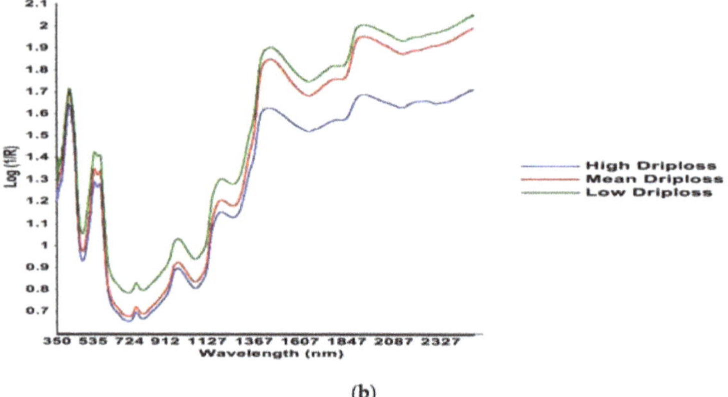

(b)

Figure 2. Average Vis–NIR spectra recorded from the LTL muscle (48 post-mortem). (**a**), Spectra of 10 high-ultimate-pH and 10 low-ultimate-pH carcass samples, (**b**) spectra of 10 high-drip-loss and 10 low-drip-loss carcass samples compared to the average spectrum of all samples.

These figures suggest that the absorbance value may be closely proportional to the drip loss, supporting the premise that Vis–NIR spectroscopy has relevance for online prediction of drip loss percentage in post-mortem beef samples

3.2. Descriptive Statistics of Beef Samples

Ranges, means, standard deviations and coefficient of variances of physico-chemical traits of all beef samples are given in Table 1. The results depicted a range of 5.16–6.91 for the ultimate pH of the beef samples stored for 48 h. An ultimate pH falling between 5.4–5.8 would be considered normal for beef, with higher values potentially corresponding to a dark, firm and dry phenotype, and lower values potentially presenting issues about their water-holding capacity [30]. Overall, however, the ultimate pH data showed a relatively low coefficient of variance (CV) of 3.23%. A relatively low variation in ultimate pH has been reported in several studies in the past by scientists who investigated the use of near-infrared spectroscopy in beef [28,31,32], chicken [33] and pork meat [34]. Here, it is possible that some outliers could contribute to the range measured, and these outliers were identified during data processing and removed from the data set. The ranges for L*, a* and b* colours were from 35.17 to 50.91, from 8.20 to 19.80 and from 6.34 to16.94, respectively. The descriptive statistical analyses of drip loss (%) and cooking loss (%) are also summarized in Table 1. The average values of drip loss and cook loss were 2.94% and 31.22%, respectively. Drip loss showed considerably higher variability (CV = 51.02%) compared to cook loss (CV = 9.55%). Data obtained from traditional laboratory methods showed that the dataset exhibited relatively low variability in certain parameters and higher variability in others. Overall, the values of the studied parameters were in line with previous findings [30].

Table 1. Ranges, mean, standard deviations and coefficient of variance of physico-chemical traits of all beef samples.

Parameter	*n*	Range	Mean	SD	CV (%)
Ultimate pH	366	5.16–6.91	5.58	0.18	3.23
Drip Loss (%)	224	0.29–9.04	2.94	1.50	51.02
Cook Loss (%)	293	16.81–38.02	31.22	2.98	9.55
Colour L*	368	35.17–50.91	42.99	2.44	5.68
Colour a*	368	8.20–19.80	14.30	1.82	12.73
Colour b*	368	6.34–16.94	11.40	1.76	15.44

n: number of samples; SD: standard deviation; CV: Coefficient of variance; L*: brightness; a*: redness; b*: yellowness.

3.3. Prediction of the pH Ultimate Values of Beef from Vis–NIR Spectra

Prediction models were developed to predict the ultimate pH (48 h post-mortem) of LTL muscles from the spectra collected from neck and rump on the day of slaughtering (1 h and 2 h post-mortem), at quartering time (24 h and 25 h post-mortem) and from LTL muscles (48 h and 49 h post-mortem). The results of prediction of the ultimate pH of beef samples, including the coefficient of determination of calibration (R^2C), the root-mean-square error of calibration (RMSEC), the coefficient of determination of cross-validation (R^2CV) and the root-mean-square error of cross validation (RMSECV), are presented in Table 2. Poor predictions were observed from the spectra collected from beef neck (1 h post-mortem: $R^2CV = 0.22$ and RMSECV = 0.18; 2 h post-mortem: $R^2CV = 0.16$ and RMSECV = 0.21) and rump (1 h post-mortem: $R^2CV = 0.04$ and RMSECV = 0.17; 2 h post-mortem: $R^2CV = 0.23$ and RMSECV = 0.20), but good predictions were recorded from the spectra collected at the time of quartering (24 h post-mortem: $R^2CV = 0.66$ and RMSECV = 0.15). It is interesting to note that when the spectral data collected at the time of quartering (24 h post-mortem) were subjected to baseline correction and SNV correction, the prediction results improved significantly ($R^2CV = 0.91$ and RMSECV = 0.17). However, these spectral corrections did not bring any improvement when applied to the spectral data collected from neck and rump (1 h and 2 h post-mortem). Vis–NIR spectra collected from LTL muscles showed good prediction of the ultimate pH (48 h post-mortem: $R^2CV = 0.67$ and RMSECV = 0.11; 49 h post-mortem: $R^2CV = 0.73$ and RMSECV = 0.11). When baseline and SNV corrections were applied to the spectral data collected from LTL muscle, they improved the PLS model ($R^2CV = 0.96$ and RMSECV = 0.25). A previous study also showed that the prediction of pH is possible using Vis–NIR spectroscopy, despite the narrow ultimate pH range of the sample. The findings of the current investigation are in line with the work of De Marchi [27], who found similar results while developing prediction models of beef quality traits using Vis–NIR spectroscopy. These results demonstrate that the prediction of the ultimate pH is possible in the meat industry immediately after quartering or for LTL muscles using Vis–NIR spectra.

Table 2. Prediction of pH ultimate of beef samples using Vis–NIR spectra.

pH Ultimate	Math Treatment	n	F	R^2C	RMSEC	R^2CV	RMSECV
Neck 1 h-PM	Log (1/R)	357	1	0.13	0.18	0.22	0.18
Neck2 h-PM	Log (1/R)	153	4	0.22	0.20	0.16	0.21
Rump 1 h-PM	Log (1/R)	358	6	0.11	0.16	0.04	0.17
Rump 2 h-PM	Log (1/R)	153	4	0.30	0.19	0.23	0.20
Quartering 24 h-PM	Log (1/R)	361	4	0.74	0.13	0.66	0.15
Quartering 24 h-PM	BS + SNV	361	6	0.92	0.34	0.91	0.17
Quartering 25 h-PM	Log (1/R)	366	9	0.36	0.14	0.22	0.16
LTL muscle 48 h-PM	Log (1/R)	223	6	0.71	0.10	0.67	0.11
LTL muscle 48 h-PM	SNV	223	5	0.96	0.33	0.96	0.25
LTL muscle 49 h-PM	Log (1/R)	191	8	0.80	0.09	0.73	0.11

PM: Post-mortem; n: number of total samples; F: number of partial least-squares (PLS) latent variables; R^2C: coefficient of determination of calibration; RMSEC: root-mean-square error of calibration; R^2CV: coefficient of determination of cross-validation; RMSECV, root-mean-square error of cross validation; Log (1/R): raw absorbance data; BS: baseline correction; SNV: standard normal variate.

3.4. Prediction of Drip Loss

Water represents about 75% of the total fresh weight of meat. Vis–NIR spectral data showed the absorbance of O–H bonds at 1450 and 1940 nm. Drip loss percentage was measured for LTL muscles 48 h post-mortem. The results of the PLS model of drip loss from the spectral data collected from various regions (neck, rump, quartered surface and LTL muscle) of beef carcass during post-slaughtering storage are plotted in Table 3. The results revealed lower coefficients of determination of cross-validation for neck (1 h post-mortem: $R^2CV = 0.20$; 2 h post-mortem: $R^2CV = 0.10$) and rump (1 h post-mortem: $R^2CV = 0.17$; 2 h post-mortem: $R^2CV = 0.09$), whilst high coefficients of determination of cross-validation were recorded for quartered surface (24 h post-mortem: $R^2CV = 0.51$) and LTL muscle (48 h post-mortem: $R^2CV = 0.99$). After applying baseline corrections and SNV, the PLS model results from the spectral data collected during quartering were improved (24 h post-mortem: $R^2CV = 0.82$). The values of RMSECV in all PLS models were between 1.12 and 1.43, which shows that the model is relatively good for the prediction of drip loss from spectral datasets. These findings have confirmed the potential of Vis–NIR spectroscopy to predict the drip loss percentage of beef during pre-rigor storage. The corollaries of the present study are in accordance with the findings of Prieto et al. [35], who investigated the capability of Vis–NIR reflectance spectroscopy for the prediction of the physical, chemical and sensory quality of beef. It was found that the best prediction model was achieved when spectral data were collected 24 h post-mortem at the quartering stage and 48 h post-mortem from the LTL muscle. This might be due to the fact that the actual drip loss of the samples was also analysed on LTL muscles. Therefore, our results indicate that the actual drip loss of LTL muscles can be predicted from the spectra obtained during quartering or from the spectra collected on the surface of LTL muscles.

Table 3. Prediction of drip loss (%) of beef samples using Vis–NIR spectra.

Drip Loss (%)	Math Treatment	n	F	R^2C	RMSEC	R^2CV	RMSECV
Neck 1 h-PM	Log (1/R)	213	4	0.24	1.32	0.20	1.36
Neck 2 h-PM	Log (1/R)	153	4	0.17	1.10	0.10	1.15
Rump 1 h-PM	Log (1/R)	212	3	0.18	1.37	0.17	1.38
Rump 2 h-PM	Log (1/R)	153	4	0.17	1.10	0.09	1.16
Quartering 24 h-PM	Log (1/R)	214	4	0.54	1.22	0.51	1.34
Quartering 24 h-PM	BS + SNV	214	3	0.82	1.44	0.82	1.43
Quartering 25 h-PM	Log (1/R)	219	4	0.22	1.32	0.17	1.37
LTL muscle 48 h-PM	Log (1/R)	224	2	0.99	0.11	0.99	1.12
LTL muscle 49 h-PM	Log (1/R)	192	7	0.43	1.12	0.32	1.24

n: number of total samples; F: number of PLS latent variables; R^2C: coefficient of determination of calibration; RMSEC: root mean square error of calibration; R^2CV: coefficient of determination of cross-validation; RMSECV, root mean square error of cross validation; Log (1/R): raw absorbance data; BS: baseline correction; SNV: standard normal variate.

3.5. Cooking Loss Measurement

The results presented in Table 4 show that the Vis–NIR spectral data provided moderate prediction accuracy for the determination of cooking loss from the PLS model. Poor coefficients of determination of cross-validation were obtained for neck (1 h post-mortem: $R^2CV = 0.19$, RMSECV = 2.70; 2 h post-mortem: $R^2CV = 0.09$, RMSECV = 3.41), rump (1 h post-mortem: $R^2CV = 0.14$, RMSECV = 2.79; 2 h post-mortem: $R^2CV = 0.28$, RMSECV = 2.99), quartered surface (24 h post-mortem: $R^2CV = 0.25$, RMSECV = 2.59; 25 h post-mortem: $R^2CV = 0.22$, RMSECV = 2.64) and LTL muscles (48 h post-mortem: $R^2CV = 0.43$, RMSECV = 2.27; 49 h post-mortem: $R^2CV = 0.45$, RMSECV = 2.23). However, it is interesting to note that a lower value of RMSECV was obtained for the PLS models, showing the potential of spectral data to predict cooking loss. Models were also built with baseline and SNV corrections, but the results did not improve significantly in terms of R^2CV. These spectral corrections had little or no effect on the regression results. According to the literature available, the prediction of cooking loss using Vis–NIR spectroscopy has not been very successful. Our findings are in accord with those of Prieto et al. [35], but nevertheless show greater accuracy than the results presented by De Marchi et al. [36].

3.6. Prediction of Colour Parameters

No significant variations were found in the colour of different analysed samples [37]. The results of the PLS models for spectral data collected from neck, rump, quartered surface and LTL muscles of beef using Vis–NIR spectroscopy and colour parameters are shown in Table 5. Higher RMSECV values for L* colour were obtained for the neck (2.20–2.30) and rump (2.18–2.26) as compared to quartered surface (1.95–1.99) and LTL muscle sections (1.76–1.80). A relatively good prediction was recorded for a* and b* colours, although with lower coefficients of determination of calibration (R^2C = 0.02–0.41 for a*; R^2C = 0.00–0.46 for b*), but good values of RMSECV (0.75–1.82 for a*; 1.38–1.83 for b*). The NIR spectra correspond to the overtones and combinations of fundamental vibrations of C–H, N–H, O–H and S–H functional groups and provide information about the chemical composition of a sample. These findings are in harmony with the work done by Williams and Norris [26], who suggested that spectral measurements along with chemometric models can be used for the prediction of colour in meat. In our research, we were able to find good prediction models based on lower RMSEC and RMSEV for colour measurement from Vis–NIR data. However, lower values for R^2C and R^2CV were obtained, which might be due to colour variation across different areas of the same slice of meat. Ideally, the sample point location in the muscle that is used for colour measurement through traditional methods would be the same as the sample point location that is used for spectral collections; this could be important in developing a regression model. However, our data were averages of three spectra and three colour measures, hence, they should be reflective of the average colour and average spectra of the piece.

Table 4. Prediction of cook loss (%) of beef samples using Vis–NIR spectra.

Cook Loss (%)	Math	n	F	R^2C	RMSEC	R^2CV	RMSECV
Neck 1 h-PM	Log (1/R)	286	8	0.34	2.44	0.19	2.70
Neck 2 h-PM	Log (1/R)	81	2	0.18	3.20	0.09	3.41
Rump 1 h-PM	Log (1/R)	285	8	0.26	2.50	0.14	2.79
Rump 2 h-PM	Log (1/R)	81	8	0.58	2.28	0.28	2.99
Quartering 24 h-PM	Log (1/R)	293	10	0.43	2.25	0.25	2.59
Quartering 25 h-PM	Log (1/R)	293	10	0.41	2.28	0.22	2.64
LTL muscle 48 h-PM	Log (1/R)	151	6	0.51	2.00	0.43	2.27
LTL muscle 49 h-PM	Log (1/R)	150	6	0.53	2.06	0.45	2.23

n: number of total samples; F: number of PLS latent variables; R^2C: coefficient of determination of calibration; RMSEC: root mean square error of calibration; R^2CV: coefficient of determination of cross-validation; RMSECV, root mean square error of cross validation; Log (1/R): raw absorbance data.

Table 5. Prediction of colour parameters of beef samples using Vis–NIR spectra.

Colour L*	Math	n	F	R^2C	RMSEC	R^2CV	RMSECV
Neck 1 h-PM	Log (1/R)	361	6	0.24	2.11	0.18	2.20
Neck 2 h-PM	Log (1/R)	155	1	0.03	2.30	0.01	2.30
Rump 1 h-PM	Log (1/R)	360	7	0.29	2.05	0.20	2.18
Rump 2 h-PM	Log (1/R)	155	9	0.37	1.88	0.11	2.26
Quartering 24 h-PM	Log (1/R)	368	9	0.42	1.85	0.33	1.99
Quartering 2 5 h-PM	Log (1/R)	368	8	0.42	1.84	0.36	1.95
LTL muscle 48 h-PM	Log (1/R)	224	8	0.60	1.55	0.49	1.76
LTL muscle 49h-PM	Log (1/R)	191	7	0.53	1.65	0.44	1.80
Colour a*							
Neck 1h-PM	Log (1/R)	361	8	0.21	1.62	0.08	1.75
Neck 2 h-PM	Log (1/R)	155	5	0.18	1.67	0.11	1.75
Rump 1 h-PM	Log (1/R)	360	1	0.02	1.81	0.09	1.82
Rump 2 h-PM	Log (1/R)	155	1	0.06	1.78	0.06	1.81
Quartering 24 h-PM	Log (1/R)	368	7	0.23	1.58	0.15	1.67
Quartering 25 h-PM	Log (1/R)	368	3	0.09	1.72	0.07	0.75
LTL muscle 48 h-PM	Log (1/R)	224	6	0.41	1.35	0.32	1.46
LTL muscle 49 h-PM	Log (1/R)	191	4	0.31	1.37	0.28	1.41
Colour b*							
Neck 1 h-PM	Log (1/R)	361	1	0.00	1.76	NA	1.79
Neck 2 h-PM	Log (1/R)	155	1	0.01	1.80	NA	1.83
Rump 1 h-PM	Log (1/R)	360	1	0.00	1.76	NA	1.79
Rump 2 h-PM	Log (1/R)	155	1	0.01	1.80	NA	1.82
Quartering 24 h-PM	Log (1/R)	368	8	0.30	1.47	0.19	1.57
Quartering 25 h-PM	Log (1/R)	368	9	0.34	1.43	0.20	1.57
LTL muscle 48 h-PM	Log (1/R)	224	6	0.46	1.30	0.39	1.39
LTL muscle 49 h-PM	Log (1/R)	191	5	0.40	1.33	0.32	1.41

n: number of total samples; F: number of PLS latent variables; R^2C: coefficient of determination of calibration; RMSEC: root mean square error of calibration; R^2CV: coefficient of determination of cross-validation; RMSECV, root mean square error of cross validation; Log (1/R): raw absorbance data.

4. Conclusions

This research work demonstrates the potential of Vis–NIR infrared spectroscopy to predict certain quality parameters of beef. The results are quite promising, showing the potential of this method for online prediction of quality traits in the meat industry. In the case of spectral pre-treatments, baseline correction (BS), SNV and their combinations served as the best pre-treatments, enhancing model accuracy in many cases. Reliable and accurate PLS models were obtained for the ultimate pH for spectra recorded immediately after slaughtering, at quartering time (24 h post-mortem) and from the LTL muscle (48 h post-mortem), whereas for drip loss, the only reliable model was achieved at quartering time (24 h post-mortem). On the other hand, the PLS models for cook loss gave moderate results for the spectra collected on the cut face of the LTL muscle 49 h post-mortem. Similarly, the cut face of the LTL muscle was also observed to be the best regions for colour measurements (48 h and 49 h post-mortem), since the accuracy of the PLS models in this muscle was significantly higher as compared to that for other regions of the carcass and time points. Taken together with findings from our earlier work [24], it can be deduced from the study that Vis–NIR spectroscopy can be a useful tool for the on-site prediction of certain meat quality parameters using different regions of the carcass and different time points, and therefore, this technology has some potential for the beef sector, both for meat management systems of processors and in livestock breeding programs.

Author Contributions: Conceptualization, P.A, T.S, A.C and R.H; Formal analysis, A.S; Funding acquisition, T.S and R.H; Investigation, A.S and J.C; Methodology, A.S, P.A, G.D and R.H; Supervision, R.H; Writing—original draft, A.S; Writing—review & editing, P.A, T.S, J.C and R.H.

Funding: This research is funded by the BreedQuality project (11/SF/311) which is supported by The Irish Department of Food, Agriculture and the Marine (DAFM) under the National Development Plan 2007–2013.

Conflicts of Interest: The authors declare no conflict of interest.

References

1. Liang, R.; Zhu, H.; Mao, Y.; Zhang, Y.; Zhu, L.; Cornforth, D.; Wang, R.; Meng, X.; Luo, X. Tenderness and sensory attributes of the longissimus lumborum muscles with different quality grades from Chinese fattened yellow crossbred steers. *Meat Sci.* **2016**, *112*, 52–57. [CrossRef] [PubMed]

2. Troy, D.; Kerry, J. Consumer perception and the role of science in the meat industry. *Meat Sci.* **2010**, *86*, 214–226. [CrossRef] [PubMed]

3. Di Luca, A.; Elia, G.; Hamill, R.; Mullen, A.M. 2D DIGE proteomic analysis of early post mortem muscle exudate highlights the importance of the stress response for improved water-holding capacity of fresh pork meat. *Proteomics* **2013**, *13*, 1528–1544. [CrossRef] [PubMed]

4. Dransfield, E.; Martin, J.-F.; Bauchart, D.; Abouelkaram, S.; Lepetit, J.; Culioli, J.; Jurie, C.; Picard, B. Meat quality and composition of three muscles from French cull cows and young bulls. *Anim. Sci.* **2003**, *76*, 387–399. [CrossRef]

5. Ripoll, G.; Alberti, P.; Panea, B.; Olleta, J.; Sanudo, C. Near-infrared reflectance spectroscopy for predicting chemical, instrumental and sensory quality of beef. *Meat Sci.* **2008**, *80*, 697–702. [CrossRef]

6. Xiaobo, Z.; Xiaowei, H.; Povey, M.J.W. Non-invasive sensing for food reassurance. *Analyst* **2016**, *141*, 1587–1610. [CrossRef]

7. Sahar, A.; Dufour, E. Classification and characterization of beef muscles using front-face fluorescence spectroscopy. *Meat Sci.* **2015**, *100*, 69–72. [CrossRef]

8. Nunes, K.M.; Andrade, M.V.O.; Filho, A.M.S.; Lasmar, M.C.; Sena, M.M. Detection and characterisation of frauds in bovine meat in natura by non-meat ingredient additions using data fusion of chemical parameters and ATR-FTIR spectroscopy. *Food Chem.* **2016**, *205*, 14–22. [CrossRef]

9. Rahman, U.U.; Sahar, A.; Pasha, I.; Rahman, S.U.; Ishaq, A. Assessing the capability of Fourier transform infrared spectroscopy in tandem with chemometric analysis for predicting poultry meat spoilage. *PeerJ* **2018**, *6*, e5376. [CrossRef]

10. Lucarini, M.; Durazzo, A.; Del Pulgar, J.S.; Gabrielli, P.; Lombardi-Boccia, G. Determination of fatty acid content in meat and meat products: The FTIR-ATR approach. *Food Chem.* **2018**, *267*, 223–230. [CrossRef]

11. Kang, Z.-L.; Li, X.; He, H.-J.; Ma, H.-J.; Song, Z.-J. Structural changes evaluation with Raman spectroscopy in meat batters prepared by different processes. *J. Food Sci. Technol.* **2017**, *54*, 2852–2860. [CrossRef] [PubMed]

12. Fowler, S.M.; Schmidt, H.; Van De Ven, R.; Hopkins, D.L. Preliminary investigation of the use of Raman spectroscopy to predict meat and eating quality traits of beef loins. *Meat Sci.* **2018**, *138*, 53–58. [CrossRef] [PubMed]

13. Xiong, Z.; Sun, D.W.; Pu, H.; Gao, W.; Dai, Q. Applications of emerging imaging techniques for meat quality and safety detection and evaluation: A review. *Crit. Rev. Food Sci. Nutr.* **2017**, *57*, 755–768. [CrossRef] [PubMed]

14. Sahar, A.; Boubellouta, T.; Lepetit, J.; Dufour, E. Front-face fluorescence spectroscopy as a tool to classify seven bovine muscles according to their chemical and rheological characteristics. *Meat Sci.* **2009**, *83*, 672–677. [CrossRef] [PubMed]

15. Allais, I.; Viaud, C.; Pierre, A.; Dufour, E. A rapid method based on front-face fluorescence spectroscopy for the monitoring of the texture of meat emulsions and frankfurters. *Meat Sci.* **2004**, *67*, 219–229. [CrossRef]

16. Aït-Kaddour, A.; Thomas, A.; Mardon, J.; Jacquot, S.; Ferlay, A.; Gruffat, D. Potential of fluorescence spectroscopy to predict fatty acid composition of beef. *Meat Sci.* **2016**, *113*, 124–131. [CrossRef]

17. Aït-Kaddour, A.; Loudiyi, M.; Ferlay, A.; Gruffat, D. Performance of fluorescence spectroscopy for beef meat authentication: Effect of excitation mode and discriminant algorithms. *Meat Sci.* **2018**, *137*, 58–66. [CrossRef]

18. Pieszczek, L.; Czarnik-Matusewicz, H.; Daszykowski, M. Identification of ground meat species using near-infrared spectroscopy and class modeling techniques—Aspects of optimization and validation using a one-class classification model. *Meat Sci.* **2018**, *139*, 15–24. [CrossRef]

19. Kamruzzaman, M.; Elmasry, G.; Sun, D.-W.; Allen, P. Non-destructive prediction and visualization of chemical composition in lamb meat using NIR hyperspectral imaging and multivariate regression. *Innov. Food Sci. Emerg. Technol.* **2012**, *16*, 218–226. [CrossRef]

20. Lohumi, S.; Lee, S.; Lee, H.; Kim, M.S.; Lee, W.-H.; Cho, B.-K. Application of hyperspectral imaging for characterization of intramuscular fat distribution in beef. *Infrared Phys. Technol.* **2016**, *74*, 1–10. [CrossRef]

21. Kamruzzaman, M.; Makino, Y.; Oshita, S. Hyperspectral imaging for real-time monitoring of water holding capacity in red meat. *LWT* **2016**, *66*, 685–691. [CrossRef]

22. Zhao, M.; Esquerre, C.; Downey, G.; O'Donnell, C.P.; Fernandez, C.A.E. Process analytical technologies for fat and moisture determination in ground beef—A comparison of guided microwave spectroscopy and near infrared hyperspectral imaging. *Food Control.* **2017**, *73*, 1082–1094. [CrossRef]

23. Rady, A.; Adedeji, A. Assessing different processed meats for adulterants using visible-near-infrared spectroscopy. *Meat Sci.* **2018**, *136*, 59–67. [CrossRef] [PubMed]

24. Cafferky, J.; Sweeney, T.; Allen, P.; Sahar, A.; Downey, G.; Cromie, A.; Hamill, R.M. Investigating the use of visible and near infrared spectroscopy to predict sensory and texture attributes of beef *M. longissimus thoracis et lumborum. Meat Sci.* **2019**. [CrossRef] [PubMed]

25. Honikel, K.O.; Hamm, R. Measurement of water-holding capacity and juiciness. In *Quality Attributes and their Measurement in Meat, Poultry and Fish Products*; Springer Science and Business Media LLC: Boston, MA, USA, 1994; pp. 125–161.

26. Williams, P.C.; Norris, K. *Near Infrared Technology in the Agricultural and Food Industries*, 2nd ed.; American Association of Cereal Chemists, Inc.: St. Paul, MN, USA, 1987.

27. De Marchi, M. On-line prediction of beef quality traits using near infrared spectroscopy. *Meat Sci.* **2013**, *94*, 455–460. [CrossRef] [PubMed]

28. Andrés, S.; Silva, A.; Soares-Pereira, A.; Martins, C.; Bruno-Soares, A.; Murray, I.; Silva, J. The use of visible and near infrared reflectance spectroscopy to predict beef *M. longissimus thoracis et lumborum* quality attributes. *Meat Sci.* **2008**, *78*, 217–224. [CrossRef] [PubMed]

29. Cozzolino, D.; Murray, I.; Paterson, R.; Scaife, J.R. Visible and near infrared reflectance spectroscopy for the determination of moisture, fat and protein in chicken breast and thigh muscle. *J. Near Infrared Spec.* **1996**, *4*, 213–223. [CrossRef]

30. Kuswandi, B.; Nurfawaidi, A. On-package dual sensors label based on pH indicators for real-time monitoring of beef freshness. *Food Control* **2017**, *82*, 91–100. [CrossRef]

31. Page, J.K.; Wulf, D.M.; Schwotzer, T.R. A survey of beef muscle color and pH. *J. Anim. Sci.* **2001**, *79*, 678–687. [CrossRef]

Foods **2019**, *8*, 525

32. Prieto, N.; Andrés, S.; Giráldez, F.; Mantecon, A.R.; Lavin, P. Ability of near infrared reflectance spectroscopy (NIRS) to estimate physical parameters of adult steers (oxen) and young cattle meat samples. *Meat Sci.* **2008**, *79*, 692–699. [CrossRef]

33. Battagin, M.; Zanetti, E.; Pulici, C.; Cassandro, M.; De Marchi, M.; Penasa, M. Feasibility of the direct application of near-infrared reflectance spectroscopy on intact chicken breasts to predict meat color and physical traits. *Poult. Sci.* **2011**, *90*, 1594–1599.

34. Liu, Y.; Chen, Y.-R. Analysis of visible reflectance spectra of stored, cooked and diseased chicken meats. *Meat Sci.* **2001**, *58*, 395–401. [CrossRef]

35. Prieto, N.; Ross, D.; Navajas, E.; Nute, G.; Richardson, R.; Hyslop, J.; Simm, G.; Roehe, R. On-line application of visible and near infrared reflectance spectroscopy to predict chemical–physical and sensory characteristics of beef quality. *Meat Sci.* **2009**, *83*, 96–103. [CrossRef] [PubMed]

36. De Marchi, M.; Berzaghi, P.; Boukha, A.; Mirisola, M.; Galol, L.; Gallo, L. Use of near infrared spectroscopy for assessment of beef quality traits. *Ital. J. Anim. Sci.* **2007**, *6*, 421–423. [CrossRef]

37. Cafferky, J.; Hamill, R.M.; Allen, P.; O'Doherty, J.V.; Cromie, A.; Sweeney, T. Effect of Breed and Gender on Meat Quality of *M. longissimus thoracis et lumborum* Muscle from Crossbred Beef Bulls and Steers. *Foods* **2019**, *8*, 173. [CrossRef] [PubMed]

 *foods*

Article

An Original Methodology for the Selection of Biomarkers of Tenderness in Five Different Muscles

Marie-Pierre Ellies-Oury [1,†], Hadrien Lorenzo [2,†], Christophe Denoyelle [3], Jérôme Saracco [4] and Brigitte Picard [1,*]

[1] Université Clermont Auvergne, INRA, VetAgro Sup, UMR Herbivores, F-63122 Saint-Genès-Champanelle, France; marie-pierre.ellies@inra.fr

[2] Univiversité de Bordeaux, Inria BSO, Inserm U1219 Bordeaux Population Health Research Center, SISTM Team, 33800 Bordeaux, France; Hadrien.Lorenzo@u-bordeaux.fr

[3] Service Qualite des Carcasses et des Viandes, Institut de l'Elevage, 69007 Lyon, France; Christophe.Denoyelle@idele.fr

[4] Inria BSO, CQFD Team, CNRS UMR5251 Institut Mathématiques de Bordeaux, ENSC Bordeaux INP, F-33400 Talence, France; jerome.saracco@math.u-bordeaux1.fr

* Correspondence: brigitte.picard@inra.fr; Tel.: +33-5-57-35-38-70

† The authors contributed equally.

Received: 24 April 2019; Accepted: 8 June 2019; Published: 11 June 2019

Abstract: For several years, studies conducted for discovering tenderness biomarkers have proposed a list of 20 candidates. The aim of the present work was to develop an innovative methodology to select the most predictive among this list. The relative abundance of the proteins was evaluated on five muscles of 10 Holstein cows: *gluteobiceps, semimembranosus, semitendinosus, Triceps brachii* and *Vastus lateralis*. To select the most predictive biomarkers, a multi-block model was used: The Data-Driven Sparse Partial Least Square. *Semimembranosus* and *Vastus lateralis* muscles tenderness could be well predicted (R^2 = 0.95 and 0.94 respectively) with a total of 7 out of the 5 times 20 biomarkers analyzed. An original result is that the predictive proteins were the same for these two muscles: µ-calpain, m-calpain, h2afx and Hsp40 measured in m. *gluteobiceps* and µ-calpain, m-calpain and Hsp70-8 measured in m. *Triceps brachii*. Thus, this method is well adapted to this set of data, making it possible to propose robust candidate biomarkers of tenderness that need to be validated on a larger population.

Keywords: predictive model; tenderness; meat; biomarker; calpain; h2afx

1. Introduction

As emphasized by many authors, tenderness is considered the most important qualitative characteristic of meat. Tenderness is also a highly variable characteristic, and this wide variability appears to be a significant reason for consumer dissatisfaction and reduction in beef consumption. Therefore, tenderness inconsistency is a priority issue for the meat industry [1].

Tenderness can be evaluated by direct methods, such as instrumental or sensorial assessments, or by indirect methods, using muscular characteristics as tenderness predictors. Sensory methods are expensive, difficult to organize and time consuming [2]. In addition, these methods are invasive, and cannot be performed early enough to allow carcasses to be adapted to markets according to their level of tenderness. Thus, there is a need for method that is available early post mortem to estimate the tenderness potential of each carcass, and among the possible methods, the abundance of certain proteins is of particular interest for predicting tenderness [3,4]. Previous works reviewed by Picard et al. [5] made it possible to identify candidate biomarkers of tenderness. They belong to numerous biological pathways: heat shock proteins, oxidative and glycolytic metabolism, oxidative stress, muscle structure and contraction and proteolysis.

Predicting the overall tenderness of the whole carcass with a reduced number of indicators could be of interest for the beef meat chain. Indeed, with such information, retailers would be able to guide meat samples either to the traditional butchery circuits (if they are of good quality) or to the boning circuit (for those with a poor level of quality), thereby meeting consumer expectations. Therefore, it is interesting to investigate whether biomarkers of one muscle could predict the tenderness of another muscle of the carcass, and thus to evaluate the possibility of predicting the tenderness of a whole carcass by sampling only a reduced number of muscles.

Thus, the aim of the present work was to identify, from among a pool of 20 biomarkers of interest measured in five different muscles, whether a combination of biomarkers could predict the tenderness of each of the five muscles. Therefore, a specific statistical methodology adapted to a low number of samples with a large number of observations [6,7] was tested.

To investigate this question, we looked at five muscles that have been described in the literature as being muscles with different levels of tenderness. These were classified, according the Warner-Bratzler measurements, as tender (3.2 < Warner-Bratzler Shear Force < 3.9 kg; *Triceps brachii* (TB) or intermediate (3.9 < Warner-Bratzler Shear Force < 4.6 kg; *gluteobiceps* (GB), *Vastus lateralis* (VL), *semimembranosus* (SM), *semitendinosus* (ST)) according to [8]. Indeed, data in the literature evaluating the correlations between the sensory tenderness of each of these five muscles and the "carcass sensory tenderness value" showed significant correlation coefficients ($p < 0.0001$) going from 0.75 for the ST muscle, to 0.81 to the VL muscle [9]. According to the literature, meat tenderness variation among muscles of the same carcass might be explained by animal genetics, feeding, handling or slaughter process [10], but also by muscle characteristics. Indeed, muscle background toughness is mainly determined by its organization and amount of connective tissue. This property is also influenced by the level of intramuscular fat, which is known to be highly variable across the muscles [11]. Then, the toughening and tenderization phases occur during postmortem storage of meat, as a result of sarcomere shortening during rigor development. The degree of contraction at which a muscle enters the state of rigor mortis is highly variable among different muscles within the carcass [12]. Thus, it might be supposed that the five muscles studied represent a diversity of physicochemical and muscular characteristics, which could be considered to be representative of the whole carcass.

2. Material and Methods

2.1. Animals

This study was conducted on 10 Holstein cows slaughtered between 29 and 90 months of age (56.5 ± 18.3 months on average) in a commercial slaughterhouse. The animals were slaughtered at an average carcass weight of 319 kg (±22 kg) and their carcasses were classified as 3 for fat score (European grade with a scale going from 1 to 5), with a conformation score going from P= to O-.

The slaughter was made in compliance with current ethical guidelines for animal welfare.

2.2. Muscle Sampling

In this experiment, we selected five muscles, which were excised for each animal and sampled for further analysis. A standardization of the sampling procedures was adopted for each muscle with a clearly identified sample location. The location of each muscle on the carcass can be seen in Figure 1.

Samples for biochemical (biomarkers and myosin heavy chains proportions) analysis were collected 15 min after slaughter, cut into small fragments, frozen in liquid nitrogen and packaged at −80 °C before grinding. Samples intended for sensory analysis were collected 24 h post mortem, and then they were vacuum-packaged and chilled for 7 days at +4 °C for ageing.

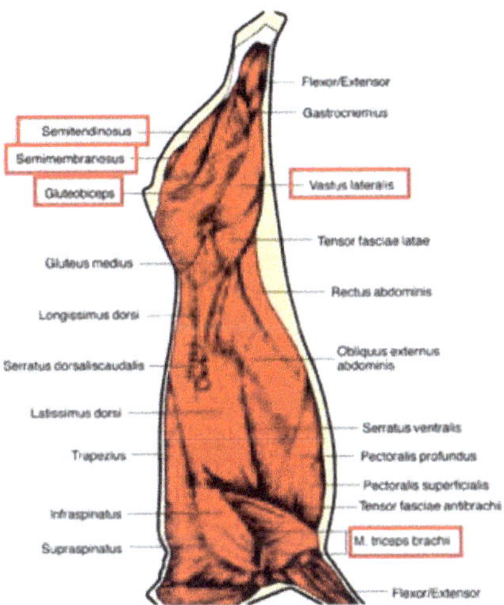

Figure 1. Location of the 5 muscles studied, the names of which are framed in red.

2.3. Biomarker Analysis

Total protein extractions were performed according to Bouley et al. [13]. The protein concentration was determined by spectrophotometry with the Bradford assay [14]. The evaluation of the relative abundance of protein biomarkers was realized by dot blot using antibodies previously validated as described in Guillemin et al. [15] by western-blotting (Table 1; [16]).

Table 1. Gene names and protein names.

Biological Functions	Protein Names	Gene Names
Heat Shock proteins	αB-crystalin	CRYAB
	Hsp20	HspB6
	Hsp27	HspB1
	Hsp40	DNAJA1
	Hsp70-8	HspA8
	Hsp70-1b	HspA1B
	GRP75	HAPS9
Oxidative resistance	Dj1	PARK7
	Cis-Peroxiredoxin	PRDX6
	Super-oxide dismutase Cu/Z	SOD1
Structure	Myosin binding protein H	MyBPH
	Myosin light chain 1F (Mylc-1f)	MYL1
	Myosin heavy chain IIx	MyH1
	α-actin	ACT1
Metabolism	β-Enolase	ENO3
	Phosphoglucomutase 1	PGM1
Apoptosis and signaling	Tumor protein p53	TP53
	H2A Histone Family Member X	H2AFX
Proteolysis	m-calpain	CAPN2
	μ-calpain	CAPN1

For dot-blot analysis, 15 μg of protein samples were spotted on a nitrocellulose membrane using Minifold I Dot-Blot apparatus. Dot-Blot membrane was air-dried during 5 min, and saturated with a 10% milk blocking buffer at 37 °C for 20 min. Then a primary antibody specific to the protein studied was incubated at 37 °C for 90 min. The anti-mouse fluorochrome-conjugated LICOR-antibody IRDye 800CW (1 mg/mL) diluted at 1/20,000 in 1% milk blocking buffer was hybridized for 30 min at 37 °C. Membranes were scanned by the Odyssey scanner (LI-COR Biosciences, Lincoln, NE, USA) and analyzed with GenePix (Axon Laboratory, Union City, CA, USA).

2.4. Myosin Heavy Chain Isoforms Proportions

The different types of myosin heavy chain (MyHC) isoforms were determined on the basis of previously determined migration patterns [17] using sodium Dodecyl Sulfate polyacrylamide gel electrophoresis (SDS-PAGE). This protocol was adapted from that of Talmadge and Roy [18]. Myofibrillar proteins were extracted from 200 mg of muscles with a buffer containing 0.5 M NaCl, 20 mM NaPPi, 50 mM Tris, 1 mM EDTA and 1mM DTT according to the protocol described in [19]. Then 5 µg of proteins were loaded in each well on a Mini-Protein II Dual Slab Cell electrophoretic system (Bio-Rad, Hercules, CA, USA). The separating gel consisted of 35% (*w/v*) glycerol, 9% (*w/v*) acrylamide-N,N′-methylenebisacrylamide (Bis) (50:1), 230 mM Tris (pH 8.8), 115 mM glycine, and 0.4% *w/v* SDS and the stacking gel of 47% (*w/v*) glycerol, 6% (*w/v*) acrylamide-Bis (50:1), 110mM Tris (pH 6.8), 6 mM EDTA, and 0.4% (*w/v*) SDS. The electrophoresis was performed at a constant voltage of 70 V for 30 h at 4 °C. At the end of migration, the gels were stained with Coomassie Blue [19] (Figure 2), and relative amounts of the different MyHC isoforms were quantified using ImageQuant TL v2003 software (Amersham Biosciences Corp., Piscataway, NJ, USA).

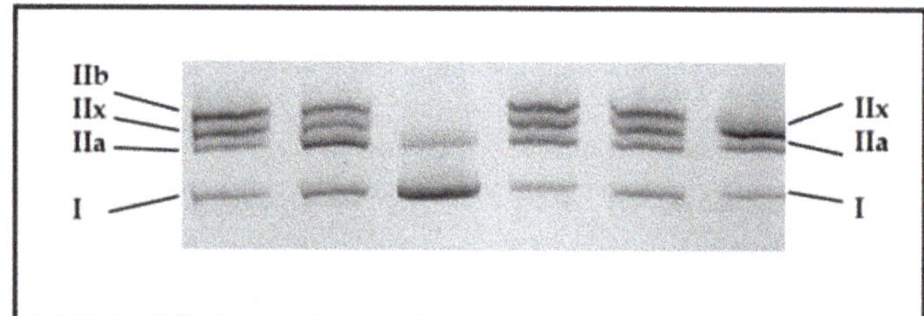

Figure 2. Illustration of the separation of the different myosin heavy chains isoforms depending on their molecular weight in different cattle skeletal muscles according to the protocol of Picard et al. [17].

2.5. Tenderness Evaluation

After ageing, the 5 muscles were trimmed and cut into 1.5 cm thick homogeneous steaks, then vacuum-packaged a second time and frozen at −20 °C until the sensory analysis. Meat in the form of 15 mm steaks was thawed and then cooked for 1 min 45 s in an Infra grill Duo Sofraca set at a temperature of 300 °C. After cooking, the steaks were cut into 20 mm cubes that were served on a plastic plate at an internal temperature of 55 °C.

Sensory assessment was conducted by a 12-member panel selected, trained and controlled according to the XP V09-503 standard. The products were evaluated according to the conventional profile method (QDA) based on NF ISO 13300 [20]. Samples were presented in a monadic mode to the panelists.

The panelists evaluated the cooked samples for tenderness. Each attribute was rated on a 100-point non-graduated scale with a score of zero, on an ascending scale of quality for each attribute, being equivalent to tough, and a score of 100 being equivalent to tender. The sessions were carried out in a sensory analysis room equipped with individual boxes, under artificial non-colored lighting (Institut de l'Elevage, Villers-Bocage, France).

Warner-Bratzler shear force on cooked samples (internal temperature of 55 °C) was measured according to the method of Schakelford et al. [21] with an INSTRON 3343 material testing machine.

For each sample (each muscle of each animal), 10 measurements of shear force were made at different thicknesses of meat cores between 0.8 and 1.2 cm (Institut de l'Elevage, Villers-Bocage, France).

2.6. Statistical Analysis

Firstly, basic statistics, such as variance analysis and mean multiple comparisons with one factor and Pearson correlations were carried out using R-software [22].

Then, a specific statistical methodology adapted to a low number of samples with a large number of observations was tested. This method is a purely geometrical dimension reduction approach, which is suitable for managing a limited sample size with numerous variables stored in blocks. This makes it possible to provide interpretable information from the available data. However, it is important to mention that one should be wary of generalizing the results too quickly. Indeed, the generalization of the results would require a validation on a larger sample. Nevertheless, the proposed method will make it possible to clearly highlight some links that exist between the biomarkers and the force/or the tenderness on the available dataset.

2.6.1. Methodology for Biomarkers Selection

Among the different methodologies that can deal with multi-block structured data sets, the Data-Driven Sparse Partial Least Square (ddsPLS) shows interest for supervised problems whether in the case of regression (i.e., when the response variables are numerical) or classification (i.e., when the response variables are categorical). The ddsPLS approach allows variable selection in the covariate and in the output parts using two parameters, which are: R, the number of components in the model, and L_0, the maximum number of variables to be selected in the final model in the covariate part. These two parameters are fixed by cross-validation according to general supervised learning problem optimization solutions. The block structure can be designed according to many known factors, which are the time or the type of variable chosen for example.

Please note that, even if there are no missing values in the data set analyzed in the present work, the ddsPLS approach was initially developed to deal with missing samples, such as entire rows of missing values in given blocks.

2.6.2. Partial Least Square (PLS)

Let $X \in \mathbb{R}^{n \times p}$ and $Y \in \mathbb{R}^{n \times p}$ be, respectively, the covariate matrix and the response matrix, describing n individuals through p resp. q variables. The *PLS* problem has been introduced by [18]. Its objective is to maximize the norm of the projected variance-covariance matrix $Y^T X \in \mathbb{R}^{q \times p}$ along the principal direction defined through $(u, v) \in \mathbb{R}^p \times \mathbb{R}^q$ with $(Yv)^T Xu \in \mathbb{R}$ being the chosen projection of individuals that must be maximized, under the unity constraints $u^T u = v^T v = 1$. Finally, the optimization problem can be written as $(u, v) = argmax_{u^T u = v^T v = 1}(Yv)^T Xu$.

This makes it possible to define u and v as the weights of the first component of the model for the X part and for the Y part respectively. The nipals algorithm [18] makes it possible to find u and v in the same time.

Those two weights make it possible to define $t = Xu \in \mathbb{R}^n$ and $s = Yv \in \mathbb{R}^n$, with the respective X and Y parts being first component scores of the model. In other words, they represent the individual positions in the dimension found.

Once that first component is defined, other dimensions are successively obtained thanks to the same kind of optimization. The deflation step makes it possible to remove the information from the current dimension to the residual data set. This step ensures that each component is orthogonal to the others.

The ddsPLS algorithm does not use the deflation step, since it has raised many questions for years [19]. It performs directly the R-dimensional Singular Value Decomposition (*SVD*) of the soft-thresholded variance covariance matrix. This solution does not permit the creation of more components than the rank of the soft-thresholded variance covariance matrix, which is itself majorized by the rank of the variance covariance matrix. In the high-dimensional context, this represents the number of variables in the Y part for multivariate regression, or the number of classes minus one in the classification cases.

2.6.3. Data-Driven Sparse Partial Square (ddsPLS)

In the multi-block framework, the covariate part is now described by T blocks (different covariates matrices) $X_t \in \mathbb{R}^{n \times p_t}, t = 1, \ldots, T$ describing the same n individuals. The ddsPLS algorithm, for a couple (R, L_0) chosen by the user, begins with the T SVD of each of the variance-covariance soft-thresholded matrices $S_\lambda\left(\frac{Y^T X_t}{n-1}\right)_+ \cdot S_\lambda : x \to sign(x)(|x| - \lambda)_+$, where $sign$ denotes the sign of a real, $|.|$ denotes the absolute value, and $(.)_+$ the max between its argument and 0. The parameter $\lambda > 0$ is chosen such as the cumulative number of non-null columns in the matrices $S_\lambda\left(\frac{Y^T X_t}{n-1}\right), t \in \{1, \ldots, T\}$ is exactly equal to L_0 and $\forall \partial\lambda \in \mathbb{R}^*_+$ the cumulative number of non-null columns in the matrices $S_{\lambda-\partial\lambda}\left(\frac{Y^T X_t}{n-1}\right)_+, t \in \{1, \ldots, T\}$ is strictly higher than $L_0 + 1$. Let $U_t \in \mathbb{R}^{p_t \times R}, t \in \{1, \ldots, T\}$, denote the respective R-dimensional weights obtained through the corresponding R-dimensional SVD.

The next step makes it possible to gather that information through the super-weights, which are symbolized by $\beta_t \in \mathbb{R}^{R \times R}$. The scaled super-weights $U_t\beta_t \in \mathbb{R}^{n \times R}$ show the effect of one variable of block t on each of the super-components.

Finally, a third step makes it possible to build a regression model based on T regression matrices B_t, such as

$$Y \approx \sum_{t=1}^{T} X_t B_t \in \mathbb{R}^{n \times q}$$

Using the Moore-Penrose pseudo-inverse, which does not imply matrix inversion, is particularly tricky in high dimensions or when $p_t > n$ in the context of regression. Let us mention that a linear discriminant analysis model is built using the *super-components* to predict new individual classes in the context of classification.

2.6.4. Model Selection

As previously mentioned, the ddsPLS approach requires two parameters to be determined by the user: L_0, the maximum number of covariates (in the X part described by the T blocks) to be included in the model, and R, the number of components (calculated in the T underlying SVD), which is the analogous of the number of components in Principal Component Analysis.

Those parameters are determined according to the minimization of a chosen error risk. For the regression framework considered in this paper, the Mean Square Error in Prediction (*MSEP*) is naturally minimized over the response variables, i.e., the five tenderness variables. The error is computed on the Leave-One-Out (LOO) cross-validation predictions, denoted $\hat{y}_{m,i}$, for muscle $m \in \{VL, ST, TB, SM, GB\}$ and for carcass $i \in [1, 10]$:

$$\forall m \in \{VL, ST, TB, SM, GB\}, MSEP(m) = \frac{1}{n} \sum_{i=1}^{10} (y_{m,i} - \hat{y}_{m,i})^2$$

Since it has been chosen to standardize the Y matrix (zero-mean and unit-variance), when MSEP values are lower than 1, this reveals that the cross-validated models are better (on average) than just predicting the average of the selected sample at each iteration of the LOO cross-validation process, and thus the ddsPLS approach succeeds in retrieving links between the blocks of covariates and the corresponding response variables. On the other hand, when MSEP values are larger than 1, this indicates that predicting to the average of the selected sample at each iteration of the LOO cross-validation process provides more accurate predictions than the underlying cross-validated models, and thus the ddsPLS approach fails to retrieve information from the blocks of covariates and the corresponding response variables.

3. Results

3.1. Data Description

The five muscles were characterized by various slow oxidative MyHC I ($p = 0.007$) and fast oxido-glycolytic MyHC IIA ($p < 0.001$) proportions, whereas MyHC IIX (fast glycolytic) proportions were not significantly different among muscles ($p = 0.129$; Table 2).

The m. *semimembranosus* (SM) appeared to be the least slow oxidative muscle, with 10.5% of MyHC I, whereas the m. *gluteobiceps* (GB) and m. *Triceps Brachii* (TB), with proportions of MyHC I twice as large (20.4 and 22.5%, respectively), were the slowest oxidative muscles. The m. *semitendinosus* (ST) might be considered to be the most glycolytic muscle; even if the proportions of MyHC IIX were not significantly different among the five muscles, this muscle had low proportions of both MyHC I and MyHC IIA.

Table 2. Muscular characteristics of the 5 muscles.

Variables	GB	SM	ST	TB	VL	SEM	p
MyHC I (%)	20.4 [b]	10.5 [a]	16.8 [ab]	22.5 [b]	15.2 [ab]	1.15	$p = 0.007$
MyHC IIA (%)	36.4 [bc]	42.9 [c]	25.8 [a]	28.8 [ab]	38.2 [bc]	1.50	$p < 0.001$
MyHC IIX (%)	43.2	46.6	57.4	48.7	46.6	1.80	$p = 0.129$
Tenderness (on 100)	33.7 [a]	42.3 [a]	43.3 [a]	57.9 [b]	41.9 [a]	1.85	$p < 0.001$
Shear force (daN)	11.0 [b]	6.4 [a]	7.4 [a]	5.2 [a]	6.8 [a]	0.39	$p < 0.001$

Different letters indicate a significant difference ($p \leq 0.05$) between the muscles. GB, *gluteobiceps*; SM, *semimembranosus*; ST, *semitendinosus*; TB, *Triceps Brachii*; VL, *Vastus lateralis*; SEM, standard error of the mean.

The TB muscle was the tenderest muscle, according the panelists. In accordance with this result, this muscle also had the lowest Warner-Bratzler shear force (Table 2). At the same time, the GB, which was the toughest muscle according to the mechanical analysis, had the lowest scores for sensory tenderness. These data are related to the significant correlation between tenderness scores and shear force values observed in this study (correlation: -0.53 when considering the five muscles together; $p < 0.001$). However, the correlation between tenderness and shear force is very muscle-dependent: for VL and ST muscles, the correlation is negative and significant (respectively -0.68 and 0.79), whereas for muscles GB, SM and TB, the correlation is non-significant.

Among the 20 analyzed biomarkers, three did not show a difference in relative abundances between the five muscles: αB-crystalin ($p = 0.179$), prdx6 ($p = 0.293$) and α-actin ($p = 0.323$) (Table 3).

The m. *Vastus lateralis* (VL) showed relative abundances significantly different from other muscles for many biomarkers. Indeed, it contained significantly less eno3, mylc1f, and Hsp40, and more m-calpain, pgm1, h2afx, and MyhcIIX than the other four muscles.

However, the VL muscle was close to the ST muscle for many biomarkers, especially Hsp70-1a, grp75 and mybph.

The SM muscle might be distinguished by a relatively high abundance of Hsp27, Hsp40, Hsp70-8, Dj1, sod1, mylc1f and μ-calpain.

The relative abundance of GB and TB muscle biomarkers was close, except for the Hsp20, which was lower in the TB muscle than in the GB muscle.

Table 3. Relative abundance of each biomarker in the five muscles.

Biomarkers	ST	SM	GB	TB	VL	SEM	*p*
Hsp20	95 [a]	150 [b]	152 [b]	99 [a]	68 [a]	7.7	<0.001
Hsp27	101 [ab]	129 [c]	118 [bc]	111 [bc]	86 [a]	3.5	<0.001
Hsp40	118 [b]	162 [c]	112 [ab]	117 [b]	99 [a]	4.0	<0.001
Hsp70-8	105 [a]	141 [bc]	123 [ab]	129 [abc]	156 [c]	4.9	=0.012
Hsp70-1b	85 [a]	188 [c]	155 [bc]	146 [b]	81 [a]	8.5	<0.001
GRP75	119 [a]	153 [b]	147 [b]	155 [b]	122 [a]	4.0	=0.002
αB-crystalin	147 [a]	195 [a]	211 [a]	153 [a]	156 [a]	10.3	=0.179
Dj1	95 [a]	125 [c]	110 [abc]	114 [bc]	101 [ab]	2.8	=0.004
PRDX6	97 [a]	110 [a]	110 [a]	110 [a]	115 [a]	2.7	=0.293
SOD1	79 [ab]	95 [a]	68 [b]	62 [b]	58 [b]	3.7	=0.008
MyBP-H	102 [a]	128 [b]	140 [b]	144 [b]	101 [a]	3.8	<0.001
Mylc-1f	78 [ab]	109 [c]	74 [a]	89 [b]	47 [d]	3.6	<0.001
MyHCIIx	27 [a]	33 [a]	28 [a]	29 [a]	45 [b]	1.3	<0.001
ENO3	103 [a]	134 [b]	121 [b]	133 [b]	83 [c]	3.5	<0.001
h2afx	137 [a]	173 [b]	127 [a]	139 [a]	226 [c]	6.4	<0.001
PGM1	121 [a]	165 [b]	132 [a]	132 [a]	306 [c]	10.8	<0.001
TP53	106 [a]	142 [b]	94 [a]	104 [a]	153 [b]	4.2	<0.001
α-actin	82 [a]	89 [a]	92 [a]	96 [a]	99 [a]	2.7	=0.323
m-calpain	106 [a]	139 [b]	120 [ab]	126 [ab]	214 [c]	6.8	=0.003
μ-calpain	123 [a]	175 [b]	136 [a]	118 [a]	134 [a]	5.3	<0.001

3.2. Links between Biomarkers and Tenderness Intra Muscles

When considering the tenderness of 1 muscle with associated biomarkers in the same muscle, Pearson correlations make it possible to highlight the following significant correlations (Figure 3):

- in the ST muscle, sensorial tenderness was negatively linked to Hsp40, Hsp27, pgm1
- in the SM muscle, sensorial tenderness was negatively linked only to mylc1f
- in the VL, GB and TB muscles, the sensorial tenderness was not significantly correlated with the abundance of any biomarkers (Figure 3).

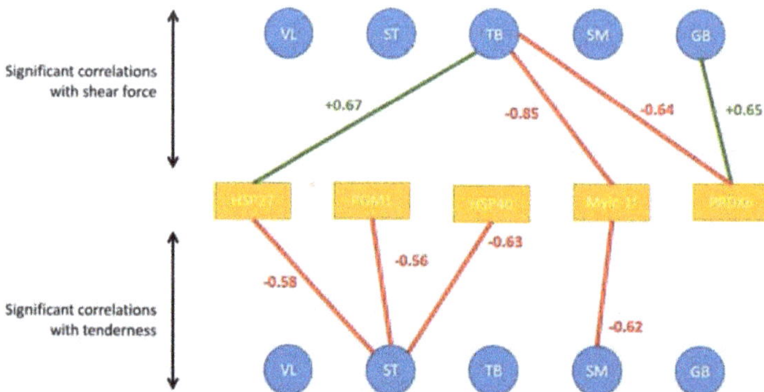

Figure 3. Significant correlations between tenderness/shear force of each muscle (represented by the blue disks) and the abundance of muscular biomarkers of the concerned muscle (represented by the orange rectangles). Red values correspond to negative correlations and green ones to negative correlations.

When considering the correlations between shear force and protein biomarkers in the same muscle, Pearson correlations make it possible to highlight the following significant correlations:

- in the GB muscle, shear force was positively linked to prdx6
- in the TB muscle, shear force was negatively linked to prdx6 and mylc1f, and positively to Hsp27;
- in the other three muscles, the shear force was not significantly correlated with the abundance of an analyzed biomarker.

3.3. Links between Biomarkers and Tenderness among Muscles

When considering the correlation that appeared between the tenderness of one muscle and the biomarker abundance of the 4 other muscles, the highest number of significant correlations were observed for VL and SM tenderness with biomarkers measured in GB and TB muscles (Figure 4).

For biomarkers measured in GB muscle, the coefficient of correlation with VL tenderness were high particularly for h2afx (+0.85), m-calpain (+0.83), μ-calpain (+0.80), Hsp40 (+0.76), Hsp70-8 (+0.65) and for m-calpain (+0.77) and μ-calpain (+0.86), Hsp70-8 (+0.71) with SM tenderness.

For the biomarkers measured in TB muscle, high correlations were observed with VL tenderness for Hsp70-8 (+0.91), m-and μ- calpains (+0.77), as observed with biomarkers measured in GB muscle. For tenderness of SM muscle, high correlations were found with m- and μ-calpains (+0.89 and +0.71, respectively), Hsp70-8 (+0.63) as observed with VL tenderness, and with α-B crystalin (+0.63).

The tenderness of these two muscles VL and SM was especially correlated with biomarkers of GB and TB muscles.

Among these correlations, it might be noted that among biomarkers, μ-calpain, m-calpain, Hsp70-1a, Hsp70-8, tp53, pgm1 and mybph are always positively linked with tenderness, as opposed to MyhcIIx and Hsp40, which can be either positively or negatively linked to meat tenderness. Some biomarkers appeared to be more frequently correlated with tenderness: m-calpain, μ-calpain and Hsp70-8. These three biomarkers, evaluated in the TB and GB muscles, might be considered interesting biomarkers to predict meat tenderness.

Some coefficients of correlations were observed between biomarkers from TB muscle and tenderness of ST, SM and VL muscles (positively). However, biomarkers from TB muscles are mostly correlated with ST and TB shear force (negatively). It should be noted that many correlations can be observed with the tenderness of ST: dj1 (the highest; +0.76), MyhcIIx (in accordance with previous results in this muscle; +0.64), pgm1 (+0.58) and a-actin (+0.59) (Figure 4).

Many correlations were noted between the proteins measured in SM muscle and GB tenderness, whereas for the other muscles, only a few correlations were significant. The highest number of correlations measured in ST muscle was observed for the tenderness of GB muscle (Figure 4).

With regard to shear force, the muscle that is the most correlated with biomarkers is the ST muscle, the shear force of this muscle being especially correlated with biomarkers from VL, TB and GB muscles (Figure 5). The biomarkers that are significantly correlated with shear force are quite different from those found for muscle tenderness even if shear force and tenderness are significantly correlated in ST and VL muscles.

The details of the correlations obtained between the either tenderness of shear force of each of the five muscles and the 20 biomarkers of each muscle have been annexed in Figure 5.

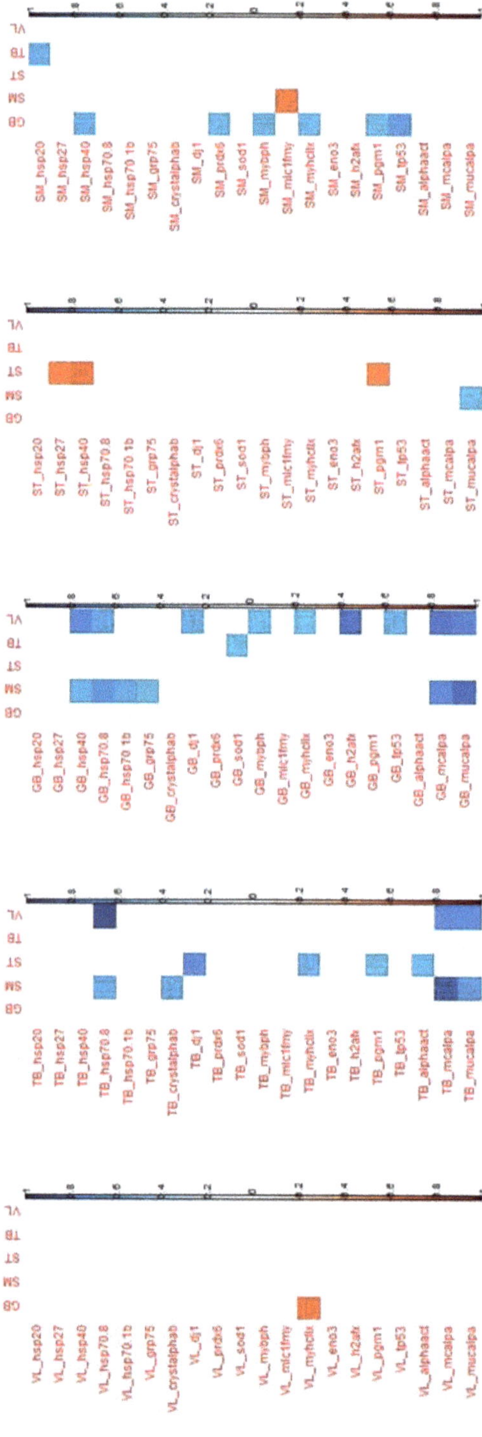

Figure 4. Correlations demonstrated between the tenderness of each of the 5 muscles (in column) and the 20 biomarkers of each muscle (in-line). Only the coefficients that were significant (corresponding to those that were superior or equal to 0.55/inferior or equal to −0.55) are indicated (positive correlations in blue and negative ones in red).

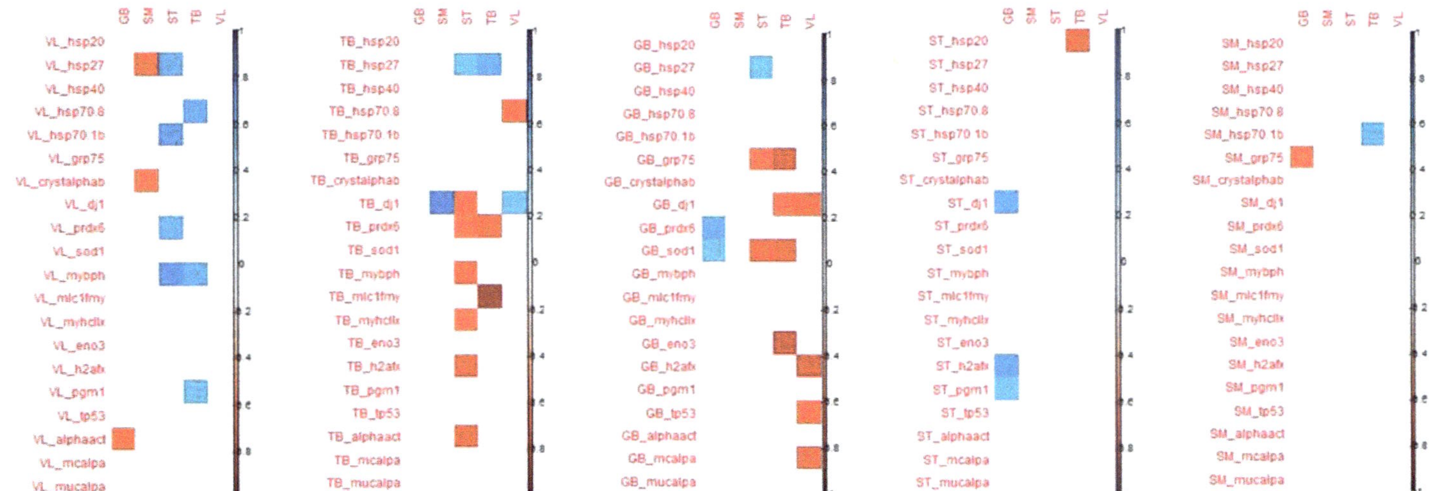

Figure 5. Correlations demonstrated between the shear force of each of the 5 muscles (in columns) and the 20 biomarkers of each muscle (in lines). Only the coefficients that were significant (corresponding to those that were superior or equal to 0.55/inferior or equal to −0.55) are indicated (positive correlations in blue and in negative ones red).

3.4. Selection of Biomarkers the Most Predictive of Tenderness

To predict tenderness, the ddsPLS multi-block approach was used for five blocks of covariates X_t, $t = 1, \ldots, 5$ (each one corresponding to the pool of 20 biomarkers for each of the 5 muscles) and for one block of response variables Y (corresponding to the five tenderness scores, one score for each muscle). Note that in the model, we entered a 6th block, X_6, consisting of performance variables (age/weight). The covariates of X_6 were never selected in tenderness prediction, indicating that the protocol is well balanced, and that the variability of the tenderness is explained neither by weight or by age, nor by the combination of those factors.

Among all the proposed models, we selected the one that minimizes the $MSEP$ (see Figure 6). Thus, the model with $L_0 = 8$ and $R = 2$ was selected.

MSEP versus regularization coefficient L_0 coefficient ddsPLS for R=2

Figure 6. Evolution of the mean square error in prediction for each tenderness muscle and for the average tenderness depending on the maximum number of biomarkers selected in the model. In the present figure, the maximum is fixed at 30 biomarkers out of 100, i.e., 5 time 20 biomarkers.

For the three muscles GB, ST and TB, the model indicated that the best way to predict tenderness is to predict the average value of the corresponding tenderness (i.e., the predicted values are equal to a constant, since there is no link with biomarkers). Hence, for these three muscles, we observe a horizontal alignment of the predicted values of tenderness, as shown in Figure 7.

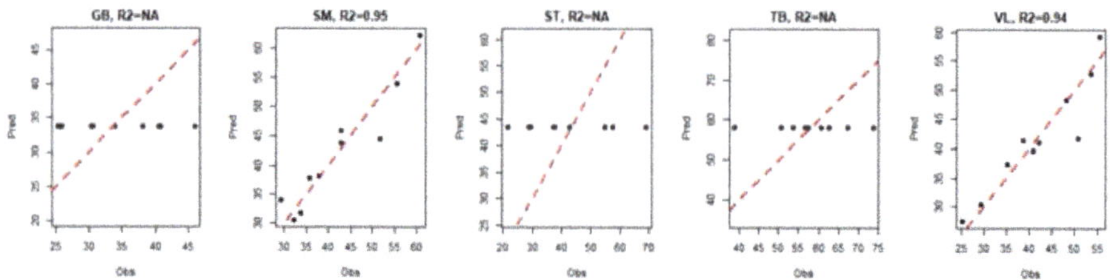

Figure 7. Evolution of predicted tenderness according to observed tenderness in each muscle.

For VL and SM muscles, the seven biomarkers that were selected by the model to predict global tenderness were derived from two different muscles: GB and TB. However, none of the selected biomarkers came from the muscle for which tenderness was predicted (Table 4). Among these seven biomarkers, two are derived from both GB muscle and TB muscle, leading to the number of distinct biomarkers solicited in the model being five.

Table 4. Biomarkers selected of tenderness prediction.

Muscle in Which Biomarkers Were Measured	Biomarkers Selected for the Prediction of Tenderness	
	SM	**VL**
GB	μ-calpain (+) m-calpain (+) h2afx (−) Hsp40 (−)	μ-calpain (+) m-calpain (+) h2afx (+) Hsp40 (+)
SM	-	-
ST	-	-
TB	μ-calpain (−) m-calpain (+) Hsp70-8 (−)	μ-calpain (+) m-calpain (+) Hsp70-8 (+)
VL	-	-
R^2	0.95	0.94

The seven selected biomarkers measured in GB or TB muscles were the same for VL tenderness prediction and for SM tenderness prediction. These proteins belong to the heat shock protein family (Hsp40 and Hsp70-8) or were associated with the proteolysis (m-calpain and μ-calpain) and to the cellular response to DNA damage during oxidative stress (h2afx). Among the biomarkers selected in the model, two were selected in both the GB and the TB muscles: μ-calpain and m-calpain. The three other biomarkers are not the same; the model selected the h2afx and Hsp40 from the GB muscle and the Hsp70-8 from the TB muscle.

It is interesting to note that three biomarkers (see Figure 8) enter with a positive impact on SM and VL tenderness prediction: μ-calpain and m-calpain of the GB muscle and m-calpain of the TB muscle. Whereas the other biomarkers are associated with tenderness prediction either positively (in the VL muscle) or negatively (in the SM muscle), depending on the muscle in which it was measured: h2afx and Hsp40 of the GB muscle, Hsp70-8 of the TB muscle.

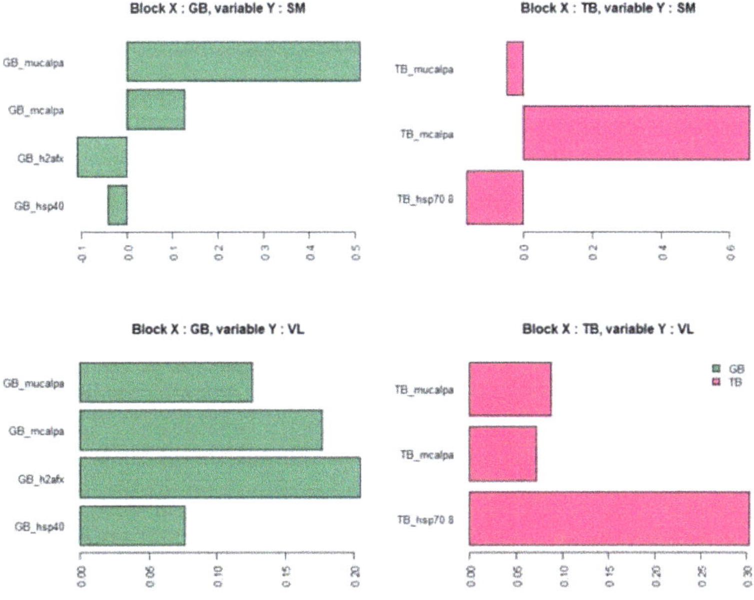

Figure 8. Coefficients associated with each biomarker in tenderness prediction models (representing the relative importance of each selected biomarker).

Thus, the contribution of biomarkers in the prediction of tenderness seems to be modulated according to the muscle.

The same analysis was performed by replacing the tenderness score with the shear force value. It appears that none of the shear force values were correctly predicted by the muscular biomarkers, regardless of the origin of the biomarkers. The model explained no more than 20% of the shear force, except for the TB muscle, where 75% of the shear force could be explained by 2 proteins: the eno3 coming from the GB muscle, and the mylc1f coming from the TB muscle. However, due to the poor quality of the model, it was difficult to draw conclusions on the relevance of these two biomarkers in a definitive way.

4. Discussion

To complete the approach presented here, several models were evaluated for various values of the parameters L_0 and R; the corresponding results are not provided here. One can observe that these models show good stability. More precisely, if we focused in the present paper on the case $L_0 = 8$ and $R = 2$, which makes it possible to minimize the mean of MSEP, we also studied cases where $L_0 \in \{4, 5, 6, 7, 8, 9\}$ (corresponding to the flat area of the MSEP curves in Figure 6). It appears that the biomarkers selected for $L_0 = 4$ are successively supplemented with 1, 2, 3 (and so on) biomarkers when selecting a maximum of 5, 6, 7 (and so on) biomarkers. Moreover, we also tried to explore the models based on a multi-block model with a block of response variables Y built only with the VL and SM tenderness, instead of the tenderness scores of the five muscles. The corresponding models selected the same biomarkers as previously indicated, and returned R^2 values close to the values indicated in Figure 7. These different elements confirm the interest and relevance of the method, accurately predicting meat tenderness. The use of a multi-block model here (based on a linear relationship) allows us to highlight significant links between the tenderness and the abundance of certain biomarkers, even if they were not significantly correlated with tenderness. It is not inconsistent that the combination of several markers (individually little correlated with tenderness) can provide a relevant prediction of tenderness. It can also be noted that several biomarkers, significantly correlated with muscle tenderness, were not selected in the predictive models; this was certainly due to there being redundant information among different biomarkers. The two approaches, correlations and predictions, are thus fully complementary for resulting in relevant conclusions.

The highest number of significant correlations was observed for tenderness of VL and SM muscles with biomarkers measured in GB and TB muscles, mainly for VL tenderness with 12 proteins significantly correlated with tenderness, and 13 in SM muscle. Some coefficients of correlation were high both with SM and VL tenderness: for Hsp70-8, m-calpain and μ-calpain measured in GB and TB muscles. High correlations were found with VL tenderness for h2afx, and Hsp40 measured in GB and TB muscles. These proteins are the same that retained in the equations of prediction of the tenderness of these two muscles with the protein measured in GB muscle. Globally, the highest coefficients were found for VL and SM tenderness for most of the correlations: Hsp20, Hsp40, Hsp70-8, prdx6, dj1, MyBP-H, MyhcIIX, eno3, h2afx, pgm1, m- and μ-calpain. This suggests an important contribution of these proteins in the tenderness of these muscles.

Two models made it possible to predict the tenderness of a muscle (SM and VL) by the biomarkers of two other muscles of the carcasses (GB and TB). The selected proteins are the same in both predicted muscles, which testifies to the importance of these proteins in tenderness construction. Nevertheless, the coefficients associated with certain biomarkers highlight a different hierarchy in the selected biomarkers. Moreover, it might be noted that some coefficients might be reversed from one muscle to another. The fact that tenderness can be predicted only in the two muscles VL and SM can be explained by the fact that these two muscles are very different from the other three in terms of abundances of the analyzed proteins. The two muscles VL and SM are opposite for some proteins such as mylc1f and Hsp40, with the highest abundances being found in SM and the lowest in VL muscle. The abundances of m-calpain and μ-calpain are particular to these two muscles, in comparison to the three others, as m-calpain abundance was the highest in VL and μ-calpain abundance was the highest in SM muscle.

Both correlation and prediction analysis highlighted five proteins among the 20 putatively analyzed as interesting candidate biomarkers of tenderness: m-calpain, μ-calpain, Hsp40, Hsp70-8 and h2afx.

Among the selected biomarkers, the abundance of the m-calpain measured in GB and TB muscles enters the model with a positive coefficient in the prediction of the tenderness of the two muscles (VL and SM) and with a high effect. That highlights a major role of this protein in tenderness independent of the type of muscle. Meanwhile, the role of μ-calpain seems to be more muscle type-dependent.

Calpains (CAPN) are a large family of intracellular Ca^{2+}-dependent cysteine-neutral proteases. The CAPN system is one of the endogenous proteolysis systems, and it plays a major role in meat tenderization [23]. Calpain's actions are mainly due to having two major isoforms: μ-calpain and m-calpain. These two isoforms require different amounts of calcium in vitro, and their names suggest that they help promote the microscopic and milli-molar concentrations of intracellular Ca^{2+} [24]. Among CAPN enzymes responsible for meat tenderization, μ-calpain, which is encoded by the CAPN1 gene, plays a predominant role [25]. As far as we know, calpains can cleave limited myofibrillar proteins such as titin, desmin and vinculin, and contribute to the improvement of tenderness, whereas, high levels of calpastatin are related to decreased proteolysis and increased meat toughness [26,27]. Furthermore, both proteases were found to interact with several proteins belonging to different pathways: homeostasis, structure, glucose metabolism, heat stress, mitochondria, apoptosis. The different implication of calpains in meat tenderness observed in the present study could be explained by a different role of each calpain in tenderization. In accordance with this hypothesis, it was shown in lamb m. *Biceps femoris* that the autolysis of μ-calpain and loss of most of the activity occur within 7 days post-mortem whereas m-calpain activity did not decrease up to 56 days post-mortem. Moreover, very little post-mortem proteolysis is observed in the muscles of μ-calpain knockout mice suggesting its predominant effect on catalysis of proteins during postmortem aging.

Thus, we confirm that these proteases might be considered good biomarkers of beef tenderness, as already shown in the literature [16,28,29]. Moreover, we also confirmed, as previously shown by Guillemin et al. [16], that the contribution of proteolysis to tenderness might be different from one muscle to another. Indeed, these authors indicated that, in the LT muscle, m-calpain and μ-calpain are positively correlated with tenderness, whereas, in the ST muscle, tenderness appears to be positively correlated with μ-calpain, but negatively with m-calpain. These authors indicated that the inverse relation between tenderness and μ-calpain from one muscle to another is coherent with the inverse correlation between contractile and metabolic proteins and tenderness in ST and LT, and thus that muscular particularities might determine calpain contribution to tenderness. Thus, our results seem to confirm this hypothesis to the extent that the contribution of μ-calpain of TB in the construction of tenderness is positive in the VL and negative in the SM. Some studies analyzing a suppression of μ-calpain expression showed an increase in cell death through an activation of apoptotic caspases and an increase in Hsp27, Hsp70, and Hsp90 expressions. These data illustrate a link between the different biomarkers involved in tenderness prediction.

Heat Shock Proteins preserve cellular proteins against denaturation and possible loss of function [30]. A large set of Hsp have been associated with meat tenderness. For example, Guillemin et al. [28] proposed the ratio of Hsp70/small Hsps as a good predictor of muscle tenderness in charolais young bulls. The results of Bernard et al. [31], on the same cattle, found Hsp40 (DNAJA1 gene) to be a positive marker of beef toughness in ST muscle. Hsp40s represent a large protein family that functions as a co-chaperone protein of Hsp70. These two families, Hsp70 and Hsp40, are often co-localized in the same subcellular compartment. Hsp40s function as ATP-independent chaperones that bind non-native polypeptides and protect cells from stress by preventing protein aggregation. The major function of Hsp40 proteins is to regulate ATP-dependent polypeptide binding by Hsp70 protein. Among the Hsp70 family, the Hsp70-8 is known to slow down the process of cellular death and to protect tissues against oxidative stress.

According to Picard et al. [29], there is an inverse relationship between tenderness and proteins from the small Hsp family according to muscle type and breed. It is thus not astonishing that the influence of these proteins on tenderness might be reversed when considering two different muscles such as SM and VL. In accordance with these authors, the results of Guillemin et al. [32] showed different relationships between proteins of the interactome of tenderness and tenderness in two different muscles such as ST and LT. The results of the present study confirm these observations.

The last protein involved in the prediction of VL and SM tenderness in the present study was h2afx. H2afx belongs to the histone h2a family involved in the cellular response to DNA damage, notably during oxidative stress [33]. It has recently been shown that h2ax specifically controls the recruitment of DNA repair proteins to the sites of DNA damage [34]. In eukaryotic cells, one of the first cellular response is the phosphorylation of h2afx within 1 to 3 min after damage. The number of H2AX phosphorylated molecules, γ-h2ax, increases linearly with the severity of the damage and is necessary for the recruitment of other factors to the sites of DNA damage [35]. During stresses, h2afx recruits metabolism enzymes to activate energy generation, enhance protein synthesis, and recruit anti-stress proteins as hspa1a to protect cells from degradation. H2afx was proposed as a potential biomarker of beef tenderness by Guillemin et al. [16], who showed that three proteins, h2afx, sumo4 and tp53, were situated at the crossroads between groups of proteins involved in tenderness, as metabolism, structure, cellular stress, oxidative stress, apoptosis, and calcium signaling proteins. As h2afx modulates different pathways and ensure genome integrity [36], it could have a crucial role in meat tenderness, which is confirmed by the results of the present study. However, why its abundance only as measured in GB muscle is explicative of tenderness of SM and VL muscles, and why the relationships with tenderness are inverted in these two muscles, remain to be understood.

In conclusion, this study made it possible to propose an original statistical methodology well adapted to this type of multi-block data set with several covariates (biomarkers) larger than the number *n* of individuals (carcasses). Moreover, seven proteins could be proposed as candidate biomarkers of tenderness, but the understanding of the biological mechanisms involved according to the muscle considered, remain to be deeply analyzed.

Author Contributions: Conceptualization: M.-P.E.-O., H.L., J.S., B.P.; Methodology: M.-P.E.-O., H.L., J.S.; Software: H.L. and J.S.; Validation: M.-P.E.-O., H.L., J.S., B.P.; Writing—Original draft preparation: M.-P.E.-O., H.L., J.S., B.P.; Writing—Review and editing: M.-P.E.-O., H.L., C.D., J.S., B.P.; Funding acquisition: B.P.

Funding: This research was funded by APIS-GENE, as a part of a French national project called Phenotend (AG-CC-V1).

Acknowledgments: The authors thank all the technicians of Inra and Idele who participated in muscle sampling, protein extraction, dot-blot analysis, sensory evaluation and measure of shear force in particular Nicole Dunoyer from INRA and Valérie Hardit from Idele.

Conflicts of Interest: The authors declare no conflict of interest.

References

1. Destefanis, G.; Brugiapaglia, A.; Barge, M.T.; Dal Molin, E. Relationship between beef consumer tenderness perception and Warner–Bratzler shear force. *Meat Sci.* **2008**, *78*, 153–156. [CrossRef]
2. Platter, W.J.; Tatum, J.D.; Belk, K.E.; Chapman, P.L.; Scanga, J.A.; Smith, G.C. Relationships of consumer sensory ratings, marbling score, and shear force value to consumer acceptance of beef strip loin steaks. *J. Anim. Sci.* **2003**, *81*, 2741–2750. [CrossRef] [PubMed]
3. Cassar-Malek, I.; Picard, B. Expression marker-based strategy to improve beef quality. *Sci. World J.* **2016**, *2016*, 2185323. [CrossRef] [PubMed]
4. Picard, B.; Lebret, B.; Cassar-Malek, I.; Liaubet, L.; Berri, C.; Le Bihan-Duval, E.; Hocquette, J.F.; Renand, G. Recent advances in omic technologies for meat quality management. *Meat sci.* **2015**, *109*, 18–26. [CrossRef] [PubMed]

5. Picard, B.; Gagaoua, M. Proteomic Investigations of Beef Tenderness. In *Proteomics in Food Science*; Elsevier: Amsterdam, The Netherlands, 2017; pp. 177–197.

6. Lorenzo, H.; Saracco, J.; Thiébaut, R. Supervised Learning for Multi-Block Incomplete Data. Available online: https://arxiv.org/abs/1901.04380 (accessed on 14 January 2019).

7. Lorenzo, H.; Razzaq, M.; Morange, P.-E.; Saracco, J.; Trégouët, D.-X.; Thiébaut, R. High-dimensional multi-block analysis of factors associated with thrombin generation potential. In Proceedings of the 32th International Symposium on Computer-Based Medical Systems, Córdoba, Spain, 5–7 June 2019.

8. Belew, J.B.; Brooks, J.C.; McKenna, D.R.; Savell, J.W. Warner–Bratzler shear evaluations of 40 bovine muscles. *Meat Sci.* **2003**, *64*, 507–512. [CrossRef]

9. Crosley, R.I.; Heinz, P.H.; De Bruyn, J.F. The relationship between beef tenderness and age classification of beef carcasses in South Africa. In Proceedings of the 8th Meat Symposium: Meat, Le Needs of the South Africaner Consumer, Pretoria, South Africa, 25 April 1995; pp. 57–66.

10. Warner, R.D.; Greenwood, P.L.; Pethick, D.W.; Ferguson, D.M. Genetic and environmental effects on meat quality. *Meat Sci.* **2010**, *86*, 171–183. [CrossRef]

11. Veiseth-Kent, E.; Pedersen, M.E.; Rønning, S.B.; Rødbotten, R. Can postmortem proteolysis explain tenderness differences in various bovine muscles? *Meat Sci.* **2018**, *137*, 114–122. [CrossRef]

12. Rhee, M.S.; Wheeler, T.L.; Shackelford, S.D.; Koohmaraie, M. Variation in palatability and biochemical traits within and among eleven beef muscles. *J. Anim. Sci.* **2004**, *82*, 534–550. [CrossRef]

13. Bouley, J.; Chambon, C.; Picard, B. Mapping of bovine skeletal muscle proteins using two-dimensional gel electrophoresis and mass spectrometry. *Proteomics* **2004**, *4*, 1811–1824. [CrossRef]

14. Bradford, M.M. A rapid and sensitive method for the quantitation of microgram quantities of protein utilizing the principle of protein-dye binding. *Anal. Biochem.* **1976**, *72*, 248–254. [CrossRef]

15. Guillemin, N.; Meunier, B.; Jurie, C.; Cassar-Malek, I.; Hocquette, J.F.; Leveziel, H.; Picard, B. Validation of a dot-blot quantitative technique for large scale analysis of beef tenderness biomarkers. *J. Physiol. Pharmacol.* **2009**, *60*, 91–97.

16. Guillemin, N.; Bonnet, M.; Jurie, C.; Picard, B. Functional analysis of beef tenderness. *J. Proteom.* **2011**, *75*, 352–365. [CrossRef]

17. Picard, B.; Barboiron, C.; Chadeyron, D.; Jurie, C. Protocol for high-resolution electrophoresis separation of myosin heavy chain isoforms in bovine skeletal muscle. *Electrophoresis* **2011**, *32*, 1804–1806. [CrossRef] [PubMed]

18. Talmadge, R.J.; Roy, R.R. Electrophoretic separation of rat skeletal muscle myosin heavy-chain isoforms. *J. Appl. Physiol.* **1993**, *75*, 2337–2340. [CrossRef]

19. Picard, B.; Barboiron, C.; Duris, M.P.; Gagniere, H.; Jurie, C.; Geay, Y. Electrophoretic separation of bovine muscle myosin heavy chain isoforms. *Meat Sci.* **1999**, *53*, 1–7. [CrossRef]

20. *NF ISO 13300. Sensory Analysis—General Guidance for the Staff of a Sensory Evaluation Laboratory*; International Organization for Standardization: Geneva, Switzerland, 2006.

21. Shackelford, S.D.; Wheeler, T.L.; Koohmaraie, M. Technical Note: Use of belt grill cookery and slice shear force for assessment of pork longissimus tenderness1. *J. Anim. Sci.* **2004**, *82*, 238–241. [CrossRef] [PubMed]

22. R: The R Project for Statistical Computing. Available online: https://www.r-project.org/ (accessed on 5 April 2019).

23. Lian, T.; Wang, L.; Liu, Y. A new insight into the role of calpains in post-mortem meat tenderization in domestic animals: A review. *Asian Aus. J. Anim. Sci.* **2013**, *26*, 443. [CrossRef]

24. Glass, J.D.; Culver, D.G.; Levey, A.I.; Nash, N.R. Very early activation of m-calpain in peripheral nerve during Wallerian degeneration. *J. Neurol. Sci.* **2002**, *196*, 9–20. [CrossRef]

25. Koohmaraie, M. Biochemical factors regulating the toughening and tenderization processes of meat. *Meat Sci.* **1996**, *43*, 193–201. [CrossRef]

26. Kemp, C.M.; Sensky, P.L.; Bardsley, R.G.; Buttery, P.J.; Parr, T. Tenderness—An enzymatic view. *Meat Sci.* **2010**, *84*, 248–256. [CrossRef]

27. Kent, M.P.; Spencer, M.J.; Koohmaraie, M. Postmortem proteolysis is reduced in transgenic mice overexpressing calpastatin. *J. Anim. Sci.* **2004**, *82*, 794–801. [CrossRef] [PubMed]

28. Guillemin, N.; Jurie, C.; Renand, G.; Hocquette, J.-F.; Micol, D.; Lepetit, J.; Picard, B. Different phenotypic and proteomic markers explain variability of beef tenderness across muscles. *Int. J. Biol.* **2012**, *4*, 26.

29. Picard, B.; Gagaoua, M.; Micol, D.; Cassar-Malek, I.; Hocquette, J.-F.; Terlouw, C.E. Inverse relationships between biomarkers and beef tenderness according to contractile and metabolic properties of the muscle. *J. Agric. Food Chem.* **2014**, *62*, 9808–9818. [CrossRef]

30. Goldberg, A.L. Protein degradation and protection against misfolded or damaged proteins. *Nature* **2003**, *426*, 895. [CrossRef] [PubMed]

31. Bernard, C.; Cassar-Malek, I.; Le Cunff, M.; Dubroeucq, H.; Renand, G.; Hocquette, J.-F. New indicators of beef sensory quality revealed by expression of specific genes. *J. Agric. Food Chem.* **2007**, *55*, 5229–5237. [CrossRef] [PubMed]

32. Guillemin, N.; Jurie, C.; Cassar-Malek, I.; Hocquette, J.-F.; Renand, G.; Picard, B. Variations in the abundance of 24 protein biomarkers of beef tenderness according to muscle and animal type. *Animal* **2011**, *5*, 885–894. [CrossRef] [PubMed]

33. Paull, T.T.; Rogakou, E.P.; Yamazaki, V.; Kirchgessner, C.U.; Gellert, M.; Bonner, W.M. A critical role for histone H2AX in recruitment of repair factors to nuclear foci after DNA damage. *Curr. Biol.* **2000**, *10*, 886–895. [CrossRef]

34. Stewart, G.S.; Wang, B.; Bignell, C.R..; Taylor, A.M.; Elledge, S.J. MDC1 is a mediator of the mammalian DNA damage checkpoint. *Nature* **2003**, *421*, 961. [CrossRef]

35. Kao, J.; Milano, M.T.; Javaheri, A.; Garofalo, M.C.; Chmura, S.J.; Weichselbaum, R.R.; Kron, S.J. γ-H2AX as a therapeutic target for improving the efficacy of radiation therapy. *Curr. Cancer Drug Targets* **2006**, *6*, 197–205. [CrossRef]

36. Dickey, J.S.; Redon, C.E.; Nakamura, A.J.; Baird, B.J.; Sedelnikova, O.A.; Bonner, W.M. H2AX: Functional roles and potential applications. *Chromosoma* **2009**, *118*, 683–692. [CrossRef]

Article

Beef Tenderness Prediction by a Combination of Statistical Methods: Chemometrics and Supervised Learning to Manage Integrative Farm-To-Meat Continuum Data

Mohammed Gagaoua [1,*], **Valérie Monteils** [1], **Sébastien Couvreur** [2] **and Brigitte Picard** [1,*]

[1] UMR Herbivores, VetAgro Sup, Université Clermont Auvergne, INRA, F-63122 Saint-Genès-Champanelle, France

[2] URSE, Ecole Supérieure d'Agriculture (ESA), Université Bretagne Loire, 55 Rue Rabelais, BP 30748, 49007 Angers, CEDEX, France

* Correspondence: mohammed.gagaoua@inra.fr or gmber2001@yahoo.fr (M.G.); brigitte.picard@inra.fr (B.P.)

Received: 3 July 2019; Accepted: 19 July 2019; Published: 22 July 2019

Abstract: This trial aimed to integrate metadata that spread over farm-to-fork continuum of 110 Protected Designation of Origin (PDO)Maine-Anjou cows and combine two statistical approaches that are chemometrics and supervised learning; to identify the potential predictors of beef tenderness analyzed using the instrumental Warner-Bratzler Shear force (WBSF). Accordingly, 60 variables including WBSF and belonging to 4 levels of the continuum that are farm-slaughterhouse-muscle-meat were analyzed by Partial Least Squares (PLS) and three decision tree methods (C&RT: classification and regression tree; QUEST: quick, unbiased, efficient regression tree and CHAID: Chi-squared Automatic Interaction Detection) to select the driving factors of beef tenderness and propose predictive decision tools. The former method retained 24 variables from 59 to explain 75% of WBSF. Among the 24 variables, six were from farm level, four from slaughterhouse level, 11 were from muscle level which are mostly protein biomarkers, and three were from meat level. The decision trees applied on the variables retained by the PLS model, allowed identifying three WBSF classes (Tender (WBSF $\leq$ 40 N/cm^2), Medium (40 N/cm^2 < WBSF < 45 N/cm^2), and Tough (WBSF $\geq$ 45 N/cm^2)) using CHAID as the best decision tree method. The resultant model yielded an overall predictive accuracy of 69.4% by five splitting variables (total collagen, μ-calpain, fiber area, age of weaning and ultimate pH). Therefore, two decision model rules allow achieving tender meat on PDO Maine-Anjou cows: (i) IF (total collagen < 3.6 μg OH-proline/mg) AND (μ-calpain $\geq$ 169 arbitrary units (AU)) AND (ultimate pH < 5.55) THEN meat was very tender (mean WBSF values = 36.2 N/cm^2, n = 12); or (ii) IF (total collagen < 3.6 μg OH-proline/mg) AND (μ-calpain < 169 AU) AND (age of weaning < 7.75 months) AND (fiber area < 3100 μm^2) THEN meat was tender (mean WBSF values = 39.4 N/cm^2, n = 30).

Keywords: beef tenderness; machine learning; farm-to-fork; carcass; rearing practices; decision trees; cows

---

1. Introduction

Among the eating qualities of meat, tenderness is often reported as one of the main drivers of beef palatability that dictates the overall liking of cooked meat or to make (re)purchasing decision [1–3]. However, it has been reviewed that for consumer confidence, there is need to guarantee consistent and high eating quality of meat [4]. From the large literature, there is a consensus that this is a challenging task to achieve consistent eating quality as meat is biochemically dynamic and susceptible to variation. Indeed, variations in beef tenderness stems from a wide range of factors which are intrinsic and extrinsic and measurable from the farm-to-fork continuum levels [5–8]. The modern beef industry seeks new strategies using the whole or part of these factors to develop management and predictive tools. These tools would provide products of consistent quality that meet consumer expectations, paying specific attention to sensory traits. Accordingly, we recently proposed a holistic approach that considers 4 levels of the farm-to-fork live period of the animals (farm level: rearing factors and animal characteristics, slaughterhouse level: carcass characteristics, muscle level: muscle characteristics and protein biomarkers, meat level: meat quality traits) to sufficiently characterize the driving factors in relation to different desirable qualities of meat, namely tenderness [8,9].

Therefore, we intend to use metadata that spread over this continuum, to identify how carcass and beef qualities can be jointly managed using rearing practices applied during the whole life of the animals or by a combination of proxies that belong to the other levels of the continuum [8]. To achieve this challenging objective, we proposed to implement various statistical strategies to analyze this metadata by defining three main purposes: (i) apply/develop appropriate statistical tools to relate accurately the different elements of the continuum; (ii) determine the most appropriate methods of rearing practices to meet the expectations of the slaughterers; and (iii) provide breeders/slaughterers with decision tools (predictive) for joint management of carcass and meat quality potential [10]. Hence, partial least squares regression (PLS) and decision trees were applied in this work to achieve the fixed objectives on PDO Maine-Anjou cull cows. Overall, the combined statistical techniques used in this trial showed the possibility to propose recommendations that would help make decisions about how joint management of the qualities of carcasses and their produced beef will help reach the targeted market specifications.

2. Materials and Methods

2.1. Experimental Design and Animal Characteristics and Rearing Factors

In this trial, we used the data of the same 110 PDO Maine-Anjou cows from previous experimental designs that are described in details by Gagaoua et al. [11] and Couvreur et al. [12]. The investigated cows are from a cooperative of livestock farmers located in the department of Maine-et-Loire, France. All the animals were collected and slaughtered following the same protocol and in the same commercial slaughterhouse (Elivia, Lion d'Angers, France). This collaboration allowed us collecting information on animals such the feeding regimen during the whole life of the animal as well as the day before slaughter, conditions of transport to the slaughterhouse and duration, conditions of resting period after arrival at the abattoir, conditions of resting with free access to water but food deprived, stunning procedure, as well the conditions of chilling and storing of the carcasses.

The rearing practices of each animal were obtained by a survey carried out by directly interviewing farmers and described in detail by Couvreur et al. [12]. The survey included 16 quantitative and qualitative questions (Table 1) subdivided into two categories:

(i)   Questions related to the finishing period: part of hay, haylage and/or grass in the finishing diet (% *w/w*); daily and global amount of concentrate (kg); fattening duration (days); physical activity (% days out)

(ii) Questions related to animal characteristics: animals with beef or dairy-ability; birth month/season; birth weight (kg); age at weaning (month); duration of the period between the last weaning and the beginning of the finishing period (days); age of first calving; number of calving; suckling value (0–10) and age at slaughter.

Table 1. Average values and variations of the data from the farm level describing the 16 variables of animal characteristics and finishing period [1].

Variables	n	Mean	SD	Min	Max
Birth weight (kg)	100	49.9	4.91	38	66
Month of birth (1–12)	110	-	-	1	12
Genetic type (0: Beef or 1: Dairy)	110	-	-	0	1
Age of weaning (month)	107	7.2	1.07	5	11
Weaning duration [2]	110	8.7	9.41	0	36
Age at first calving (month)	110	32.4	4.09	18	43
Number of calving	110	3	2.05	1	9
Suckling score (0–10)	103	5.9	1.36	3	9
Fattening duration (day)	110	98.6	29.96	37	203
Haylage diet (%)	110	27.8	36.98	0	100
Hay diet (%)	110	48.2	37.39	0	100
Grass diet (%)	110	24	32.1	0	100
Daily concentrate diet (kg)	110	7.7	2.13	2	13
Global concentrate diet (kg)	110	738	244	178	1330
Activity (%)	110	54	46.21	0	100
Age at slaughter (month)	110	67.5	24.79	34	120

[1] These data were obtained for each individual cow following the survey described in the questionnaire conducted by Couvreur et al. [12] including information about the finishing period and animal characteristics. [2] Represent the period between the last weaning and the beginning of the fattening period (days).

2.2. Slaughtering, Carcass Characteristics and Muscle Sampling

All the cows were slaughtered using captive bolt pistol prior to exsanguination. They were dressed following the standard commercial practices in compliance with the French welfare and EU regulations (Council Regulation (EC) No. 1099/2009). The carcasses were not electrically stimulated. We chilled the carcasses during 24 h *p-m* (post mortem) at 2–3 °C. After slaughter, the carcasses were characterized and graded using the European beef grading system (CE 1249/2008). A total of 8 carcass characteristics (Table 2) were recorded: the carcass weight (kg), conformation score (1–15 scale), weight of the 5th ribeye, muscle carcass weight (g) of the 5th rib, fat carcass weight (g) of the 5th rib, fat-to-muscle ratio in the 5th rib (% *w/w*), color score of the carcass (1–5 scale) and tenderness score of the carcass (1–5 scale).

Table 2. Average values and variations of the data from the slaughterhouse level describing the eight carcass characteristics.

Variables	n	Mean	SD	Min	Max
Carcass weight (kg)	110	438.2	36.09	380	553
Conformation score (1–15 scale) [1]	107	7.8	0.82	6	10
5th rib weight (g)	110	3079	638	1793	5640
Muscle carcass weight (g) [2]	110	1882	403	1145	3478
Fat carcass weight (g) [2]	110	582	190	216	1338
Fat-to-muscle ratio in the 5th rib (% *w/w*)	110	31.3	10.17	16	85
Color score of the carcass (1–5) [3]	105	2.9	0.38	2	4
Tenderness score of the carcass (1–5) [4]	105	3.4	0.65	2	5

[1] EUROP classification grid for carcass conformation scores from P− = 1 to E+ = 15. [2] Muscle and fat carcass weights were estimated after dissection of the 5th rib as Gagaoua et al. [7]. The equations used are described in detail in the study by Couvreur et al. [12]. [3] Visual score assessing meat color from 1–5 was evaluated by the same experts familiar with the EUROP grid according to the PDO Maine-Anjou agreement. [4] Palpation of the 5th rib allowed determining on 1–5 scale the tenderness potential of the steaks.

The *Longissimus thoracis* (LT, mixed fast oxido-glycolytic) muscle samples were removed from the right-hand side of each animal carcass 24 h *p-m* from as detailed in Gagaoua et al. [11]. Briefly, from the four parts that were taken, one was frozen in liquid nitrogen and kept at −80 °C until analyzed for muscle biochemistry by the quantification of fiber area, the percentages of myosin heavy chains isoforms (MyHC), the activities of metabolic enzymes describing both the glycolytic and oxidative pathways, biomarkers of beef tenderness quantified by the immunobased Dot-Blot technique. The second part was cut into pieces of 1–2 cm cross-section, vacuum packed and stored at −20 °C until analyzed for intramuscular fat content and intramuscular connective tissue. The third part was used to evaluate meat color coordinates and ultimate pH. The fourth part was cut into 20 mm thick steaks and vacuum packed in sealed plastic bags for 14 days ageing at 4 °C. For these aged meat samples, the steaks were frozen and stored at −20 °C until shear force measurements of tenderness.

2.3. Muscle Characteristics Determination

There were 30 muscle characteristics quantified from the muscle level (Table 3). The parameters corresponded to myosin fibers describing the contractile properties, oxidative and glycolytic metabolic enzyme activities to define the metabolic properties of the muscles; intramuscular connective tissue properties by collagen contents and beef tenderness protein biomarkers by their abundance [6,11].

Table 3. Average values and variations of the data from the muscle level describing the 30 quantified characteristics in *Longissimus thoracis* muscle including protein biomarkers for the 110 cows.

Variables	Mean	SD	Min	Max
a. Contractile properties by myosin fibers characterization				
Fiber area. μm^2	2906	646	1762	5203
MyHC-I, %	31.2	7.37	15.22	69
MyHC-IIa, %	56.6	12.78	23.76	84.78
MyHC-IIx/b, %	12.2	14.03	0	53.91
b. Metabolic properties by metabolic enzyme activities				
LDH ($\mu mol \cdot min^{-1} \cdot g^{-1}$)	1.05	0.33	0.31	2.26
ICDH ($\mu mol \cdot min^{-1} \cdot g^{-1}$)	703	109	491	939
c. Intramuscular connective tissue properties				
Total collagen μg OH-prol·mg^{-1} DM	3.1	0.42	2.08	4.06
Insoluble collagen μg OH-prol·mg^{-1} DM	2.4	0.33	1.61	3.26
Soluble collagen %	20.8	2.94	14.85	26.58
d. Protein biomarkers quantified by Dot-Blot (in arbitrary units)				
Heat shock proteins				
CRYAB	226.4	83.96	59.04	576.89
Hsp20	164.8	45.45	59.84	306.74
Hsp27	79.7	19.83	36.88	134.56
Hsp40	130.5	20.97	96.09	280.56
Hsp70-1A	111.4	24.81	61.29	180.36
Hsp70-1B	120.1	26.16	70.38	187.36
Hsp70-8	184.5	49.43	50.12	432.19
Hsp70-Grp75	144.5	30.5	87.12	213.24
Metabolism				
Enolase 3 (ENO3)	144.3	36.22	78.74	258.12
Phosphoglucomutase 1 (PGM1)	101	27.26	46.88	254.36
Structure				
α-Actin	122.7	40.37	56.99	266.14
Myosin binding protein H (MyBP-H)	90.2	27.49	42.05	184.32
Myosin light chain 1F (MyLC-1F)	63.8	12.91	33.23	91.06
Mysoin heavy chain IIx (MyHC-IIx)	124.9	18.55	80.91	182.28
Oxidative stress				
Superoxide dismutase [Cu-Zn] (SOD1)	101.5	37.92	23.95	167.44
Peroxiredoxin 6 (PRDX6)	106.2	17.41	73.78	163.74
Protein deglycase (DJ1)	90.6	13.9	58.12	146.92
Proteolysis				
μ-calpain	151.7	38.24	75.28	281.08
m-calpain	96.1	12.62	64.69	124.75
Apoptosis and signaling				
Tumor protein p53 (TP53)	118.3	22.31	78.36	175.78
H2A Histone Family Member X (H2AFX)	98.7	19.01	58.72	153.83

For metabolic muscle type, we measured the activities of isocitrate dehydrogenase (ICDH; EC 1.1.1.42) and lactate dehydrogenase (LDH; EC 1.1.1.27) [13]. Both enzymes are representative of main steps of the oxidative and glycolytic pathways, respectively and are routinely used to determine the metabolic types of beef muscles [14].

The contractile properties were determined by the determination of the percentages of myosin heavy chains (MyHC) isoforms using an adequate mini-gel electrophoresis protocol [15]. Controls of bovine muscle containing 3 (MyHC-I, IIa and IIx) or 4 (MyHC-I, IIa, IIx and IIb) muscle fibers were run at the extremities of each gel [16]. Thus, 3 isoforms of MyHC isoforms were quantified. The quantification of the bands revealed no existence of MyHC-IIb isoform in PDO Maine-Anjou breed, therefore only MyHC-I, IIa and IIx isoforms are reported.

The muscle mean cross sectional fiber area for all the animals was determined on 10-μm thick sections cut perpendicular to the muscle fibers with a cryotome [13]. Between 100 and 200 fibers in each of the two different locations in the muscle were used to determine the mean fiber area (in μm^2) by computerized image-analysis.

For total, insoluble collagen and percentage of soluble collagen, we used the frozen muscle. First, it was homogenized in a household cutter, freeze-dried for a period of 48 h before pulverization in a horizontal blade mill. Afterward, it was stored at +4°C in stopper plastic flasks until analyses. For total collagen and following our previously described protocol, about 250 mg of muscle powder were weighed and acid hydrolysed with 10 mL of 6 N HCl overnight at 110 °C in a screw-capped glass tube. The acid hydrolysate was diluted 5 times in 6 N HCl and the subsequent procedure used was as Dubost et al. [17]. For soluble/insoluble collagen, muscle powder was solubilised and hydrolysed according to the same method as for total collagen. For total and insoluble collagen, each sample was weighed and measured in duplicate and data were expressed in mg of hydroxyproline per g of dry matter (mg OH-Pro·g^{-1} DM). From the average values of these parameters and for each sample, the solubility of collagen was calculated as their ratio as following:

$$Soluble\ collagen = \frac{Total\ collagen - Insoluble\ collagen}{Total\ collagen} \times 100\%$$

The relative abundances of the 21 beef tenderness biomarkers were determined as cited above using Dot-Blot [16,18]. The quantified biomarkers belong to 6 different but interacting biological pathways:

1. heat shock proteins (αB-crystallin, Hsp20, Hsp27, Hsp40, Hsp70-1A, Hsp70-1B, Hsp70-8 and Hsp70-Grp75);
2. metabolism (Enolase 3 (ENO3) and Phosphoglucomutase 1 (PGM1));
3. structure (α-actin, Myosin binding protein H (MyBP-H), Myosin light chain 1F (MyLC-1F) and Mysoin heavy chain IIx (MyHC-IIx));
4. oxidative stress (Superoxide dismutase [Cu-Zn] (SOD1), Peroxiredoxin 6 (PRDX6) and Protein deglycase (DJ1));
5. proteolysis (μ-calpain and m-calpain);
6. apoptosis and signaling (Tumor protein p53 (TP53) and H2A Histone (H2AFX)).
7. The conditions retained and suppliers for all primary antibodies dilutions and details of the protocol are exactly the same of our previous work using the same data [11]. The relative protein abundances of the biomarkers were based on the normalized volume and expressed in arbitrary units (A.U).

2.4. Meat Quality Traits

At the meat level, 6 eating quality were evaluated (Table 4).

Table 4. Descriptive statistics of the 6 variables from the meat level corresponding to meat quality traits measured in *Longissimus thoracis* muscle.

Variables	n	Mean	SD	Min	Max
Warner-Bratzler shear force (N/cm^2)	110	44.6	11.21	23.55	81.49
Intramuscular fat (IMF) content (% *w/w*)	110	16.3	6.18	6.15	40.34
Ultimate pH (pHu)	107	5.6	0.1	5.34	6.22
Lightness (L*)	110	39.7	2.3	34.36	46.84
Redness (a*)	110	8.8	1.24	4.17	11.77
Yellowness (b*)	110	7.4	1.43	4.02	11.42

Ultimate pH (pHu) was evaluated at 24 h *p-m* in each muscle sample using a Hanna pH meter (HI9025, Hanna Instruments Inc., Woonsocket, RI, USA) suitable for meat penetration. The measurements were done by inserting a glass electrode between the 6th and 7th rib. The pH meter was calibrated at chilling temperature using standard pH 4 and pH 7 buffers.

For surface fresh meat color determination, a portable colorimeter (Minolta CR400, Konica Minolta, Japan) was used to measure L*, a* and b* coordinates as described by Gagaoua et al. [7].

For intramuscular fat (IMF) content, a Dionex ASE 200 Accelerated Solvent Extractor (Dionex Corporation, Sunnyvale, CA, USA) was used. Briefly, muscle dry matter was assayed gravimetrically after drying at 80 °C for 48 h. Then, total lipids were extracted by mixing 6 g of muscle powder with chloroform-methanol according to the method of Folch et al. [19]. Each sample was measured in triplicate and data were expressed in g per 100 g of dry matter (g/100 g DM).

For objective tenderness, Warner-Bratzler shear force (WBSF) was measured according to Lepetit and Culioli [20] using a Warner-Bratzler shear device (Synergie200 texturometer, MTS, Eden Prairie, MN, USA). After thawing 48 h at 4 °C, the steaks were placed for 4 h in a thermostated bath at 18 °C [12]. Then, they were cooked using an Infra Grills E (Sofraca, Athis-Mons, France) set at 300 °C until the temperature at the heart of the steak reached 55 °C, a usual temperature in France [21]. From 3–5 test pieces (1 × 1 × 4 cm) were taken from the heart of the steak in the direction of the fibers and 3–4 repetitions per test tube were carried out. A 1 kN load cell and a 60 mm/min crosshead speed were used (universal testing machine, MTS, Synergie 200H). The peak load (N) and energy to rupture (J) of the muscle sample were determined.

2.5. Statistical Analyses

The data analyses were carried out following 4 main steps as described in Figure 1 using the following statistical software: XLSTAT 2018.2 (AddinSoft, Paris, France) and SAS 9.3 (SAS Institute INC, Cary, NC, USA). A total of 60 variables (q) at 4 levels of the continuum: (i) farm ($q_X = 16$), (ii) slaughterhouse ($q_X = 8$), (iii) muscle ($q_X = 30$) and iv) meat ($q_X = 5/ q_Y = 1$) were integrated in this trial (Tables 1–4). Smirnov-Grubb's outlier test at a significance level of 5% was first applied for the whole data to check any entry errors or outliers. Subsequently, Shapiro-Wilk tests were applied to determine the normality of data distribution. Descriptive analyses for all the variables were computed (Tables 1–4). For modeling and before Partial Least Squares (PLS) analyses on $q_X = 59$ variables to explain WBSF ($q_Y = 1$), the data were standardized by computing Z-scores. Z-scores are deviation of each observation relative to the mean of each individual (cow) and amongst each rearing practice [11,22]. PROC STANDARD of SAS was used to standardize the whole data to a mean of 0 and standard deviation of 1. This step allowed removing the effects of rearing practices and variability in the units as well as in the scales among the different variables. It is worthwhile to note that our previous work using the same data showed that rearing practices had no effect on tenderness assessed by trained panelist or instrumental measure by WBSF [11,12].

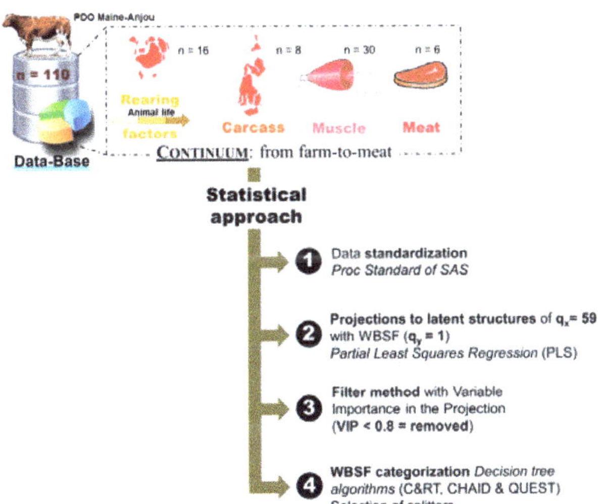

Figure 1. Summary of the statistical approach highlighting the four main statistical steps followed in this study for the selection of best variables from the 59-continuum data from farm-to-meat and then Warner-Bratzler Shear force (WBSF) prediction/categorization into different classes using 3 decision tree algorithms to select the best method.

PLS was then used to identify how the set of explanatory variables (q_X = 59) was associated to WBSF (instrumental beef tenderness, q_Y = 1) and to select the main driving factors (variables) from each level of the continuum. Briefly, this method consists of relating two data matrices X and Y to each other. In our case, X consists of continuum data except WBSF (X-matrix, 59 variables) and Y is instrumental tenderness measured by WBSF (Y-matrix, 1 variable). The filter method with the variable importance in the projection (VIP) was subsequently used to select the most important variables in the model [23,24]. Thus, the variables with a VIP < 0.8 were all eliminated and the retained variables were used to build decision trees based on the frequently used decision tree algorithms [25]: C&RT (classification and regression tree); QUEST (quick, unbiased, efficient regression tree) and CHAID (Chi-squared Automatic Interaction Detection). This step intends to validate the main variables allowing splitting into three tenderness categories (Tender, Medium and Tough) the beef cuts according to their WBSF values using the whole retained variables in the PLS model. The same criteria of accuracy, sensitivity and specificity used by Gagaoua et al. [25] were applied in this data to choose the best decision tree method. Therefore, the best decision tree was obtained by CHAID method. The identified tenderness groups were further separated by variance analysis using the PROC GLM of SAS on each splitter retained in the decision tree. Variables were considered significantly different among the tenderness classes at the significance level of p < 0.05 using Tukey's test.

3. Results and Discussion

The descriptive analyses of the data (mean, SD, and minimum and maximum ranges) at each level of the continuum are given in Tables 1–4. Warner-Bratzler shear force (WBSF) is a routine instrumental measure used as a proxy for sensory testing for meat tenderness. The WBSF values ranged from 23.55–81.49 N/cm^2, with an average of 44.6 N/cm^2 (SD 11.21 N/cm^2). The coefficient of variation was, therefore, 25.1%. This indicates a high variability in tenderness of the population of PDO Maine-Anjou (Figure 2). This was reported by previous studies [26–30] and is a common result in the field of meat texture quality. However, there is scarcity in the studies available on meat tenderness of PDO Maine-Anjou breed rather than this database to perform any comparisons.

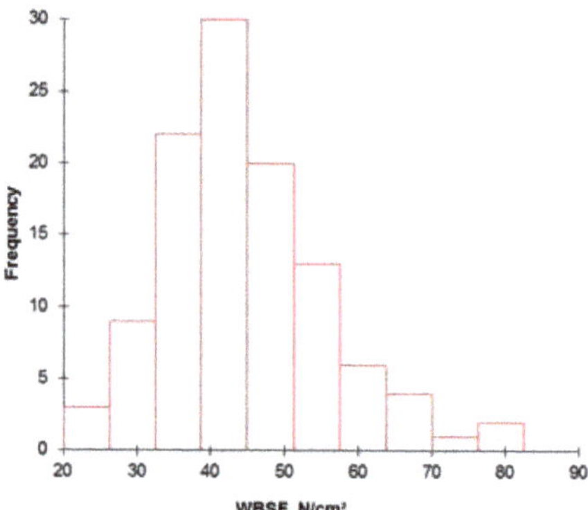

Figure 2. Histogram highlighting the relative frequency for meat tenderness assessed by WBSF on the 110 PDO Maine-Anjou cows.

The best WBSF PLS model retained 24 variables to explain tenderness variability (Table 5). From the whole 59 explanatory (independent) variables included in the PLS model, 35 had variable importance in the projection (VIP < 0.80) and were removed based on the filter method (Figure 1). This step improved the variation explained in the second model (R^2X: from 0.17–0.31) and the powerful of the link (R^2Y: from 0.37–0.64) with the dependent variable that is WBSF. The final model explained 75% of the variability of WBSF (Table 5). Among the 24 variables, six were from farm level (age of weaning; grass diet, %; haylage diet, %; birth month; type of animal (meat or dairy) and physical activity at farm, %), four from slaughterhouse level (color and tenderness scores of the carcasses; ribeye weight and EUROP conformation score), 11 were from the muscle level and mostly they were tenderness protein biomarkers (fiber area, μ-calpain, m-calpain, SOD1, ICDH, DJ-1, PGM1, HSP70-8, LDH, and total and insoluble collagen) and three were from meat level (pHu, redness (a*) and yellowness (b*)). The ranks of each of the 24 variables in the model based on their VIP are further given in Table 5.

The objective with respect to implementing decision tools is to propose a model that accurately explains beef tenderness by learning simple decision rules inferred from the individual values of WBSF using the retained potential predictors from Table 5. Accordingly, the best decision tree built using the retained variables in the PLS model was that of CHAID method (Figure 3). CHAID is a recursive partitioning based on the χ^2-test, which is used to select the best split at each step [31]. Briefly, a CHAID tree is a decision tree that is built by repeatedly splitting subsets of the space into two or more child nodes, beginning with the first whole dataset [32]. To define the best split at any node of decision tree, any allowable pair of categories of the independent variables is merged until there is no statistically significant difference within the pair with respect to the target variable. CHAID has also the peculiarity to proceed stepwise, thus being more adept at handling interactions between explanatory variables, which are available from an examination of the tree. Then, the final nodes of the tree identify subcategories determined by different sets of explanatory variables. The WBSF CHAID decision tree in this trial (Figure 3A) has 4 levels and 11 nodes out of which 6 are considered as terminal (they do not split further). From the 6 terminal nodes, the tree allowed identifying 3 different classes of WBSF (differing in their tenderness: Tender (2 nodes), Medium (1 node) and Tough (3 nodes)) using 5 splitting variables only. The resultant model yielded an overall predictive accuracy of 69.4% compared to the raw data of each individual observation of WBSF.

Table 5. Best Partial Least Squares (PLS) model of Warner-Bratzler Shear force (WBSF) showing the ranking of the 24 retained variables from the continuum data from farm-to-meat and their variable importance in the projection (VIP) values.

Variables of the Continuum from Farm-To-Meat Data	Rank	VIP
Farm level: rearing factors and animal characteristics		
Age of weaning, month	3	1.99
Grass diet, %	10	1.31
Haylage diet, %	14	1.12
Birth month	15	1.11
Type of animal (meat or dairy)	16	0.97
Physical activity at farm, %	24	0.84
Slaughterhouse level: carcass characteristics		
Color score, 1–5 scale	5	1.8
Carcass tenderness score, 1–5 scale	21	0.9
Ribeye weight, g	20	0.94
EUROP Conformation score, 1–15 scale	23	0.87
Muscle level: protein biomarkers		
Fiber area, μm^2	2	2.01
SOD1, AU	4	1.94
m-calpain, AU	6	1.64
ICDH, $\mu mol \cdot min^{-1} \cdot g^{-1}$	7	1.57
Protein deglycase (DJ-1), AU	9	1.51
PGM1, AU	11	1.27
Insoluble collagen, μg OH-proline/mg DM	13	1.18
HSP70-8, AU	17	0.97
μ-calpain, AU	18	0.96
Total collagen, μg OH-proline/mg DM	19	0.96
LDH, $\mu mol \cdot min^{-1} \cdot g^{-1}$	22	0.89
Meat level: meat quality traits		
pHu	1	3.29
Redness (a*)	8	1.53
Yellowness (b*)	12	1.27

From the five splitters, three were the first drivers of the PLS model (highlighted in bold character in Table 5). The first splitter was total collagen and generated two groups. As expected, the 15 steaks of the first node (right of decision tree) with total collagen ≥ 3.6 µg OH-proline/mg had the highest WBSF values (mean value = 50.3 N/cm^2) and considered as Tough meat [27,33]. After that, the second group (n = 95) was clustered by µ-calpain at a threshold of 169 AU. The group on the right (n = 26) was then separated by ultimate pH at a threshold of 5.55 into 14 medium steaks (WBSF mean value = 43.4 N/cm^2) and 12 very tender steaks (WBSF mean value = 36.2 N/cm^2). The group on the left (n = 69) was separated by the age of weaning of the animals into a final tough group (WBSF ≥ 45, n = 18) and a medium group of 51 steaks, which were then categorized by fiber area at a threshold of 3100 µm^2 into 30 tender (WBSF mean value = 39.4 N/cm^2) and 21 tough steaks (WBSF mean value = 48.1 N/cm^2). The mean values of WBSF among the three tenderness categories are 38.9 ± 8.1 N/cm^2, 43.4 ± 6.6 N/cm^2 and 49.4 ± 12.0 N/cm^2 for tender, medium and tough steaks, respectively. Therefore, the CHAID decision tree could simply and easily apply the discrimination rule based on five splitters to identify these three tenderness categories.

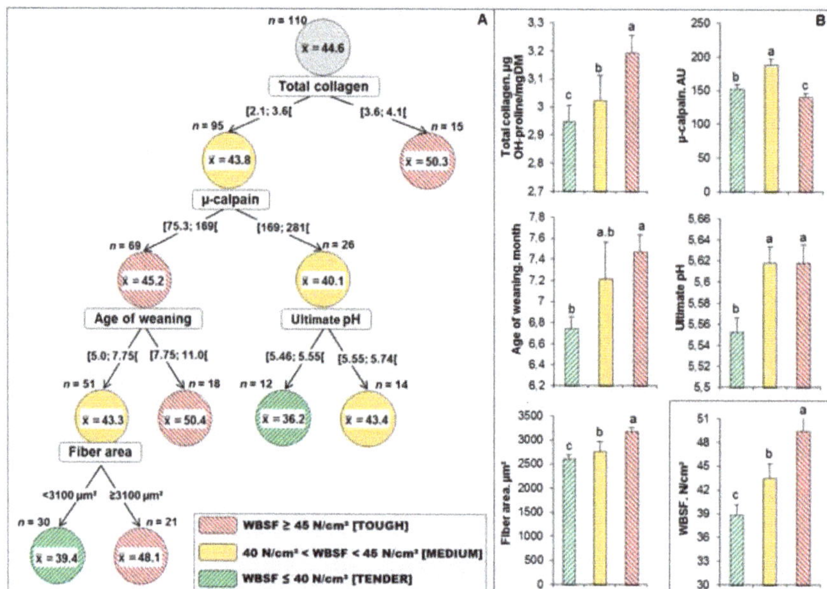

Figure 3. Categorization of the 110 steaks into different tenderness categories based on the decision tree. (**A**) Best decision tree obtained by the Chi-squared Automatic Interaction Detection (CHAID) method built using the list of variables retained in Table 5 to predict correctly 69.4% of WBSF values into three tenderness categories (Red: Tough meat; Orange: Medium meat; Green: Tender meat). The distribution of the animals in each WBSF cluster was used for accuracy measurement. At the beginning of the decision tree, all of the data (*n* = 110) are concentrated at a root node located at the top of the tree. This was then divided into two child nodes on the basis of an independent variable (1st splitter = total collagen), that creates the best homogeneity. The cut-off value of each dividing splitter was calculated from the data of all the subjects. Therefore, the data in each child node are more homogenous than those in the upper parent node. This process is continued repeatedly for each child node until all of the data in each node have the greatest possible homogeneity. This node is called a terminal node and no more branches are possible. (**B**) Variance analysis on the variables retained by the CHAID decision tree among the three WBSF (tenderness) categories that were all significant at *p* < 0.05 (Tukey's test). The mean values of WBSF among the three tenderness categories were further given at the bottom right of the graph B. Least-square means in the same graph with different superscript letters (a–c) are significantly different (*p* < 0.05).

It seems from the whole data of the continuum based on the PLS model and CHAID decision tree that muscle characteristics (from the muscle level) namely total collagen content, μ-calpain and fiber area, are the main potential discriminators/predictors of tenderness of PDO Maine-Anjou (Figure 3B). Collagen is well known to be associated with the background of meat toughness [34,35]. In a meta-analysis on the main parameters affecting collagen amount and its heat-solubility, Blanco et al. [36] hypothesized that total collagen content is different among muscles but with high amounts at birth, thereafter decreasing from birth to puberty as a function of muscle growth [35]. They further linked their model to proteolysis, in line to the involvement of μ-calpain, cross-sectional area of the fibers and age of weaning of the animal. In a recent metadata of 308 young bulls that were categorized into three tenderness classes, total collagen was a discriminator of the built tenderness classes with a mean value for the tough class of 3.51 μg OH-prol·mg⁻¹ DM. These findings agree with the negative link of collagen with tenderness evaluated by trained panelist or instrumental, as reported in numerous studies [13,14,17,37] and reviewed by Lepetit [38,39].

The two other variables are related to farm level (rearing practices) by age of weaning or at the meat level by ultimate pH. Accordingly, the decision tree (Figure 3A) allowed to identify that a steak of the PDO Maine-Anjou was considered tender (lowest WBSF: $< 40\text{N/cm}^2$) if it matched the following two rules:

(i) IF (total collagen < 3.6 µg OH-proline/mg) AND (µ-calpain ≥ 169 AU) AND (ultimate pH < 5.55) THEN meat was very tender (mean WBSF values $= 36.2$ N/cm^2, $n = 12$); or

(ii) IF (total collagen < 3.6 µg OH-proline/mg) AND (µ-calpain < 169 AU) AND (age of weaning < 7.75 months) AND (fiber area < 3100 µm^2) THEN meat was tender (mean WBSF values $= 39.4$ N/cm^2, $n = 30$).

In addition to the predictive rules allowed by the statistical approach applied in this study, possible biological mechanisms behind meat tenderness determinism of PDO Maine-Anjou breed are further revealed. It seems that the final tenderness of this breed is mainly related to the extent breakdown of structural properties in the muscle and to the background toughness related to connective tissue. This is in agreement with the recent findings of a proteomic study based on a sub dataset of eight PDO Maine-Anjou cows categorized into four tender (WBSF < 31 N/cm^2) and 4 tough (N > 60 N/cm^2) samples [40]. In this study, the authors identified eight potential biomarkers explaining differences in tenderness, of which six are structural proteins. The interesting links allowed by the decision tree within pH drop presented by ultimate pH and proteolysis by the abundance of µ-calpain support these findings. Furthermore, these are consistent with several studies from the large literature [41–44]. For example, an earlier work by Dransfield et al. [45] reported strong relationship between glycolysis and µ-calpain activity with a strong effect on final tenderness of meat.

The involvement of age at weaning and the cross-sectional area of the fibers in the prediction of tenderness may be explained by two reasons. First, it is known that animal growth affect muscle development and consequently its composition including connective tissue [46], evaluated in this study by collagen content and solubility. Second, animal growth had important consequences on protein turnover know to influence the final qualities of meat [47]. Indeed, previous studies have reported significant relationships between the speed of development of the animal and therefore its carcass composition with tenderness of cows according to age at weaning [48]. Overall, these results allow us to propose the PLS—CHAID decision trees as an interesting tool for validation on other types of animals and other qualities of meat for use by the farmers as well as the slaughterers in order to classify (or predict) the potential quality of carcasses soon after slaughter.

4. Conclusions

The purpose of this trial was to investigate the usefulness of combining chemometrics and machine learning tools to predict tenderness of PDO Maine-Anjou cows. First, we analyzed the potential of Partial Least Squares to select the main variables (or variables of interest) from the continuum to explain WBSF of ribeye steaks. The filter method allowed retaining 24 variables from 59 to explain WBSF variability. Second, using the CHAID decision tree as the best algorithm method among others, the 110 steaks were categorized into three tenderness classes using five splitters: total collagen, µ-calpain, fiber area, age of weaning and ultimate pH. Three of these potential predictors belong mainly to the muscle level and the two last other predictors to the farm and meat levels. The original statistical approach applied in this trial allowed us to properly group steaks for their tenderness potential using variables of the farm-to-meat continuum data. In the future, this proposed approach would be combined with sensory scores of tenderness (using trained panelists or consumers) to improve the prediction power and accuracy as well the validation of the retained splitters as predictors of PDO Maine-Anjou tenderness. Finally, the proposed tool would be further adopted for validation on other animal types and will be proposed for use by the beef sector to accurately categorize carcasses according to their tenderness potential. This would be beneficial at both the economic and consumer levels.

Foods **2019**, *8*, 274

Author Contributions: Conceptualization, M.G., V.M., S.C. and B.P.; methodology, M.G., V.M. and B.P.; validation, V.M. and B.P.; formal analysis, M.G.; investigation, S.C. and M.G.; resources, V.M., S.C. and B.P.; data curation, M.G., S.C., and B.P.; writing—original draft preparation, M.G.; writing—review and editing, V.M., S.C. and B.P.; visualization, M.G.; funding acquisition, V.M., S.C. and B.P.

Funding: This research was funded by the INRA and Pays de la Loire Region (PSDR grant PA–PEss-AOCMA_2009). The grant of project S3-23000846 given to M.G was funded by Région Auvergne-Rhône-Alpes and Fonds Européens de Développement Régional (FEDER).

Acknowledgments: Authors thank the PDO Maine-Anjou association, ADEMA and the Elivia slaughterhouse for their help in the surveys, the farm selection and the sample collection. The authors thank the many colleagues involved in this project, namely, Alberic Valais, Guillain Le Bec and Ghislain Aminot from SICA Rouge des Preés and Didier Micol from INRA for their assistance in data collection, the management and slaughter of animals, muscle sampling, and biochemical measurements. The authors convey special thanks to Nicole Dunoyer and Jeremy Huant for their technical assistance concerning the biomarkers quantifications by Dot-Blot and David Chadeyron for myosin heavy chains quantification by electrophoresis.

Conflicts of Interest: The authors declare no conflict of interest.

References

1. Miller, M.F.; Carr, M.A.; Ramsey, C.B.; Crockett, K.L.; Hoover, L.C. Consumer thresholds for establishing the value of beef tenderness. *J. Anim. Sci.* **2001**, *79*, 3062–3068. [CrossRef] [PubMed]

2. Henchion, M.; McCarthy, M.; Resconi, V.C.; Troy, D. Meat Consumption: Trends and Quality Matters. *Meat Sci.* **2014**, *98*, 561–568. [CrossRef] [PubMed]

3. Troy, D.J.; Kerry, J.P. Consumer perception and the role of science in the meat industry. *Meat Sci.* **2010**, *86*, 214–226. [CrossRef] [PubMed]

4. McCarthy, S.N.; Henchion, M.; White, A.; Brandon, K.; Allen, P. Evaluation of beef eating quality by Irish consumers. *Meat Sci.* **2017**, *132*, 118–124. [CrossRef] [PubMed]

5. Ferguson, D.M.; Bruce, H.L.; Thompson, J.M.; Egan, A.F.; Perry, D.; Shorthose, W.R. Factors affecting beef palatability - farmgate to chilled carcass. *Aust. J. Exp. Agric.* **2001**, *41*, 879–891. [CrossRef]

6. Gagaoua, M.; Monteils, V.; Picard, B. Data from the farmgate-to-meat continuum including omics-based biomarkers to better understand the variability of beef tenderness: An integromics approach. *J. Agric. Food Chem* **2018**, *66*, 13552–13563. [CrossRef] [PubMed]

7. Gagaoua, M.; Picard, B.; Monteils, V. Associations among animal, carcass, muscle characteristics, and fresh meat color traits in Charolais cattle. *Meat Sci.* **2018**, *140*, 145–156. [CrossRef] [PubMed]

8. Gagaoua, M.; Picard, B.; Monteils, V. Assessment of cattle inter-individual cluster variability: The potential of continuum data from the farm-to-fork for ultimate beef tenderness management. *J. Sci Food Agric.* **2019**, *99*, 4129–4141. [CrossRef]

9. Gagaoua, M.; Picard, B.; Soulat, J.; Monteils, V. Clustering of sensory eating qualities of beef: Consistencies and differences within carcass, muscle, animal characteristics and rearing factors. *Livest. Sci.* **2018**, *214*, 245–258. [CrossRef]

10. Gagaoua, M.; Picard, B.; Monteils, V. Beef quality management based on the continuum data from farmgate-to-meat: Which statistical strategies for meat science metadata analyses? In Proceedings of 24. Rencontres autour des recherches Ruminants (3R), Paris, France, 5–6 December 2018; pp. 1–5.

11. Gagaoua, M.; Monteils, V.; Couvreur, S.; Picard, B. Identification of Biomarkers Associated with the Rearing Practices, Carcass Characteristics, and Beef Quality: An Integrative Approach. *J. Agric. Food Chem.* **2017**, *65*, 8264–8278. [CrossRef]

12. Couvreur, S.; Le Bec, G.; Micol, D.; Picard, B. Relationships Between Cull Beef Cow Characteristics, Finishing Practices and Meat Quality Traits of *Longissimus thoracis* and *Rectus abdominis*. *Foods* **2019**, *8*, 141. [CrossRef] [PubMed]

13. Jurie, C.; Picard, B.; Hocquette, J.F.; Dransfield, E.; Micol, D.; Listrat, A. Muscle and meat quality characteristics of Holstein and Salers cull cows. *Meat Sci.* **2007**, *77*, 459–466. [CrossRef] [PubMed]

14. Gagaoua, M.; Terlouw, E.M.C.; Micol, D.; Hocquette, J.F.; Moloney, A.P.; Nuernberg, K.; Bauchart, D.; Boudjellal, A.; Scollan, N.D.; Richardson, R.I.; et al. Sensory quality of meat from eight different types of cattle in relation with their biochemical characteristics. *J. Integr. Agric.* **2016**, *15*, 1550–1563. [CrossRef]

15. Picard, B.; Barboiron, C.; Chadeyron, D.; Jurie, C. Protocol for high-resolution electrophoresis separation of myosin heavy chain isoforms in bovine skeletal muscle. *Electrophoresis* **2011**, *32*, 1804–1806. [CrossRef] [PubMed]

16. Gagaoua, M.; Terlouw, E.M.C.; Picard, B. The study of protein biomarkers to understand the biochemical processes underlying beef color development in young bulls. *Meat Sci.* **2017**, *134*, 18–27. [CrossRef]

17. Dubost, A.; Micol, D.; Meunier, B.; Lethias, C.; Listrat, A. Relationships between structural characteristics of bovine intramuscular connective tissue assessed by image analysis and collagen and proteoglycan content. *Meat Sci.* **2013**, *93*, 378–386. [CrossRef] [PubMed]

18. Guillemin, N.; Meunier, B.; Jurie, C.; Cassar-Malek, I.; Hocquette, J.F.; Leveziel, H.; Picard, B. Validation of a Dot-Blot quantitative technique for large scale analysis of beef tenderness biomarkers. *J. Physiol Pharm.* **2009**, *60* (Suppl. 3), 91–97.

19. Folch, J.; Lees, M.; Sloane Stanley, G.H. A simple method for the isolation and purification of total lipides from animal tissues. *J. Biol. Chem.* **1957**, *226*, 497–509.

20. Lepetit, J.; Culioli, J. Mechanical properties of meat. *Meat Sci.* **1994**, *36*, 203–237. [CrossRef]

21. Gagaoua, M.; Terlouw, C.; Richardson, I.; Hocquette, J.F.; Picard, B. The associations between proteomic biomarkers and beef tenderness depend on the end-point cooking temperature, the country origin of the panelists and breed. *Meat Sci.* **2019**, *157*, 107871. [CrossRef]

22. Picard, B.; Gagaoua, M.; Al Jammas, M.; Bonnet, M. Beef tenderness and intramuscular fat proteomic biomarkers: Effect of gender and rearing practices. *J. Proteom.* **2019**, *200*, 1–10. [CrossRef] [PubMed]

23. Mehmood, T.; Liland, K.H.; Snipen, L.; Sæbø, S. A review of variable selection methods in Partial Least Squares Regression. *Chemom. Intell. Lab. Syst.* **2012**, *118*, 62–69. [CrossRef]

24. Chong, I.-G.; Jun, C.-H. Performance of some variable selection methods when multicollinearity is present. *Chemom. Intell. Lab. Syst.* **2005**, *78*, 103–112. [CrossRef]

25. Gagaoua, M.; Monteils, V.; Picard, B. Decision tree, a learning tool for the prediction of beef tenderness using rearing factors and carcass characteristics. *J. Sci. Food Agric.* **2019**, *99*, 1275–1283. [CrossRef] [PubMed]

26. Belew, J.B.; Brooks, J.C.; McKenna, D.R.; Savell, J.W. Warner–Bratzler shear evaluations of 40 bovine muscles. *Meat Sci.* **2003**, *64*, 507–512. [CrossRef]

27. Shackelford, S.D.; Morgan, J.B.; Cross, H.R.; Savell, J.W. Identification of Threshold Levels for Warner-Bratzler Shear Force in Beef Top Loin Steaks. *J. Muscle Foods* **1991**, *2*, 289–296. [CrossRef]

28. Wheeler, T.L.; Shackelford, S.D.; Koohmaraie, M. Sampling, cooking, and coring effects on Warner-Bratzler shear force values in beef2. *J. Anim. Sci.* **1996**, *74*, 1553–1562. [CrossRef]

29. Magnabosco, C.U.; Lopes, F.B.; Fragoso, R.R.; Eifert, E.C.; Valente, B.D.; Rosa, G.J.M.; Sainz, R.D. Accuracy of genomic breeding values for meat tenderness in Polled Nellore cattle1. *J. Anim. Sci.* **2016**, *94*, 2752–2760. [CrossRef] [PubMed]

30. Jerez-Timaure, N.; Huerta-Leidenz, N.; Ortega, J.; Rodas-González, A. Prediction equations for Warner–Bratzler shear force using principal component regression analysis in Brahman-influenced Venezuelan cattle. *Meat Sci.* **2013**, *93*, 771–775. [CrossRef]

31. Kass, G.V. An Exploratory Technique for Investigating Large Quantities of Categorical Data. *J. R. Stat. Soc. Ser. C (Appl. Stat.)* **1980**, *29*, 119–127. [CrossRef]

32. Ture, M.; Tokatli, F.; Kurt, I. Using Kaplan–Meier analysis together with decision tree methods (C&RT, CHAID, QUEST, C4.5 and ID3) in determining recurrence-free survival of breast cancer patients. *Expert Syst. Appl.* **2009**, *36*, 2017–2026. [CrossRef]

33. Nian, Y.; Zhao, M.; O'Donnell, C.P.; Downey, G.; Kerry, J.P.; Allen, P. Assessment of physico-chemical traits related to eating quality of young dairy bull beef at different ageing times using Raman spectroscopy and chemometrics. *Food Res. Int.* **2017**, *99*, 778–789. [CrossRef] [PubMed]

34. Purslow, P.P. New developments on the role of intramuscular connective tissue in meat toughness. *Annu. Rev. Food Sci. Technol.* **2014**, *5*, 133–153. [CrossRef]

35. Purslow, P.P. Contribution of collagen and connective tissue to cooked meat toughness; some paradigms reviewed. *Meat Sci.* **2018**, *144*, 127–134. [CrossRef] [PubMed]

36. Blanco, M.; Jurie, C.; Micol, D.; Agabriel, J.; Picard, B.; Garcia-Launay, F. Impact of animal and management factors on collagen characteristics in beef: A meta-analysis approach. *Animal* **2013**, *7*, 1208–1218. [CrossRef] [PubMed]

37. Dransfield, E.; Martin, J.-F.; Bauchart, D.; Abouelkaram, S.; Lepetit, J.; Culioli, J.; Jurie, C.; Picard, B. Meat quality and composition of three muscles from French cull cows and young bulls. *Anim. Sci.* **2003**, *76*, 387–399. [CrossRef]

38. Lepetit, J. A theoretical approach of the relationships between collagen content, collagen cross-links and meat tenderness. *Meat Sci.* **2007**, *76*, 147–159. [CrossRef] [PubMed]

39. Lepetit, J. Collagen contribution to meat toughness: Theoretical aspects. *Meat Sci.* **2008**, *80*, 960–967. [CrossRef]

40. Beldarrain, L.R.; Aldai, N.; Picard, B.; Sentandreu, E.; Navarro, J.L.; Sentandreu, M.A. Use of liquid isoelectric focusing (OFFGEL) on the discovery of meat tenderness biomarkers. *J. Proteom.* **2018**, *183*, 25–33. [CrossRef]

41. Gagaoua, M.; Terlouw, E.M.; Micol, D.; Boudjellal, A.; Hocquette, J.F.; Picard, B. Understanding Early Post-Mortem Biochemical Processes Underlying Meat Color and pH Decline in the *Longissimus thoracis* Muscle of Young Blond d'Aquitaine Bulls Using Protein Biomarkers. *J. Agric. Food Chem.* **2015**, *63*, 6799–6809. [CrossRef]

42. Kendall, T.L.; Koohmaraie, M.; Arbona, J.R.; Williams, S.E.; Young, L.L. Effect of pH and ionic strength on bovine m-calpain and calpastatin activity. *J. Anim. Sci.* **1993**, *71*, 96–104. [CrossRef] [PubMed]

43. Hwang, I.H.; Thompson, J.M. The interaction between pH and temperature decline early postmortem on the calpain system and objective tenderness in electrically stimulated beef *Longissimus dorsi* muscle. *Meat Sci.* **2001**, *58*, 167–174. [CrossRef]

44. Ouali, A.; Gagaoua, M.; Boudida, Y.; Becila, S.; Boudjellal, A.; Herrera-Mendez, C.H.; Sentandreu, M.A. Biomarkers of meat tenderness: Present knowledge and perspectives in regards to our current understanding of the mechanisms involved. *Meat Sci.* **2013**, *95*, 854–870. [CrossRef] [PubMed]

45. Dransfield, E.; Etherington, D.J.; Taylor, M.A.J. Modelling post-mortem tenderisation—II: Enzyme changes during storage of electrically stimulated and non-stimulated beef. *Meat Sci.* **1992**, *31*, 75–84. [CrossRef]

46. Listrat, A.; Gagaoua, M.; Picard, B. Study of the Chronology of Expression of Ten Extracellular Matrix Molecules during the Myogenesis in Cattle to Better Understand Sensory Properties of Meat. *Foods* **2019**, *8*, 97. [CrossRef] [PubMed]

47. Moloney, A.P.; McGee, M. Factors Influencing the Growth of Meat Animals. In *Lawrie's Meat Science, Eight Edition*; Toldrá, F., Ed.; Woodhead Publishing: Sawston, UK, 2017; pp. 19–47.

48. Meyer, D.L.; Kerley, M.S.; Walker, E.L.; Keisler, D.H.; Pierce, V.L.; Schmidt, T.B.; Stahl, C.A.; Linville, M.L.; Berg, E.P. Growth rate, body composition, and meat tenderness in early vs. traditionally weaned beef calves1,2. *J. Anim. Sci.* **2005**, *83*, 2752–2761. [CrossRef] [PubMed]

Article

New Approach Studying Interactions Regarding Trade-Off between Beef Performances and Meat Qualities

Alexandre Conanec [1,2,3], Brigitte Picard [1], Denis Durand [1], Gonzalo Cantalapiedra-Hijar [1], Marie Chavent [2,3], Christophe Denoyelle [4], Dominique Gruffat [1], Jérôme Normand [4], Jérôme Saracco [2,5] and Marie-Pierre Ellies-Oury [1,*]

1   Universite Clermont Auvergne, INRA, VetAgro Sup, UMR Herbivores, F-63122 Saint-Genes-Champanelle, France; alexandre.conanec@agro-bordeaux.fr (A.C.); brigitte.picard@inra.fr (B.P.); denis.durand@inra.fr (D.D.); gonzalo.cantalapiedra@inra.fr (G.C.-H.); dominique.gruffat@inra.fr (D.G.)
2   Contrôle de Qualité et Fiabilité Dynamique (CQFD) team, Inria BSO, F-33400 Talence, France; marie.chavent@math.u-bordeaux.fr (M.C.); jerome.saracco@math.u-bordeaux.fr (J.S.)
3   Universite de Bordeaux, IMB, UMR 5251, F-33400 Talence, France
4   Institut de l'Elevage, Service Qualite des Carcasses et des Viandes, 69007 Lyon, France; christophe.denoyelle@idele.fr (C.D.); jerome.normand@idele.fr (J.N.)
5   ENSC Bordeaux INP, IMB, UMR 5251, F-33400 Talence, France
*   Correspondence: marie-pierre.ellies@inra.fr; Tel.: +33-5-57-35-38-70

Received: 2 May 2019; Accepted: 4 June 2019; Published: 7 June 2019

Abstract: The beef cattle industry is facing multiple problems, from the unequal distribution of added value to the poor matching of its product with fast-changing demand. Therefore, the aim of this study was to examine the interactions between the main variables, evaluating the nutritional and organoleptic properties of meat and cattle performances, including carcass properties, to assess a new method of managing the trade-off between these four performance goals. For this purpose, each variable evaluating the parameters of interest has been statistically modeled and based on data collected on 30 Blonde d'Aquitaine heifers. The variables were obtained after a statistical pre-treatment (clustering of variables) to reduce the redundancy of the 62 initial variables. The sensitivity analysis evaluated the importance of each independent variable in the models, and a graphical approach completed the analysis of the relationships between the variables. Then, the models were used to generate virtual animals and study the relationships between the nutritional and organoleptic quality. No apparent link between the nutritional and organoleptic properties of meat ($r = -0.17$) was established, indicating that no important trade-off between these two qualities was needed. The 30 best and worst profiles were selected based on nutritional and organoleptic expectations set by a group of experts from the INRA (French National Institute for Agricultural Research) and Institut de l'Elevage (French Livestock Institute). The comparison between the two extreme profiles showed that heavier and fatter carcasses led to low nutritional and organoleptic quality.

Keywords: trade-off; meat quality; beef performances; modeling

---

1. Introduction

The meat sector is facing various challenges in France, due to a current context of high quality demand and competiveness. Indeed, consumers expect homogeneous organoleptic quality, and more recently, consumers' expectations in terms of healthiness and the nutritional quality of food are increasing. The beef sector is obviously concerned, due to the lower meat consumption in the past few years [1]. In addition, farmers and retailers are facing low margins in a competitive sector. Therefore, to preserve the beef cattle industry in the territory, the sustainability of these main actors has to be enhanced by rearing more efficient animals to decrease the cost of production and provide standardized carcasses of high quality. Despite the stakes, few studies have been carried out to tackle the problem as a whole, describing the relation between these four parameters of interest: animal performances, carcass properties, nutritional qualities, and organoleptic qualities. Among these studies, a creative approach proceeding to a double dimension reduction (via a clustering of variables followed by a principal component analysis) on a dataset made of multiple variables measuring carcass value, fatty acid profile, and sensory tasting descriptors concluded that there is no relationship between the nutritional and sensory quality of the meat produced by young bulls [2]. Furthermore, other studies have highlighted some correlations (positive and negative) between these two qualities. For instance, it has been reported that a high amount of intramuscular fat enhances the flavor, the juiciness, and the tenderness of the meat [3,4], but reduces the proportion of polyunsaturated fatty acid (PUFA), whereas saturated (SFA) and monounsaturated fatty acids (MUFA) proportions increase [5]. The first objective of the present work was to integrate these four parameters of interest to analyze the relationships between them. A second aim was to set up a method of managing the trade-off to help the beef cattle industry design specifications that take the possible interaction between the nutritional and organoleptic qualities of the meat into account.

2. Materials and Methods

The whole method is based on a dataset of 30 animals, which will be described further. First, 62 variables measured on those animals were pre-treated to reduce the redundancy between them and keep the best indicators to evaluate the parameters of interest. From the output variables of this pre-treatment, as many models as variables were created to explain each variable by the other ones. Then, those models were evaluated with statistical tools, and the relationships that were observed with these models were compared to the two-by-two relationships described in the literature. After evaluating the modeling part, models were used to create a virtual population of animals to study the relationships between the modeled variables with more accuracy. To compare the nutritional and sensory quality of the meat, a set of importance weights was proposed to aggregate the variables evaluating those qualities and create a synthetic index for each parameter. The correlation between both of those indexes was calculated to study their relation. Finally, the 30 best and worst profiles (based on an objective of balance between nutritional and organoleptic quality) were studied to compare their main trait differences.

2.1. Data

2.1.1. Animals

Data were collected in a research project named "Lipivimus" (ANR-06-PNRA-018) investigating the effects of several lipids sources in the animal feeding on the nutritional and organoleptic qualities of the meat. Experimental procedures and animal-holding facilities respected French animal protection legislation, including the licensing of experimenters. They were controlled and approved by the French Veterinary Services (the abattoir and cattle experimental facilities license numbers were #63 345 01 and #63 345 17, respectively). A total of 30 Blonde d'Aquitaine heifers, which were homogeneous in terms of initial age (28 ± 1.3 months), live weight (540 ± 13 kg), and body condition score (2.2 ± 0.05) were raised during a finishing period of 100 days. Animals were assigned at random to one of the four rations that were i) isoenergetic on a net energy basis at a level of 1.71 Mcal of net energy/kg dry matter (DM) and ii) isonitrogenous on a metabolizable protein basis expressed in Protein Digestible in the small Intestine (PDI) units (defined by the French National Institute for Agricultural Research, or INRA, tables [6]). A lipid treatment (40 g/kg DM) was added to the basal diet made of molasses straw (30%) and concentrate (70%). The control group ($n = 8$) received the basal diet without any lipid source supplements, the flaxseed group ($n = 8$) received the basal diet added with extruded flaxseed, the rapeseed group ($n = 6$) received a mixture composed of extruded flaxseed (1/3) and rapeseed (2/3), and the palmitostearate group ($n = 8$) received palmitostearate. Two of the eight initial heifers of the rapeseed diet group are missing in the database: one for health problems was removed from the group, and the other one had not been tasted by the panel of degustation; thus, numerous data were missing for this animal. Average daily gain (ADG) was calculated based on the difference between the end and the beginning weights of the fattening period. The dry matter intake was measured daily for each pen to calculate the feed conversion ration per pen (FCR), based on a mean ADG of the animals of each pen.

2.1.2. Animal Slaughtering Process

Animals were slaughtered at the slaughterhouse of the Société Vitréenne d'Abattage (SVA) in Vitré, France, at the mean live weight of 693 kg (± 10 kg), with a body fat score of three (± 0.1) on a scale varying from one (very lean) to five (very fat). Overall, average daily gain was 1528 g/day (± 53 g/day) for the 100 day-finishing period with no significant variations between diets. Slaughtering was performed in compliance with French welfare regulations. The carcasses were not electrically stimulated, and they were chilled and stored at 4 °C until 24 h post mortem. Ultimate pH was recorded between the sixth and seventh rib using a pH meter equipped with a glass electrode at 24 h post mortem. Cold carcass weight, fat score, and conformation were measured at slaughter based on the EUROP grading system [7]. The fat, muscle, and bone proportions were estimated from the regression equations base on the dissection of the sixth rib [8]. Various fat tissues (kidney fat, pelvic fat, trimming fat) were also weighed. Professional experts from the Institut de l'Elevage scored each carcass for the intermuscular and intramuscular fat developments of the *Longissimus thoracis* muscle (LT) on two scales varying either from zero (absence of intermuscular fat) to five (strong intermuscular fat) based on an Institut de l'Elevage grading system, or from three (no intramuscular fat visible) to 12 (high intramuscular fat) [9], respectively. Meat color was monitored at the slaughterhouse on the LT muscle, at the sixth rib level twice, with a visual evaluation based on an Institut de l'Elevage grade varying from 1 (meat very light) to 4 (meat very dark) and using a portable colorimeter (CR400 MINOLTA) and D_{65} as the illuminant, because it closely approximates daylight [10]. Color coordinates were calculated in the CIELAB system [11]: L^* (lightness), a^* (redness) and b^* (yellowness). The chroma (as $C^* = (a^{*2} + b^{*2})^{1/2}$) and hue angle ($h^* = \arctan(b^*/a^*)$) were also calculated. Measurements were taken at three locations on each steak and averaged.

2.1.3. Nutritional Quality Measurement

Samples of LT muscles (100–120 g) were collected one day post mortem at the level of the 9th and 10th ribs, cut into small pieces, frozen in liquid N_2, and stored at −80 °C. Just before analysis, the frozen samples were mixed in liquid N_2 in an analytical mill (modelM-20, IKA-Werke, Stouten, Germany) to obtain fine powder. Muscle lipids and fatty acids were extracted and quantified as previously described by Habeanu et al. [12]. Briefly, total lipids were extracted according to the method of Folch et al. [13] by mixing the LT muscle powder with a 2/1 chloroform/methanol mixture (*v/v*) and quantified by gravimetry. Fatty acid extraction and transmethylation into fatty acid methyl esters (FAME) were subsequently performed according the methods of Bauchart et al. [14]. Fatty acid methyl ester analysis was performed with gas liquid chromatography using a Peri 2100-chromatography system (Perichrom Society, Saulx-les-Chartreux, France) fitted with a CP-Sil 88 glass capillary column (Varian, Palo Alto, CA, USA; length = 100 m; diameter = 0.25 mm). The carrier gas was H_2, and the oven and flame ionization detector temperatures described by Scislowski et al. [15] were used. Total fatty acid (FA) was quantified using C19:0 as an internal standard. The identification of each individual FAME and the calculation of the response coefficients for each individual FAME were performed using the quantitative mix C4–C24 Fame (Supelco, Bellafonte, PA, USA).

2.1.4. Organoleptic Quality Measurement

At 48 h post mortem, the *Longissimus thoracis* muscle samples that were used for organoleptic evaluation were removed from carcasses (7^e and 8^e rib), placed in sealed plastic bags under vacuum, and kept at 4 °C without light for aging for 10 days. Then, each sample was frozen and stored at −20 °C awaiting organoleptic evaluation. Samples were thawed, without stacking or overlapping, for 24 h before cooking and organoleptic assessment. Then, they were cut into mini-roasts (50-g cubes, 4 cm thick). Each mini-roast was baked (thermolyne muffle furnace, Model 6000, Bioblock Scientific, Illkirch Graffenstaden, France) without dressing. There were heated to 310 °C for seven minutes, and cut into four bites that were immediately served in another porcelain plate to 12 trained panelists. The panelists rated the sample on a 10-cm unstructured line scale (from 0 to 100) measured in mm for the following texture attributes: global tenderness defined as the ease of chewing the sample between teeth: from extremely tough (0) to extremely tender (100), juiciness defined as the amount of moisture released in the mouth: from extremely dry (0) to extremely juicy (100), and global intensity of flavor from low intensity (0) to very high intensity (100). Height descriptors of flavor were also used, from low intensity (0) to very high intensity (100): sweet flavor, acid flavor, bitter flavor, fatty flavor, metallic flavor, rancid flavor, fish flavor, and blood flavor. The sessions were carried out in a room equipped with individual booths under artificial red light to reduce the influence of the appearance of the samples. At each session, a monadic presentation of a maximum of 12 samples was done, with each sample being selected in random order.

2.2. Modelization

2.2.1. Pre-Treatment of the Variables

This experiment resulted in the collection of 62 variables (described in Appendix A), which were centered and reduced by the standard deviation. The variables were classified into one of the four parameters of interest (i.e., animal performances, carcass properties, nutritional quality, and organoleptic quality of meat). Then, a clustering of the variables was performed with the variables evaluating each parameter to reduce the redundancy among the variables and only keep a small set of indicators to evaluate those parameters. Clusters of correlated variables were formed with this statistical tool [16], and a criteria was calculated to evaluate the cohesiveness of those clusters (the closer the criteria was to 100%, the more correlated the variables that were gathered into each clusters were to each other). Then, the authors decided to resume the clusters by either a linear combination of the variables of the cluster (first component of the principal component analysis (PCA) on the variables of the cluster) or by one of the variables from the cluster. These choices are assumed to be subjective, but were based on the literature and the main indicators that were used to describe the carcass traits [17], the nutritional quality [18–20], and the organoleptic quality [21].

2.2.2. Variable Relations Modeling

To study the relations between the four parameters of interest, each variable was modeled with the set of all the variables that were evaluating another parameter of interest other than the one that the variable was to model. To clear the notation, let's define M_{var} to denote the model built to explain the dependent variable var.

To determine the most relevant model as possible, several types of regression models (linear model (lm), random forest (rf), sliced inverse regression (sir), ridge regression, and partial least square regression (plsr)) were compared using modvarsel R package [22]. The underlying methodology used in this package is entirely computational, generating a training sample several times ($n = 50$), to build models that were then evaluated with a mean square error (MSE) criteria calculated on the corresponding test dataset sample. This package also provides a tool to measure the importance of each independent variable in the corresponding regression model, and then select the best ones to build the most accurate model. The quality of the model was evaluated with the adjusted coefficient of determination (adjusted coefficient of determination (adjR^2) is a R^2 penalized by the number of independent variables into the model) calculated twice: first on the training data, which were used to build the model, and then a second time with a method developed by Harrell et al. [23]

based on bootstrap resampling (B = 100) to evaluate without overfitting the accuracy of the model. The combination of both methods enables estimating model overfitting.

2.2.3. Study of the Relationships Modeled

Two strategies were set up to analyze the relationships between the dependent variable and the independent variables. As mixed parametric and non-parametric models complicate the quantification of the importance of each independent variable in the prediction models, a common method was used to evaluate this importance, mobilizing the decomposition of the variance [24] in the sensitivity R package [25] by calculating a sensitivity index (Si ∈ [0, 1]). A sensitivity index was assigned to each indicator into the model: the most influential ones were associated with a value close to one and the less influential was assigned a value close to zero. The second strategy to analyze the model behavior was to visualize the variations of the model predictions where all the (centered and reduced) independent variables were set to their mean zero, except for one that was taking a range of values in $0 \pm 2\sigma$ (i.e., [−2, 2] as $\sigma = 1$). This procedure was repeated as many times as the number of independent variables of the model.

2.3. Trade-Offs Methodology

2.3.1. Aggregation of the Variable Evaluating the Parameters of Interest

Several variables were used to evaluate each parameter of interest. Therefore, to compare the nutritional and organoleptic quality, a synthetic index was created for both. Those indexes were calculated by a linear combination of the variables evaluating the parameters of interest weighted by their importance. The importance weights (Table 1) were proposed by a set of experts from the INRA (French National Institute for Agricultural Research) and Institut de l'Elevage (French Livestock Institute) based on their knowledge and the literature. For organoleptic quality, the weights were mainly based on the Meat Quality 4 (MQ4) [21] but for the nutritional quality, even if there are some considerations about the studied indicators in the literature, no objective equation has been created so far [18–20].

Table 1. Importance weights related to each variable evaluating the nutritional and organoleptic qualities.

Parameter of Interest NQ	Weights	Parameter of Interest OQ	Weights
lipid content	−0.15	tenderness	+0.425
long FA	+0.05	juiciness	+0.150
C16:0/C18:0 ratio	−0.15	flavor intensity	+0.175
n-6/*n*-3 ratio	−0.25	bitter flavor	−0.025
PUFA/MUFA ratio	+0.25	rancid and fish flavors	−0.125
CLA	0.1	fatty vs. metal	−0.050
trans FA	−0.05	blood and acid flavors	−0.050
Total (in absolute value)	1	Total (in absolute value)	1

CLA: conjugated linoleic acids, FA: fatty acid, MUFA: monounsaturated fatty acids, PUFA: polyunsaturated fatty acid, NQ: nutritional quality, OQ: organoleptic quality.

2.3.2. Generation of Virtual Animals

Facing a low number of animals, the models of prediction were used to create virtual animals ($n = 500$). The aim of this approach was to study the interactions between the parameters of interest based on the models and not the real animals (which have only been used to create the models). However, it was assumed that the virtual animals were built to be realistic (i.e., those animals could be real), even if the bias of the models has to be taken into account.

The (simplified) algorithm to generate virtual animals is given in Appendix B. First, one of the parameters of interest was chosen to set the value of its variables. Each variable could vary from −2 to 2 ($0 \pm 2\sigma$) with a step of 0.1. To prevent unrealistic combinations of the variables from that chosen parameter of interest, simple linear regressions were computed between these variables. A confidence interval at 95% based on those regression were decided if the combination were realistic or not. Then, all the variables that were not in the chosen parameter of interest were estimated through the prediction models. Since the models were dependent all together and the estimation of the variables made successively, all the estimation were initially set to 0. The estimations of all the variables were made several times ($n = 10$), with an order of the estimation randomly assigned, to converge to a stable state.

2.3.3. Statistical Analysis

After their generation, the virtual animals were aggregated following the procedure described in Section 2.3.1. The correlation between the two synthetic indexes was computed to evaluate the link between the nutritional and the organoleptic qualities.

The trade-off method consisted of picking the best animal based on its performances regarding both of the qualities studied. To determine the best animals, a targeted point was set, and the 30 closest animals to this point were selected. The targeted point was set at the coordinate (NQ = 2, OQ = 2) corresponding to two standard deviations of the synthetic indexes (which means that only a few animals would reach such performances). The same procedure was applied to select the 30 worst animals close to the opposite targeted point (NQ = −2, OQ = −2).

Then, the best and worst selected virtual animals were analyzed. The distribution of their traits was compared by using the Wilcoxon test (robust non-parametric test which does not need to verify the normality assumption) on each variable.

3. Results

3.1. Pre-Treatment of the Variables

The selection of the relevant variables evaluating the parameters of interest are explained in this section. The cluster affectation of each variable's information and the correlation with the linear combination of all the variables of the cluster are given in Appendix A. A clustering of variables was performed for each parameter of interest, except for Animal Performances (APs) since the (only) three variables evaluating this parameter were uncorrelated.

For Carcass Properties (CP), six clusters were created, which had good cohesiveness (63.7%). Intermuscular fat score and carcass fat development were associated in the same cluster, whereas the amount of the different fats (such as abdominal, fifth quarter, etc.) were all associated in another cluster. Thus, intermuscular fat appears to be only weakly linked to removed fat. This distribution of fat variables is consistent with results of Yang et al. [26], who indicated that intermuscular fat is independent from the other fat deposits. Therefore, two linear combinations were chosen to represent these two clusters (Table 2). Logically, the color variables of a^*, b^*, and C^* were positively linked, confirming the correlations already established by Mancini and Hunt [27]. The clustering of h^* and L^* color variables with fat proportion might be explained by the impact of fat on muscle color perception. Two linear combinations were also computed to represent both of these color aspects. Let us also mention that cluster CP1 was gathering carcass weight and bone proportion (logically negatively correlated) and pH. Since pH is an important indicator for the meat quality, it was proposed to separate the pH from the two other variables, which were informing more on the carcass weight. Finally, to summarize this cluster, pH and the carcass weight variables were kept as output variables of the pre-treatment operation.

The Nutritional Quality (NQ) of the meat was evaluated by 28 variables: the lipid content and 27 fatty acid measures or the ratio/sum of them. Therefore, the decision was made to withdraw the lipid content from the analysis (because the information of this indicator is also important to compare with intramuscular fat, for example), and perform the clustering of variables only on the fatty acid variables. The cohesiveness of the six clusters was higher than that for CP (around 72%). Two clusters were not explicitly clear. The NQ3 cluster regrouped variables describing the opposition between n-3 and *n*-6 fatty acids. The presence of SFA and C16:0 in this cluster could be explained by a link between the low amount of *n*-3 in proportion when SFA (mainly C16:0) is high. For this cluster, it was proposed to take the *n*-6/*n*-3 ratio, which is a strong nutritional indicator, into the ANSES recommendations [28]. The NQ4 cluster was also not clear, but on the whole, it appeared to be describing the opposition between MUFA and PUFA. For this cluster, the linear combination of the variable was taken and renamed as the ratio PUFA/MUFA.

Organoleptic Quality (OQ) is usually described in the literature with three main indicators: tenderness, juiciness, and flavor. The sensory variables available in this experiment gave a larger analysis of the flavor aspect. Therefore, the clustering of variables was performed only with these variables. Five clusters were formed with a cohesiveness equal to 63.6%. The synthetic index that was chosen was strongly correlated with the input variables, meaning that a small amount of information was lost in the dimension reduction step.

Table 2. Output variables of the pre-treatment operation describing the four parameters of interest (PI).

PI	Cluster Codification	Output Variables Name/Abbreviation	Output Variables Description
Animal Performances (AP)	-	Slaughter weight	Slaughter weight
	-	ADG	Average daily gain during the finishing period
	-	FCR	Feed conversion ratio (ADG/feed intake (DM))
Carcass properties (CP)	CP1	Carcass weight	Carcass weight
		pH	Ultimate pH at 24 h post mortem
	CP2	Conformation	Carcass conformation
	CP3	h*L*	Aggregation of hue (h^*), luminosity (L^*) of carcass lean, and expert evaluation of carcass muscle color
	CP4	Fat development	Aggregation of several fat tissues (in the fifth quarter, carcass fat, etc.)
	CP5	Fat proportion	Fat percentage relative to the other components of the carcass (bone and muscle), aggregate with intermuscular and intramuscular fat score
	CP6	$a^*b^*C^*$	a^*, b^*, and chroma (C^*) of carcass lean
Nutritional Quality of meat (NQ)	-	Lipid content	Lipid content
	NQ1	Long FA	Long-chain fatty acid amount and proportion
	NQ2	C16:0/C18:0	C16:0/C18:0 ratio
	NQ3	*n*-6/*n*-3	*n*-6/*n*-3 ratio
	NQ4	PUFA/MUFA	Polyunsaturated fatty acids/Monounsaturated fatty acids ratio
	NQ5	CLA	Conjugated linoleic acids
	NQ6	Trans FA	Trans fatty acids
Organoleptic Quality (OQ)	-	Tenderness	Tenderness
	-	Juiciness	Juiciness
	OQ1	Flavor intensity	Flavor intensity
	OQ2	Bitter flavor	Bitter flavor
	OQ3	Rancid fish	Flavors rancid and fish flavors
	OQ4	Fatty vs metal	Fatty versus metallic flavors
	OQ5	Blood acid flavors	Blood acid flavors

ADG: average daily gain, FCR: feed conversion ration per pen.

3.2. Choice and Quality of the Prediction Models

A synthesis of the 24 models' quality is represented in Figure 1. Random forest was chosen most of the time (18 green intervals) before ridge regression (five red intervals) and the linear regression (one blue interval). Regarding the accuracy of the models, the adjusted R^2 (adjR^2) values were higher than 0.8 in most of the prediction models. However, some models were characterized by low or very low adjR^2 values, especially M_{ADG}, M_{pH}, M_{CLA}, and $M_{juiciness}$, which denote unreliable models.

The correction of the adjusted R^2 (cor_adjR^2) enabled taking the overfitting of the models into account, and thus estimating the true accuracy of the model on unknown data. For example, the cor_adjR^2 of the M_{FCR} model was only around 0.33 when the adjR^2 was higher than 0.9. Similarly, the $M_{a^*b^*C^*}$ and $M_{long\ FA}$ appeared to be less accurate when considering their cor_adjR^2 (0.35 and 0.38, respectively) compared to their adjR^2 (0.95 and 0.90, respectively). It is interesting to notice that the models with high overfitting were all random forest models.

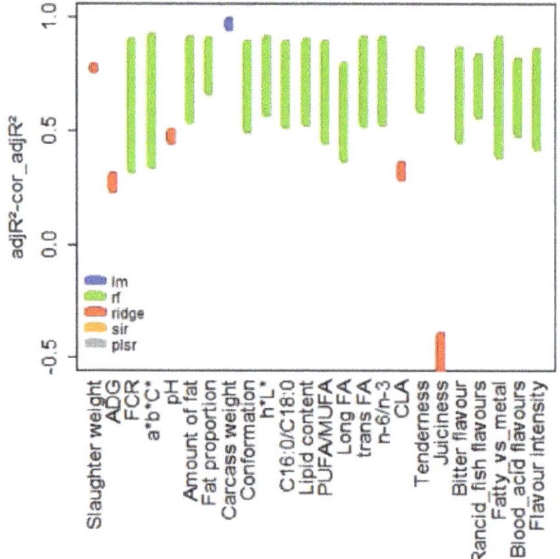

Figure 1. Quality of the 24 prediction models based on the adjusted coefficient of determination calculated twice, represented by vertical segments. The upper (respectively lower) limit of the interval corresponds to the adjusted R^2 (calculated on the training data) (respectively corrected adjusted R^2 estimate with a bootstrap approach [23]). The color indicates the model selected from the five competing regression models (lm = linear model, rf = random forest, ridge, sir = slice inverse regression, plsr = partial least square regression).

3.3. Examination of Models Behavior

Some of the 24 multivariate regression models that were built were analyzed in this section. The aim of this analysis was to show the main relations between the variables to compare and discuss them further with the current knowledge (Section 4.2.). All the relations are summarized in Figure 2 by the sensitivity index (Si) of every dependent variable into each model. Only the selected variables with the modvarsel method were used to build the model and thus have a corresponding square on the row of the model. It would be too long to describe all the results in Figure 2 one by one, but to take an example, the M_{FCR} is mainly influenced by the pH (Si = 0.20), the conformation (Si = 0.20), and the CLA (Si = 0.22). The rest of the independent variables seem to be not strongly linked to the feed efficiency.

The second method that was used to describe the models' behavior gave complementary information to the sensitivity indexes. The $M_{fat\ proportion}$ seemed to be mainly influenced by the lipid content (Si = 0.90), but the graphical approach also showed that the ADG significantly influenced the model predictions (Figure 2a), even though the sensitivity index was low (Si = 0.05). As observed in Figure 2a, a low growth rate was associated with a high portion of fat and a high amount of intramuscular and intermuscular fat deposits. This example highlighted a contradiction between the two methods, which will be discussed further (Section 4.1.).

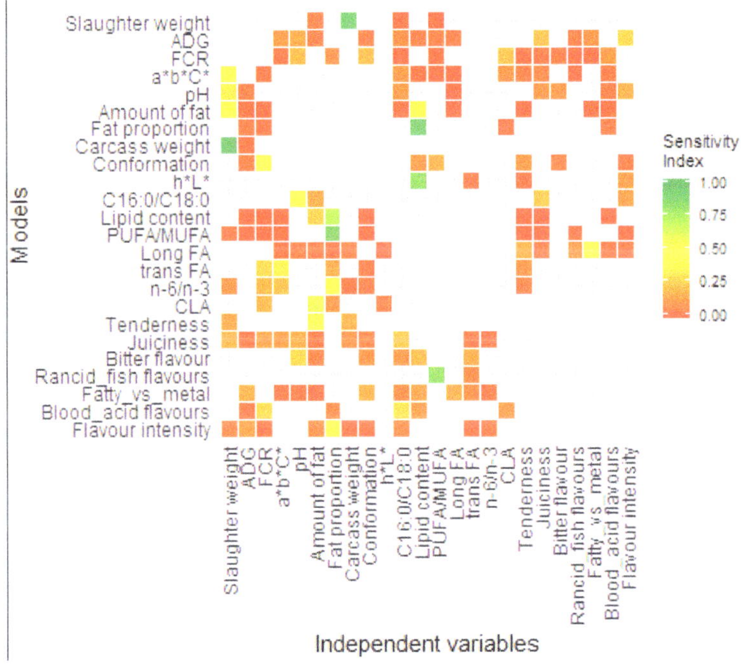

Figure 2. Heat map of the sensitivity index for each variable used (column) in each model (row). Green indicates values that are higher in the sensitivity index. If there is no square, the variable was not used in the corresponding model.

The graphical approach was also a tool to describe the shape of the variations of the prediction. $M_{PUFA/MUFA}$ was quite exclusively influenced by fat proportion (Si = 0.86), even if nine other variables were selected as predictors in model construction. However, it was only thanks to the graphical approach that the relation appeared to be clearly negative. Combining the previous information, it was possible to conclude that a thin carcass was characterized by a higher portion of PUFA (relatively to MUFA) than fatter carcasses. Nevertheless, this relation seemed to not be linear in the model (Figure 2b).

Curious variations were sometimes observed. The $M_{n-6/n-3}$ was built with seven independent variables, including the FCR (Si = 0.20) and the fat proportion (Si = 0.42), which were the most influential. The *n-6/n-3* prediction was low for the intermediate values of FCR and fat proportions, and higher when the variables were getting close to the edges of the values tested (Figure 2c).

$M_{Tenderness}$ appeared to be mainly linked to the amount of fat (Si = 0.55) and the weight of the animal (Si = 0.15) or of its carcass (Si = 0.19). The sum of all the independent variables was not exactly equal to one. This observation was almost always true for all the model analyses, and will therefore be discussed further (Section 4.1.). The influence of fat development was not very clear in Figure 2d. The highest tenderness seems to be reached for intermediate fat development. The tenderness seemed to be also very low when the animals were very fat.

Although some of the models were difficult to interpret with biological certitude, there were models where the conclusions were simpler. For instance, the $M_{rancid\ and\ fish\ flavors}$ value was mostly affected by the PUFA/MUFA (Si = 0.81). In Figure 2e, it is clearly shown that after reaching a certain proportion of PUFA (relatively to MUFA), the level of unwanted flavors dramatically increased.

Likewise, the $M_{flavor\ intensity}$ was highly impacted by fat proportion (Si = 0.48) with a high intensity when the measure was high (Figure 2f). Therefore, the flavor intensity was easily linked to the intermuscular and the intramuscular fat, which partly composed the cluster and thus the linear combination calculated from it.

3.4. Global Relation between Nutritional and Organoleptic Quality

The Organoleptic Quality (OQ) and the Nutritional Quality (NQ) indexes of the virtual animals (points) are displayed in Figure 3, and overlapped by the 30 real animals (triangles) used to build the models. A slight negative correlation was observed between NQ and OQ ($r = -0.17$) for the indexes calculated with the virtual animals. The same correlation with the real animals was a little bit higher ($r = -0.25$), but stayed very low, indicating that these two parameters are only weakly linked. The correlation test indicates that the correlation between the synthetic indexes calculated with the virtual animals was significantly different from zero. This test does not show the strength of the correlation and is highly influenced by the number of observations, as it can be observed with a higher p value for the real animal, even though the correlation seemed to be stronger.

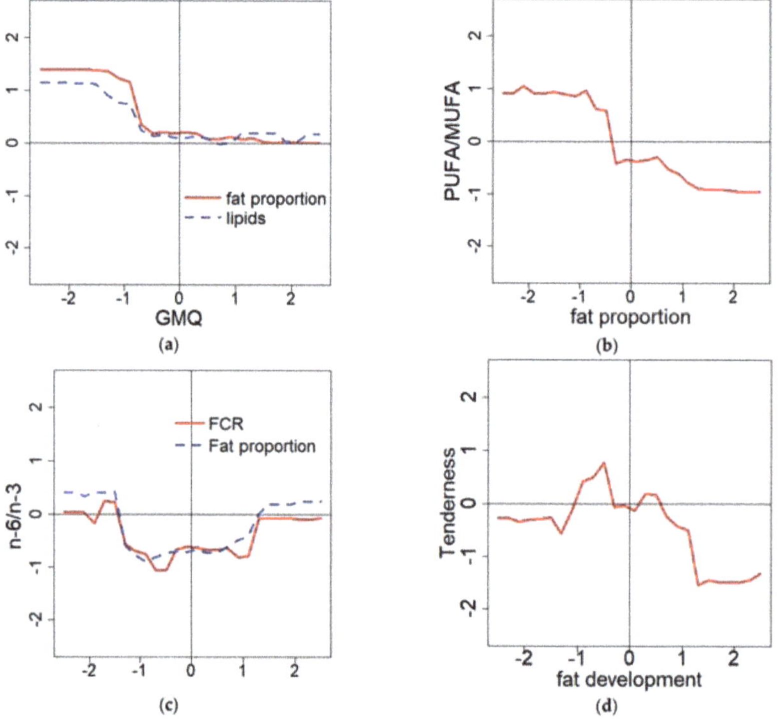

Figure 3. *Cont.*

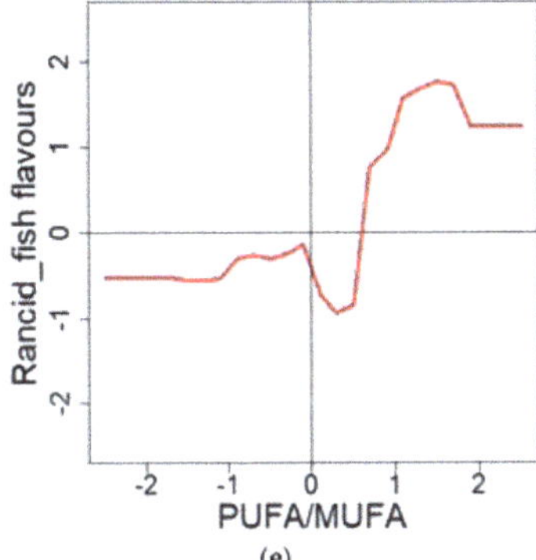
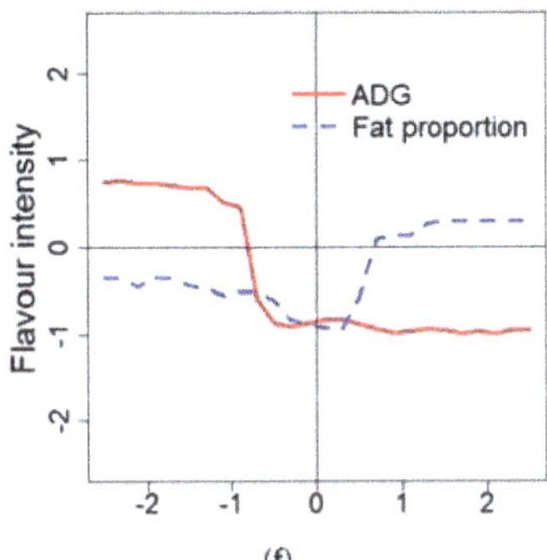

Figure 3. (**a**) Prediction of the fat proportion and the lipid content versus the ADG (average daily gain) variations; (**b**) Prediction of the PUFA/MUFA (polyunsaturated fatty acid/monounsaturated fatty acid)ratio versus the fat proportion variations; (**c**) Prediction of the *n-6/n-3* ratio versus the FCR (feed conversion ratio) and fat proportion variations; (**d**) Prediction of the tenderness versus the fat development variation; (**e**) Prediction of the unwanted rancid-fish flavor versus the PUFA/MUFA ratio variations; (**f**) Prediction of the flavor intensity versus the ADG and the fat proportion variations.

3.5. Comparison between the Best and the Worst Profile

Among the virtual animals (Figure 4), the traits of the 30 best (blue points) and worst (red points) animals were compared in Figure 5. Overlapped on the boxplots, three real animals for each category (best and worst) were also selected (black triangles) in Figure 4 and compared with virtual ones in Figure 5 (blue and red triangles). Since they were selected for their nutritional and organoleptic quality, the best animals naturally have better traits regarding these qualities.

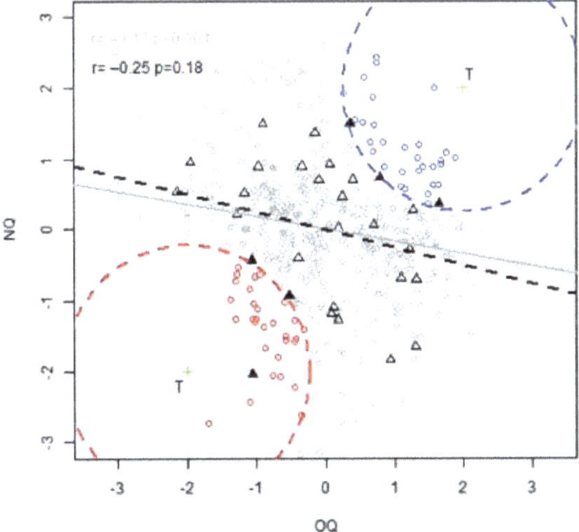

Figure 4. Nutritional and organoleptic indexes of the virtual (points) and real (triangles) animals. The correlation of the virtual (respectively real) animals is given in grey (respectively in black) in the top left corner. Regression (NQ~OQ) line for both virtual and real animals is added to visualize the correlation. Two targeted green crosses T were set to select the closest best (blue) and worst (red) virtual animals (and real animals filled in black for both categories).

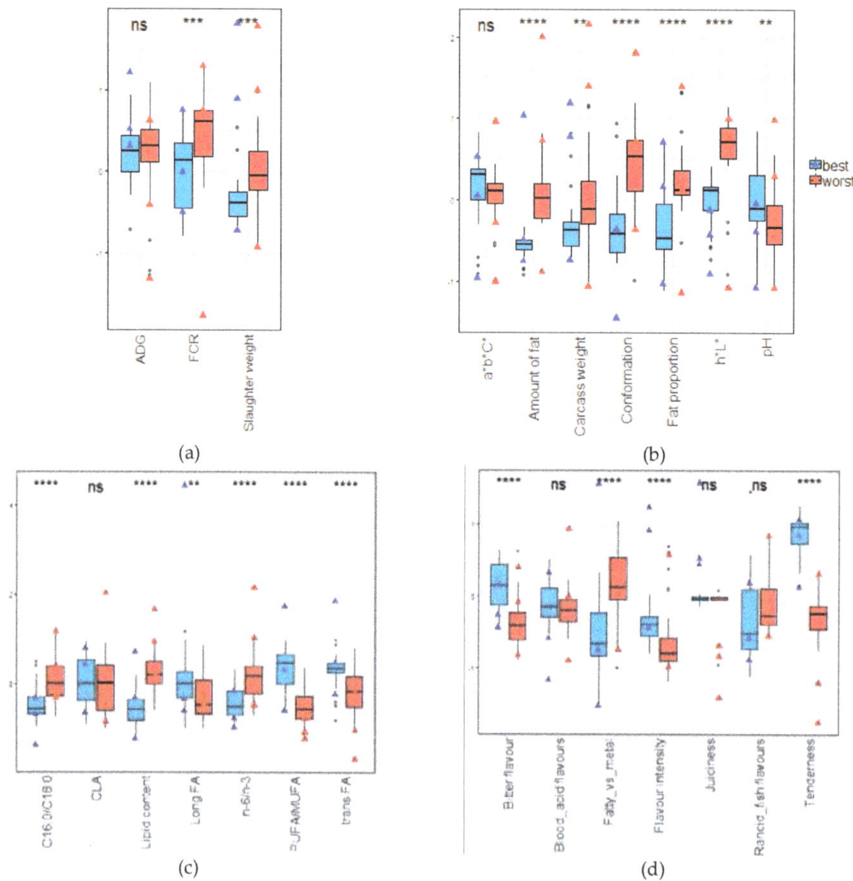

Figure 5. Comparison (boxplots) of the virtual animals traits between the worst (red) and the best (blue) profiles selected. The three best and worst real animals are added on the boxplots with triangles. For each variable, the result of a Wilcoxon test is provided (ns: no significant, **: *p* value < 0.01, ***: *p* value < 0.001, ****: *p* value < 0.0001). (**a**) animal performances; (**b**) carcass properties; (**c**) nutritional quality; (**d**) organoleptic quality.

Regarding the organoleptic quality (Figure 4d), the best animals had a better tenderness and flavor intensity than the worst animals. Even though the models were useless for the juiciness (predicting only the mean juiciness, because the model does not fit the variation of the juiciness at all), the best real animals (triangles) seemed to have a better juiciness than the worst real animals selected. On the abnormal flavor side, the best animals seemed to be more bitter than the worst animals, but the fatty/metal taste was stronger for the worst animals. No significant difference was reported for the rancid/fish and blood/acid flavor between the two extremes.

Regarding nutritional quality, the best animals had lower lipid content (Figure 5c). Logically, the C16:0 proportion was lower, meaning probably that the SFA content was also lower. In addition, PUFA/MUFA was higher, which also seems logical when the SFA is low. Among the higher proportion of PUFA, the *n*-3 fatty acid seemed to have a higher proportion in comparison to the *n*-6. Associated with higher PUFA proportions, *trans* fatty acids were also higher. No significant difference between CLA proportions was observed, but the proportion of long fatty acids was higher for the best animals.

Although not selected for this parameter of interest, the worst animals seemed to have fatter carcasses that were heavier and had conformations (Figure 5b). Those fatter carcasses seemed to have an impact on the luminescence (L^*) of the meat, but not on the $a^*b^*C^*$ indicator, where no difference was observed. The pH of the virtual animals seemed higher for the best virtual animals, even though this result was in contradiction with the six real animals selected.

Related to these observations on the carcass traits, the slaughter weight of the best animals was lighter. To produce fatter carcasses, it seems that the worst animals also had a lower feed efficiency (Figure 5a). No difference between the average daily gain was observed even if best real animals tended to be better.

4. Discussion

4.1. Modeling and Analytic Choices

All the approaches were based on a reliable modeling of the relations between the variables evaluating the parameters of interest. Some of the models were clearly not reliable. The worst case was that of the juiciness, which had very low adjusted R^2 values, and seemed not able to predict anything other than the juiciness mean (as can be seen in Figure 5d).

The results coming out the modvarsel package will also be discussed, especially regarding the type of model selection. Among the five considered regression models in the modvarsel R package, two model types have never been selected: plsr and sir (see Figure 1). This could be explained by a lack of (linear) structure regarding the variables and the relatively low number of observations (animals) compared to the relatively high number of independent variable candidates. Thus, the dimension reduction techniques were affected and not better than the other competitive fully parametric or non-parametric regression models. Linear regression was selected only once. In contrast, random forest (a purely non-parametric approach) gave a good accuracy in this kind of situation, which explained the high rate of random forest model selected by modvarsel.

Linear regression was selected to predict the carcass weight, which was strongly correlated with the animal's body weight before slaughter. Curiously, the slaughter weight was not fitted by a linear regression model, in which carcass weight would be the main independent variable in return. Indeed, it appears that a higher number of variables was selected by modvarsel to fit the slaughter weight (carcass weight, fat development, C16:0/C18:0 ratio, and PUFA/MUFA ratio) than to fit the carcass weight (slaughter weight and ADG). Some slight correlations between the four variables were observed, such as between carcass weight and fat development ($r = 0.67$, p value < 0.001) for instance, which might explain the use of penalized regression to build a more stable model, as aimed by the modvarsel algorithm. The low number of observations ($n = 30$) with influential points could also be the source of instability (high variance of the parameters' estimation) in the linear model fitting slaughter weight.

As it was noticed in the results (Section 3.2.), all 18 random forest models had high overfitting. This is related to the non-parametric nature of the algorithm, which is known to produce high overfitting when the number of observations is low compared to the number of independent variable [29]. For instance, it was interesting to notice that the $M_{n\text{-}6/n\text{-}3}$ variations were characterized by a kind of "wave", where the ratio was favorable for intermediate values of FCR and fat proportion. These relations did not have a clear biological explanation. Nevertheless, increasing the size of the training dataset might prevent the overfitting problem, with the robustness of the non-parametric models being based on the law of large numbers.

As noticed in Section 3.3, when analyzing the models, contradictions appeared between the two methods (i.e., the sensitivity indexes (Si) and the graphical approach). For instance, regarding the $M_{\text{fat proportion}}$ value, the ADG had a low sensitivity index, but was responsible for a significant part of the variations with the graphical method. In general, those contradictions appeared when the number of independent variables into the model was important, and the accuracy of the model was low. When both of these conditions were met, it could lead to difficulties regarding obtaining a reliable sensitivity index. Furthermore, the graphical approach was based on the variation of only one independent variable, and the other one was set to the mean (i.e., 0). Therefore, interaction phenomena were factored out of the analysis, which could also explain the opposition between both of the methods' results.

4.2. Outcome from the Holistic Approach

First, this original approach has shown that there is not a strong antagonism ($r = -0.17$) between nutritional and organoleptic quality. Moreover, the best profiles selected had very satisfying meat quality regarding the quality based on the initial population of 30 animals. This observation consolidated a previous work [2], where Ellies-Oury et al. indicated that these two parameters were orthogonal to each other and thus linearly independent. These results were obtained in a population of heterogeneous animals, which came from three different breeds (Angus, Limousine, Blonde d'Aquitaine), whereas in the present work, the results were obtained on a homogenous population of Blonde d'Aquitaine females. The lack of a strong linkage between the two parameters of interest (NQ and OQ) means that a trade-off is not fully necessary to maximize NQ and OQ simultaneously.

Going deeper in the best profiles analysis, the thin carcasses with a low proportion of fat seemed to drive to high PUFA and w3 proportions in the meat, as well as a higher intensity of flavor and tender meat. In contrast, the fattest carcasses seemed to produce meat with high lipids content that lacked PUFA and had an imbalanced n-6/n-3 ratio. The models explain these results. In Section 3.2, a negative relation was shown between the fat proportion and the PUFA/MUFA ratio. This result support those from Warren et al. [30] indicating that muscles with higher lipid content have a lower proportion of PUFA compared to MUFA and SFA. These could be explained by a higher *de novo* synthesis of fatty acid, which was mostly saturated and monounsaturated. In leaner meat, the triglycerides/phospholipids ratio was higher, and it is well known that PUFA mainly esterified on phospholipids [31].

In the literature, studies have shown a positive relation between fat and tenderness [32]. However, the model built in this study was different. As noticed in Figure 2d, the tenderness optimum is reached when the fat development is between -1 and 0 (which means that the animal is close to the mean or lower by one standard deviation), and the worst tenderness predicted was obtained with an upper fat development that was one standard deviation over the mean. This result was also observed in the best and worst profiles. The contradiction between this model and the literature cannot be explained by a low model quality, because the accuracy of the $M_{tenderness}$ is quite satisfying (adjR^2 = 0.85 and cor_adjR^2 = 0.60). However, the model was fit on a specific population of 30 Blonde d'Aquitaine heifers. This breed is well known to produce very lean meat in comparison to the other suckling breeds [33], which could explain the different relationship between this fat indicator and the tenderness of the meat.

As previously shown, the best animals had higher PUFA/MUFA in the meat. However, high PUFA content can also lead to the risk of unwanted flavor (e.g., the fish flavor indicated in Figure 2e). This effect seems to be related to PUFA oxidation [5]. Nevertheless, the higher level of PUFA in the best profile meat did not cause this kind of abnormal flavor.

Finally, these results confirmed that meat with higher intramuscular and intermuscular fat increases the flavor intensity [4], indicating that intramuscular fat is a precursor of many aromatic compounds formed, notably, during cooking processes.

4.3. Limits and Perspectives of the Trade-Off Method

Generating virtual observations is unusual in the animal research field, and might raise interrogations and doubts regarding the results for the reader. However, there are no statistical aberrations, because this study did not try to show the impact of one factor (the diet effect for instance) on the beef performances or meat qualities or the strength of relation between two variables. All the probabilistic tests that were performed in this paper (the Wilcoxon test of comparison between the best and worst profiles traits) should not be interpreted literally; they only demonstrate that with the relation modeled in this study, the best profiles are significantly different than the worst ones regarding various traits.

Aside from this innovative approach, the trade-off method seemed to be highly sensitive to the weighting granted for each variable by the experts. For instance, tenderness (whose weighting is significant) was the highest among the best profiles (and lowest among the worst profiles). Then, the result of optimizing was clearly dependent on the expertise provided for the relative importance of each variable evaluating the parameters of interest. Therefore, the choice of variable weights has to be carried out carefully and wisely to manage the trade-offs as efficiently as possible.

The weighting dependence can also be related to the method used to aggregate the variables, considering its numerous disadvantages. The linear combination calculated here to aggregate the variables led to compensation between them. For instance, a high proportion of PUFA would compensate for a high *trans* fatty acid proportion. No threshold was taken into account as a way to prefer a more balanced animal compared to an animal performing well on only a few indicators. More complex methods exist to resolve these problems such as outranking methods, but suffer from less clarity and are difficult to assess [34]. In addition, the weight-related techniques need variables varying in the same scale and the same direction. In the case of some of them, such as the carcass weight, the optimum value is not a maximum or a minimum weight, but rather a homogeneous weight distributed around a target value.

5. Conclusions

The present paper has exposed a very innovative approach to assess the trade-off between parameters of interest in the beef cattle industry. It showed that there is no antagonism between organoleptic quality and nutritional quality. The modeling approach has also been very interesting to highlight the relation between the variables and show that they are interconnected, and not only by two-by-two relations. This knowledge should be taken into account when designing new product brand or label expectations.

However, the very strong influence of the weight setting and the aggregation method on the results of the trade-off was observed. This should encourage the meat beef industry to clarify their own expectations regarding these two parameters of interest in order to set consensual indicators with associated hierarchy to measure the organoleptic and nutritional quality.

Moreover, the significance of these results is only relevant in the particular context of the data in terms of the animals' breed, age, diets, fattening duration, muscle, etc. Thus, the extrapolation of the present results has to be done with caution.

Author Contributions: Methodology, A.C.; formal analysis, C.A., M.-P.E.-O., B.P., J.S., M.C.; data curation, J.N., C.D.; writing—original draft preparation, A.C.; writing—review and editing, C.A., M.-P.E.-O., G.C.-H., B.P., D.D., D.G.; visualization, A.C.; supervision, M.-P.E.-O., B.P., J.S., M.C., D.D., D.G., J.N.

Funding: This research used data which were funded by National agency of research (Agence Nationale de la Recherche, ANR) under the "National program for research in human alimentation and nutrition", project "ANR-06-PNRA-018-03".

Acknowledgments: We would like to thank the technician and administrative staff of the experimental farm of Mauron, the French Livestsock Institute, and of the National Institute of Agricultural Research, which contributed to this work with material and time support. We thank all of the people involved in this project for their assistance in data collection, the management and slaughtering of animals, and muscle sampling. We especially thank Daniel Le Pichon (Brittany Regional Chamber of Agriculture), Didier Bastien, Jérôme Normand, Aude Bertout (French Livestock Institute), Dominique Bauchart, Denys Durand and Mylène Delosière (INRA UMRH, Theix) for the collection and provision of data.

Conflicts of Interest: The authors declare no conflict of interest.

Appendix A

Table A1. Variables in the initial dataset with the clustering assignment and the associated coefficient of correlation between the variables and the synthetic index (calculated as the first component of the principal analysis component (PCA) with all the variables in the cluster).

Variable	Precision and Units	Cluster Codification	Correlation with the Cluster Synthetic Index
Slaughter weight	Slaughter day weight	-	-
ADG	Average daily gain (kg/day) during the fattening period (100 days)	-	-
FCR	Feed conversion ratio (DM intake/ADG)	-	-
Carcass weight	Carcass weight (kg)	CP1	−0.96
Conformation	EUROP conformation (from 1 (P−) to 15 (E+))	CP2	1
Body fat score	Body fat score at slaughter (from 1 to 5)	CP3	−0.74
Pelvis fat	(kg)	CP4	0.79
Kidney fat	(kg)	CP4	0.92
Trimming fat	(kg)	CP4	0.84
Total fat	(kg)	CP4	0.99
Eliminated fat	(kg)	CP4	0.98
Intramuscular fat score	Score from 3 (no trace of fat) to 12 (extremely fat)	CP5	−0.75
Intermuscular fat score	Score from 0 (without intermuscular fat) to 5 (high intermuscular fat)	CP5	−0.87
pH	pH	CP1	0.81
Meat color score	Score from 1 (very light) to 4 (very red)	CP3	0.74
L^*	from black (0) to white (100) of the CIE $L^*a^*b^*$ coordinates system	CP3	−0.81
a^*	from green (−) to red (+) of the CIE $L^*a^*b^*$ coordinates system	CP6	0.99
b^*	from blue (−) to yellow (+) of the CIE $L^*a^*b^*$ coordinates system	CP6	0.98
C^*	The chroma of the CIE $L^*C^*h^*$ coordinates system	CP6	1
h^*	The hue angle of the CIE $L^*C^*h^*$ coordinates system	CP3	−0.85
Muscle proportion	Muscle proportion into the carcass	CP5	0.91
Fat proportion	Fat proportion into the carcass	CP5	−0.95
Bone proportion	Bone proportion into the carcass	CP1	0.97
Lipid content	Lipid content into LT muscle	-	-
C16:0	(%/total FA)	NQ3	−0.88
C18:0	(%/total FA)	NQ2	0.95
SFA	(%/total FA)	NQ3	-0.86
C18:1 cis9	(%/total FA)	NQ4	−0.94
MUFA cis	Sum of all MUFAcis (%/total FA)	NQ4	−0.93
C18:1 tr9	(%/total FA)	NQ5	0.61
C18:1 tr10	(%/total FA)	NQ6	0.98
C18:1 tr11	(%/total FA)	NQ5	0.92
MUFA trans	Sum of all MUFA trans (%/total FA)	NQ6	0.98
Total MUFA	Sum of all MUFA (%/total FA)	NQ4	−0.92
LA	Linoleic acid C18:2 *n*-6 en (%/total FA)	NQ4	0.95
C20:4 *n*-6	C20:4 *n*-6 (%/total FA)	NQ4	0.80
Total *n*-6	Sum of all PUFA *n*-6 (%/total FA)	NQ4	0.97
ALA	Linolenic acid C18:3 *n*-3 (%/total FA)	NQ3	0.92
EPA	C14:0 (%/total FA)	NQ4	0.93
DPA	C14:0 (%/total FA)	NQ4	0.89
Total *n*-3 LC	Sum of all long-chain FA *n*-3 (%/total FA)	NQ4	0.91
Total *n*-3	Sum of all FA *n*-3 (%/total FA)	NQ3	0.90
LC FA content	Content of all the long-chain FA	NQ1	0.98
LC FA prop	Sum of all the long chain FA (%/total FA)	NQ1	0.98
CLA	Conjugated linoleic acid C18:2 9cis 11trans (%/total FA)	NQ5	0.97
total CLA	Sum of all the conjugated FA (%/total FA)	NQ5	0.96
Total PUFA	Sum of PUFA (%/total FA)	NQ4	0.94
n-6/*n*-3	*n*-6 PUFA/*n*-3 PUFA ratio (%/total FA)	NQ3	−0.89
LA/ALA	C18:2 *n*-6/C18:3 *n*-3 ratio (%/total FA)	NQ3	−0.87
PUFA /SFA	PUFA/SFA ratio (%/total FA)	NQ4	0.93
C16:0/C18:0	C16:0/C18:0 ratio (%/total FA)	NQ2	−0.95
Tenderness intensity	Score from 1 (very hard) to 100 (very tender)	-	-
Juiciness intensity	Score from 1 (very dry) to 100 (very juicy)	-	-
Flavor intensity	Score from 1 (low intensity) to 100 (high intensity)	OQ1	−0.82
Sweet	Score from 1 (not sweet) to 100 (very sweet)	OQ1	−0.82
Acid	Score from 1 (not acid) to 100 (very acid)	OQ5	−0.85
Bitter	Score from 1 (no bitter) to 100 (very bitter)	OQ2	1
Metallic taste	Score from 1 (low metallic taste) to 100 (strong metallic taste)	OQ4	0.80
Rancid taste	Score from 1 (low rancid taste) to 100 (strong rancid taste)	OQ3	0.88
Fat taste	Score from 1 (low metallic fat) to 100 (strong fat taste)	OQ4	−0.80
Fish taste	Score from 1 (low fish taste) to 100 (strong fish taste)	OQ3	0.88
Blood taste	Score from 1 (low blood taste) to 100 (strong blood taste)	OQ5	−0.85

CP$_i$: carcass property cluster i; NQ$_i$: nutritional quality cluster i; OQ$_i$: organoleptic quality cluster i.

Appendix B

```
 1   Initiation
 2
 3       load(Models) #A set of 24 model of prediction for each variable
 4       Threshold <- 0.5
 5
 6       #Initialize a virtual animal with 0 and choose the PI to set
 7       virtual_animal <- rep(value = 0, time = 24) #[0,0,...0]
 8       PI <- [AP=1:3, CP=4:10, NQ=11:17, OQ=18:24][sample(1:4, 1)]
 9
10   Treatment
11
12       #set the variable values of a randomly chosen PI
13       Do
14           virtual_animal[PI] <- rep(value = rnorm(n = 1, mean = 0, sd = 2), time = length(PI))
15       while (checkIfRealistic*(virtual_animal[PI]))
16
17       #use models to predict the rest of the virtual_animal variables
18       Do
19           For (i in sample((1:24)[-PI]))
20               virtual_animal_prec <- virtual_animal
21               virtual_animal[i] <- predict(Models[i], newdata = virtual_animal)
22           endFor
23       while (sum(virtual_animal_prec - virtual_animal) > threshold)
24
25   Output
26
27       virtual_animal
28
29   *this function check if the set values respect the linear relation between the
30   variables based on linear model and confidense interval at 95%.
```

Figure A1. Algorithm to generate a virtual animal. The pseudo code was inspired from the R code.

References

1. FranceAgriMer. *Les Filières Animales Terrestres et Aquatiques Bilan 2013 Perspectives 2014*; FranceAgriMer: Paris, France, 2014.
2. Ellies-Oury, M.P.; Cantalapiedra-Hijar, G.; Durand, D.; Gruffat, D.; Listrat, A.; Micol, D.; Ortigues-Marty, I.; Hocquette, J.F.; Chavent, M.; Saracco, J. An innovative approach combining animal performances, nutritional value and sensory quality of meat. *Meat Sci.* **2016**, *122*, 163–172. [CrossRef] [PubMed]
3. Jeremiah, L.E.; Dugan, M.E.R.; Aalhus, J.L.; Gibson, L.L. Assessment of the relationship between chemical components and palatability of major beef muscles and muscle groups. *Meat Sci.* **2003**, *65*, 1013–1019. [CrossRef]
4. Thompson, J.M. The effects of marbling on flavour and juiciness scores of cooked beef, after adjusting to a constant tenderness. *Austral. J. Exp. Agric.* **2004**, *44*, 645–652. [CrossRef]
5. Wood, J.D.; Enser, M.; Fisher, A.V.; Nute, G.R.; Sheard, P.R.; Richardson, R.I.; Hughes, S.I.; Whittington, F.M. Fat deposition, fatty acid composition and meat quality: A review. *Meat Sci.* **2008**, *78*, 343–358. [CrossRef] [PubMed]
6. Tables Inra. *Alimentation des Bovins, Ovins et Caprins. Besoin des Animaux—Valeurs des Aliments*; Quae: Versailles, France, 2007.
7. FranceAgriMer. *Pesée/Classement/Marquage*; Guide Technique et Reglementaire; FranceAgriMer: Paris, France, 2010.
8. Robelin, J.; Geay, Y.; Jailler, R.; Cuylle, G. Estimation de la composition des carcasses de jeunes bovins à partir de la composition d'un morceau monocostal prélevé au niveau de la 11e côte. I.—Composition anatomique de la carcasse. *Ann. Zootech.* **1975**, *24*, 391–402. [CrossRef]
9. Wellington, G.H.; Stouffer, J.R. *Beef Marbling: Its Estimation and Influence on Tenderness and Juiciness, 941*; Cornell University Agricultural Experiment Station: Ithaca, NY, USA, 1959.
10. Hunt, M.C.; Acton, J.C.; Benedict, R.C.; Calkins, C.R.; Cornforth, D.P.; Jeremiah, L.E.; Olson, D.G.; Salm, C.P.; Savell, J.W.; Shivas, S.D. *Guidelines for Meat Color Evaluation*; American Meat Science Association, Kansas State University: Manhattan, KS, USA, 1991; pp. 1–17.
11. McLaren, K. An introduction to instrumental shade passing and sorting and a review of recent developments. *J. Soc. Dye. Colour.* **1976**, *92*, 317–326. [CrossRef]
12. Habeanu, M.; Thomas, A.; Bispo, E.; Gobert, M.; Gruffat, D.; Durand, D.; Bauchart, D. Extruded linseed and rapeseed both influenced fatty acid composition of total lipids and their polar and neutral fractions in longissimus thoracis and semitendinosus muscles of finishing Normand cows. *Meat Sci.* **2014**, *96*, 99–107. [CrossRef]

13. Folch, J.; Lees, M.; Stanley, G.H.S. A simple method for the isolation and purification of total lipides from animal tissues. *J. Biol. Chem.* **1957**, *226*, 497–509.

14. Bauchart, D.; Gobert, M.; Habeanu, M.; Parafita, E.; Gruffat, D.; Durand, D. Influence des acides gras polyinsaturés *n*-3 et des antioxydants alimentaires sur les acides gras de la viande et la lipoperoxydation chez le bovin en finition. *Cah. Nutr. Diététique* **2010**, *45*, 301–309. [CrossRef]

15. Scislowski, V.; Durand, D.; Gruffat, D.; Bauchart, D. Dietary linoleic acid-induced hypercholesterolemia and accumulation of very light HDL in steers. *Lipids* **2004**, *39*, 125–133. [CrossRef]

16. Chavent, M.; Kuentz-Simonet, V.; Liquet, B.; Saracco, J. ClustOfVar: An R Package for the Clustering of Variables. *J. Stat. Softw.* **2012**, *50*, 1–16. [CrossRef]

17. Monteils, V.; Sibra, C.; Ellies-Oury, M.P.; Botreau, R.; de la Torre, A.; Laurent, C. A set of indicators to better characterize beef carcasses at the slaughterhouse level in addition to the EUROP system. *Livest. Sci.* **2017**, *202*, 44–51. [CrossRef]

18. Pereira, P.M.C.C.; Vicente, A.F.R.B. Meat nutritional composition and nutritive role in the human diet. *Meat Sci.* **2013**, *93*, 586–592. [CrossRef]

19. Williams, C.M. Dietary fatty acids and human health. *Ann. Zootech.* **2000**, *49*, 165–180. [CrossRef]

20. Williams, C.M. Nutritional composition of red meat. *Nutr. Diet.* **2007**. [CrossRef]

21. Polkinghorne, R.; Philpott, J.; Gee, A.; Doljanin, A.; Innes, J. Development of a commercial system to apply the Meat Standards Australia grading model to optimise the return on eating quality in a beef supply chain. *Austral. J. Exp. Agric.* **2008**, *48*, 1451–1458. [CrossRef]

22. Ellies-Oury, M.P.; Chavent, M.; Conanec, A.; Bonnet, M.; Picard, B.; Saracco, J. A relevant way based on variable importance for biomarkers selection to predict meat tenderness. *Sci. Rep.* **2019**. under review.

23. Harrell, F.; Lee, K.; Mark, D. Multivariable prognostic models: issues in developing models, evaluating assumptions and adequacy, and measuring and reducing errors. *Stat. Med.* **1996**, *15*, 361–387. [CrossRef]

24. Sobol, I. Global sensitivity indices for nonlinear mathematical models and their Monte Carlo estimates. *Math. Comput. Simul.* **2001**, *55*, 271–280. [CrossRef]

25. Iooss, B.; Janon, A.; Pujol, G. Sensitivity: Global Sensitivity Analysis of Model Outputs. R Package Version 1.15.2. Available online: https://CRAN.R-project.org/package=sensitivity (accessed on 6 June 2019).

26. Yang, X.J.; Albrecht, E.; Ender, K.; Zhao, R.Q.; Wegner, J. Computer image analysis of intramuscular adipocytes and marbling in the longissimus muscle of cattle. *J. Anim. Sci.* **2006**, *84*, 3251–3258. [CrossRef]

27. Mancini, R.A.; Hunt, M. Current research in meat color. *Meat Sci.* **2005**, *71*, 100–121. [CrossRef]

28. ANSES. *Actualisation des Apports Nutritionnels Conseillés Pour les Acides Gras*; 2006-SA-0359; ANC AG; ANSES: Maisons-Alfort, France, 2006.

29. Breiman, L. Random forests. *Mach. Learn.* **2001**, *45*, 5–32. [CrossRef]

30. Warren, H.E.; Scollan, N.D.; Enser, M.; Hughes, S.I.; Richardson, R.I.; Wood, J.D. Effects of breed and a concentrate or grass silage diet on beef quality in cattle of 3 ages. I: Animal performance, carcass quality and muscle fatty acid composition. *Meat Sci.* **2008**, *78*, 256–269. [CrossRef] [PubMed]

31. De Smet, S.; Raes, K.; Demeyer, D. Meat fatty acid composition as affected by fatness and genetic factors: A review. *Anim. Res.* **2004**, *53*, 81–98. [CrossRef]

32. Bonny, S.P.F.; Hocquette, J.-F.; Pethick, D.W.; Farmer, L.J.; Legrand, I.; Wierzbicki, J.; Allen, P.; Polkinghorne, R.J.; Gardner, G.E. The variation in the eating quality of beef from different sexes and breed classes cannot be completely explained by carcass measurements. *Animal* **2016**, *10*, 987–995. [CrossRef]

33. Maniaval, O. Une Nouvelle Utilisation Zootechnique de L'échographie: Estimation de L'état Corporel des Bovins; Application sur Quarante Blondes D'aquitaine en Période D'engraissement. Ph.D. Thesis, Ecole Nationale Vétérinaire de Toulouse—ENVT, Toulouse, France, 2008.

34. Roy, B. The outranking approach and the foundations of ELECTRE methods. In *Readings in Multiple Criteria Decision Aid*; Springer: London, UK, 1990; pp. 155–183. ISBN 978-3-642-75935-2.

Communication

Assessment of RNAlater® as a Potential Method to Preserve Bovine Muscle Proteins Compared with Dry Ice in a Proteomic Study

Yao Zhu [1,2], Anne Maria Mullen [1], Dilip K. Rai [1], Alan L. Kelly [2], David Sheehan [3], Jamie Cafferky [1] and Ruth M. Hamill [1,*]

[1]   Teagasc Food Research Centre, Ashtown, Dublin D15KN3K, Ireland; yao.zhu@teagasc.ie (Y.Z.);
      Anne.mullen@teagasc.ie (A.M.M.); dilip.rai@teagasc.ie (D.K.R.); jamie.cafferky@teagasc.ie (J.C.)
[2]   School of Food and Nutritional Sciences, University College Cork, Cork T12 K8AF, Ireland; a.kelly@ucc.ie
[3]   Department of Chemistry, Khalifa University, Abu Dhabi PO Box 127788, UAE; david.sheehan@ku.ac.ae
*   Correspondence: ruth.hamill@teagasc.ie; Tel.: +353-(0)18059933

Received: 9 January 2019; Accepted: 2 February 2019; Published: 5 February 2019

Abstract: RNAlater® is regarded as a potential preservation method for proteins, while its effect on bovine muscle proteins has rarely been evaluated. Bovine muscle protein samples (n = 12) collected from three tender (Warner–Bratzler shear force: 30.02–31.74 N) and three tough (Warner–Bratzler shear force: 54.12–66.25 N) Longissimus thoracis et lumborum (LTL) samples, preserved using two different sampling preservation methods (RNAlater® and dry ice), at two *post mortem* time points (day 0 and day 14), were characterized using one-dimensional electrophoresis. Fourteen bands with molecular weights ranging from 15 to 250 kDa were verified, both in the dry ice and RNAlater® storage groups, at each time point, using image analysis. A shift from high to low molecular weight fragments, between day 0 and day 14, indicated proteolysis of the muscle proteins during post mortem storage. Liquid chromatography-tandem mass spectrometry (LC-MS/MS) analyses and database searching resulted in the identification of 10 proteins in four bands. Protein profiles of muscle preserved in RNAlater® were similar to those of muscle frozen on dry ice storage, both at day 0 and day 14. The results demonstrate that RNAlater® could be a simple and efficient way to preserve bovine muscle proteins for bovine muscle proteomic studies.

Keywords: muscle proteins; one-dimensional electrophoresis; bovine proteomics; LC-MS/MS; sample preparation

---

1. Introduction

Meat proteomics research critically depends on the reliability of tissue samples. It is essential to avoid proteolysis and enzymatic activity in sample preservation, so as to conserve the structural integrity of muscle proteins. Snap freezing of samples in liquid nitrogen, at $-196\ ^\circ$C, or on dry ice, at $-79\ ^\circ$C, is an efficient method to stabilize samples. Nonetheless, they can be difficult to carry out in-factory. RNAlater®, an aqueous solution, is extensively used to preserve RNA in fresh tissue and cell samples for clinical genomic and transcriptomic studies [1]. The main compounds in RNAlater® are quaternary ammonium sulphates and cesium sulphate, which denature proteins, including RNases, DNases and proteases, thereby stabilizing RNA, DNA and protein content [2].

RNAlater® has been investigated as an effective potential preservation method for proteins originating from bacteria, plants, blood vessels and human colon mucosal biopsies [2–4]. However, comparative proteomic studies between frozen and RNAlater® preserved tissues have shown some differences across tissues, for example, more proteins were detected in mouse leukocytes preserved using RNAlater® [5], compared to frozen tissue, and three of 20 analysed proteins were more abundant in snap-frozen mouse liver [6], whereas one was more abundant in RNAlater® preserved tissue. Furthermore, it was observed that 84 soluble proteins and 120 membrane-bound proteins were expressed differently in RNAlater® fixed samples compared with liquid nitrogen [7].

To date, little attention has been paid to the impact of RNAlater® on proteomic profiles from muscle tissue. In a study on gulf killifish preserved in liquid nitrogen and RNAlater®, it was demonstrated that, in contrast to their findings in other tissues, for skeletal muscle and brain, snap freezing and preservation in RNAlater® showed similar protein abundances [8]. Large biobanks of samples stored in RNAlater® exist for muscle transcriptomics, which could serve as an important resource for meat proteomics if muscle protein abundance profiles post mortem were similar to those in snap frozen muscle. However, no studies are available which assess the potential differential effect of storage in RNAlater® versus freezing on dry ice on post mortem proteolytic patterns in bovine muscle.

The objective of this study was to compare proteome patterns of *post mortem* bovine muscle stored in RNAlater® or dry ice by using one-dimensional sodium dodecyl sulphate-polyacrylamide gel electrophoresis (1D SDS-PAGE). To the best of our knowledge, this is the first systematic study of the effect of RNAlater® on bovine muscle proteins, with a view to investigate its potential as a specimen-stabilization solvent for beef proteomics.

2. Materials and Methods

2.1. Tissue Sampling

Twenty-eight Limousin bulls were slaughtered in an EU-licensed abattoir under standard conditions. *Longissimus thoracis et lumborum* (LTL) samples (2 g) were collected approximately 1 h post-slaughter (day 0), from the region of the 10th rib of each carcass on the right side, in the chill room. Muscle (LTL) tissue samples (1 g) were finely macerated and stored at 4 °C in 5 mL RNAlater® for 24 h and, then, RNAlater® was removed by plastic pipette and the sample was transferred to −80 °C. In parallel, 1 g tissue was frozen on dry ice, and transferred to a freezer at −80 °C for downstream analysis. Two steaks (2.54 cm thick) from each LTL was excised on day 2 post mortem, vacuum packaged and stored at 4 °C, and one was frozen at day 14 for shear force analysis. The other was sampled for proteomics, applying the same preservation methods as on day 0.

2.2. Warner–Bratzler Shear Force Measurement

Steaks from each of the 28 animals were subjected to Warner–Bratzler shear force in accordance with American Meat Science Association (AMSA) guidelines [9]. Steaks (2.5 cm) were thawed in a circulating water bath at 20 °C, cooked in a circulating water bath (Grant Instruments Ltd., Cambridge, UK) at 72 °C, until an internal temperature of 69 °C was achieved. After cooking, the samples were stored at 4 °C overnight.

Eight cores (1.25 cm diameter) were taken from each steak parallel to fibre direction. Cores were sheared at the central point on a Warner–Bratzler device, attached to an Instron Universal testing machine with a 50 mm/min crosshead speed. Highest and lowest shear values were excluded for each sample and the mean values for 6 cores were reported.

2.3. Extraction of Muscle Proteins

The three most tender and three toughest animals, based on the shear force values (Section 2.2), were selected for proteomic analysis. All tissue samples (100 mg), RNAlater®-preserved and dry ice-preserved, at day 0 and day 14, from each of these six animals (n = 24 samples in total), were homogenized in 1 mL of 8.3 M urea, 2 M thiourea, 1% Dithiothreitol, 2% 3-[(3-cholamidopropyl) dimethylammonio]-1-propanesulfonate and 2% Immobilized pH gradient (IPG) buffer pH 3–10 (GE Healthcare). Homogenates were incubated with shaking for 30 min on ice, followed by a 30 min centrifugation at 10,000× g in order to remove unextracted cellular components, high molecular weight protein complexes and insoluble proteins. Protein concentrations of the supernatant were analysed at

595 nm using a microplate reader (BGM LABTECH, Ortenberg, Germany) based on the method of Bradford [10]. Bovine serum albumin (BSA) was used as the standard and the concentration of protein was shown to be between 3 to 6 mg/mL.

2.4. Proteomic Analysis

Protein extracts were separated by 1D SDS-PAGE [11] on commercial Mini-PROTEAN® TGX™ precast gradient gels of 8.6 × 6.7 × 0.1 cm and 4%–20% polyacrylamide (Bio-Rad Laboratories, Deeside, UK). The tissue sample protein lysis liquid was mixed 1:1 with Laemmli sample buffer (Bio-Rad Laboratories, Deeside, UK). Twenty µL (5 µg protein) was loaded in each gel lane. Each of the 24 samples was loaded in duplicate in contiguous lanes over 6 gels. Ten µL of Precision Plus Protein™ (Bio-Rad Laboratories, Deeside, UK) was included as a standard in one lane per gel. All gels were further replicated giving a total of four lanes per sample (96 lanes) with each sample run on two separate gels. Gels were run under constant voltage at 90 V for 45 min, then at 120 V for 50 min. Subsequently, gels were stained with 50 mL Coomassie stain (Bio-Rad Laboratories, Deeside, UK) with gentle shaking for 1 h and then destained with distilled water.

2.5. Image Analysis

Gel images were acquired using a GS-800 densitometer (Bio-Rad Laboratories, Deeside, UK) and analysed by Quantity One software (Bio-Rad). Bands were detected by optical density and quantified by integrating the area under the curve of pixel intensity and band width (trace quantity × mm). For each sample, each of the four replicate lanes were scanned and averaged to give the sample mean abundance for each animal × treatment × time point. Statistical analysis of the sample mean abundance was undertaken using Genstat (Release 14.1, VSN international, London, UK). The analysis was a 2 × 2 × 2 factorial with terms for Quality (Tender, Tough), Treatment (Dry Ice, RNAlater®) and Time point (Day 0/14). Tukey's test was used to compare means. A value of $p < 0.05$ was considered statistically significant.

2.6. Mass Spectrometry Analysis

Gels were transferred on to a glass plate and the protein bands of interest were excised with a sterile scalpel. Then, the gel pieces were spun down on a bench-top vortex. Bands were subjected to liquid chromatography-mass spectrometry at a core facility (King's College, Aberdeen, UK) according to the method of Wilm [12].

3. Results and Discussion

As shown in Figure 1, 14 bovine muscle protein bands with molecular weights ranging from 250 kDa to 15 kDa (B1-B14) were detected using quantitative image analysis. Fourteen bands were detected in each lane for both RNAlater®-preserved and dry ice-preserved samples.

No significant difference in optical density was detected for any band between the protein profiles of muscle preserved in RNAlater® and those of muscle samples frozen on dry ice, both at day 0 and day 14 (Table 1). These results are consistent with findings of Abbaraju [7], which showed that RNAlater® had no impact on protein patterns in skeletal muscle of gulf killifish compared with the snap-freezing method, and suggests we can extend this finding to the mammalian muscle context.

In relation to tenderness, no significant differences in protein optical density were identified between tender and tough samples, either on day 0 or at day 14 *post mortem* (Table 1). However, highly significant differences were observed for band 7 ($p = 0.01$) and band 9 ($p < 0.001$), between day 0 and day 14 (marked bold in Table 1). As shown in Table 2, there was a decrease in intensity of band 7 and an increase in intensity in band 9, from day 0 to day 14 *post mortem*.

As bands 7 and 9 showed significant differences during *post mortem* aging, they were further analysed by Liquid chromatography-tandem mass spectrometry. Bands 11 ($p = 0.08$) and 5 ($p = 0.08$) tended towards a treatment effect and an interaction effect of quality × treatment, respectively, so were also analysed by LC-MS/MS. A total of 10 proteins were identified in the four selected bands. Many were enzymes of carbohydrate metabolism, while some were structural proteins (Table S1). Troponin and glyceraldehyde-3-phosphate dehydrogenase were both found in both band 7 and 9 (Table S1).

The increase in density of band 9 (~30 kDa), from day 0 to day 14, could be explained partly by the appearance of polypeptides migrating at approximately 30 kDa during meat aging, which is in accordance with previous studies of meat tenderization [13,14]. MS of band 9 identified troponin-T, suggesting that this protein may be a proteolyzed breakdown product from band 7. Therefore, we compared the sequence of the observed troponin peptides between band 7 and band 9, results suggested that the KPLN IDHLSEDKLR sequence was not present in band 9 (Figure S1). These findings point to the increased degradation of troponin-T with increasing *post mortem* aging time, which is consistent with reports in the literature [15].

The ability to observe proteolytic abundance changes between these two bands has important implications for deducing the relationship between proteolysis of troponin-T and the emergence of a 30 kDa protein post-mortem. It has previously been shown that purified bovine troponin-T can be degraded by μ-calpain in vitro to produce polypeptides in the 30 kDa region [16]. In addition, 30 kDa products of troponin-T have been previously identified by Western blotting [15]. Interestingly, a noteworthy observation to emerge from the data comparison was that samples stored in RNAlater® retained a similar protein profile to samples preserved in dry ice in band 7 and band 9 during the aging time. Taken together, these results show that bovine muscle proteins preserved in RNAlater® present a consistent pattern with those frozen on dry ice and stored at −80 °C, when studied using proteomic approaches.

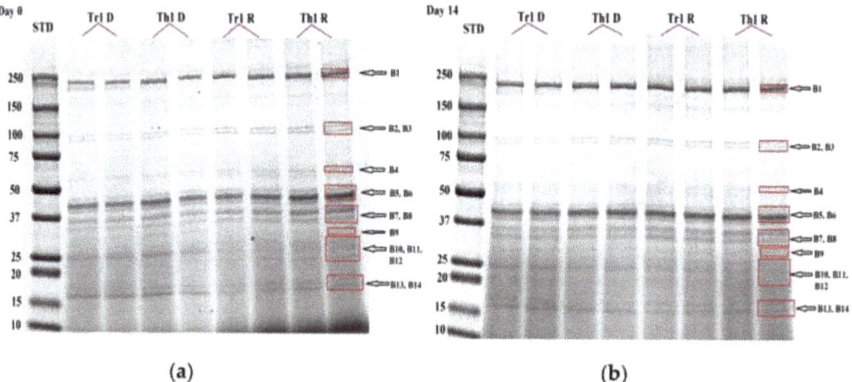

(a)   (b)

Figure 1. Four replicate lanes were run for each sample across 12 gels. Typical Coomassie-stained one-dimensional sodium dodecyl sulphate-polyacrylamide gel electrophoresis (1D SDS-PAGE) gels from the study, illustrating treatment, time point and quality samples run in duplicate, here data from tender sample 1 (Tr1) and tough sample 1 (Th1) are shown. The images above show proteins from *Longissimus thoracis et lumborum* (LTL) tender sample 1 in dry ice (Tr1 D) and RNAlater® (Tr1 R) and tough sample 1 in dry ice (Th1 D) and RNAlater® (Th1 R), at time point day 0 (**a**) and day 14 (**b**). Protein Plus protein standards (Bio-Rad Laboratories) comprising a wide range of molecular weights (10–250 kDa) were included in every gel for molecular weight determination (STD). The bands, as defined for quantitative image analysis are indicated with arrows (e.g., B1–B14).

Table 1. The *p*-values for effects of quality (tender/tough), treatment (dry ice/RNAlater®) and time point (day 0/14) on protein band abundance, and their two-way interactions.

Bands No.	Quality	Treatment	Time Point	Quality × Treatment	Quality × Time Point	Treat × Time Point
1	0.93	0.90	0.17	0.52	0.50	0.94
2	0.58	0.25	0.42	0.57	0.67	0.60
3	0.51	0.51	0.52	0.22	0.61	0.24
4	0.91	0.78	0.93	0.69	0.77	0.71
5	0.34	0.86	0.15	0.08 †	0.81	0.91
6	0.51	0.83	0.87	0.13	0.74	0.87
7	0.74	0.23	0.01	0.43	0.31	0.39
8	0.98	0.12	0.20	0.94	0.72	0.63
9	0.30	0.33	<001	0.46	0.38	0.88
10	0.87	0.81	0.61	0.84	0.21	0.79
11	0.67	0.08 †	0.69	0.87	0.71	0.77
12	0.83	0.61	0.86	0.91	0.62	0.54
13	0.64	0.60	0.45	0.45	0.68	0.68
14	0.63	0.21	0.66	0.78	0.96	0.88

p-values in bold are significant at the 0.05 level; † denotes a tendency with $p < 0.1$.

Table 2. Mean protein abundance value of band 7 and band 9.

	Dry Ice						RNAlater®					
	Day 0	Day 14	e.s.e.	Tender	Tough	e.s.e.	Day 0	Day 14	e.s.e.	Tender	Tough	e.s.e.
Band 7	0.076 ab	0.052 a	0.008	0.066	0.061	0.008	0.101 b	0.056 ab	0.014	0.072	0.085	0.014
Band 9	0.005 a	0.021 b	0.002	0.012	0.015	0.002	0.003 a	0.020 b	0.002	0.011	0.012	0.002

Values within a row that do not share a common superscript differ significantly from each other at the 0.05 level; e.s.e—standard errors of means.

4. Conclusions

This study demonstrates that, based on SDS-PAGE and MS, RNAlater® is a reliable storage agent for bovine muscle tissue that preserves proteins for proteomic analysis in a similar way to freezing in dry ice. It is concluded that use of RNAlater® is suitable for meat proteomic experiments where snap-freezing may not be a viable option for sample stabilization. Further research could focus on different tissue types or other red meat species to establish the wider relevance of our results.

Supplementary Materials: The following are available online at http://www.mdpi.com/2304-8158/8/2/60/s1, Table S1: Protein identifications from band 5, 7, 9, 11 of SDS-PAGE Gel by LC-MS, Figure S1: Amino acid sequences of peptides of Troponin T in band 7 and 9 identified by mass spectrometry are underlined. The peptide sequence KPLNIDHLSEDKLR (196-210) was detected only in band 7.

Author Contributions: Conceptualization, R.M.H. and A.M.M.; methodology, R.M.H., D.K.R. and Y.Z.; lab work, Y.Z., J.C.; resources, R.M.H.; data curation, D.K.R. and Y.Z.; draft preparation, Y.Z.; writing—review and editing, A.L.K., D.S., A.M.M., D.K.R. and R.M.H.; project administration, R.M.H. and A.M.M.; funding acquisition, R.M.H., A.M.M., D.K.R. and D.S.

Funding: This research was funded by Teagasc and the Walsh Fellowship programme, project number NFFQ0017.

Acknowledgments: We acknowledge the Irish Cattle Breeding Federation for access to samples.

Conflicts of Interest: The authors declare no conflicts of interest.

References

1. Saito, M.A.; Bulygin, V.V.; Moran, D.M.; Taylor, C.; Scholin, C. Examination of Microbial Proteome Preservation Techniques Applicable to Autonomous Environmental Sample Collection. *Front. Microbiol.* **2011**, *2*, 1–10. [CrossRef]
2. Manoj, C.R.; Douglas, S.M.; Rebecca, A.R.; Kathleen, M.E. A Method for the Extraction of High-Quality RNA and Protein from Single Small Samples of Arteries and Veins Preserved in RNAlater. *J. Pharmacol. Toxicol. Methods* **2002**, *47*, 87–92.

3. Bennike, T.B.; Kastaniegaard, K.; Padurariu, S.; Gaihede, M.; Birkelund, S.; Andersen, V.; Stensballe, A. Comparing the Proteome of Snap Frozen, RNAlater Preserved, and Formalin-Fixed Paraffin-Embedded Human Tissue Samples. *EuPA Open Proteom.* **2016**, *10*, 9–18. [CrossRef]

4. Rader, J.S.; James, P.M.; Julia, G.; Petra, G.; Rebecca, A.B.; Loan, N.; Dan, L.C. A Unified Sample Preparation Protocol for Proteomic and Genomic Profiling of Cervical Swabs to Identify Biomarkers for Cervical Cancer Screening. *Proteomics* **2008**, *2*, 1658–1669. [CrossRef]

5. Lenchik, N.I.; Desiderio, D.M.; Gerling, I.C. Two-Dimensional Gel Electrophoresis Characterization of the Mouse Leukocyte Proteome, Using a Tri-Reagent for Protein Extraction. *Proteomics* **2005**, *5*, 202–209. [CrossRef] [PubMed]

6. Barclay, D.; Ruben, Z.; Andres, T.; Rajaie, N.; David, S.; Yoram, V. A Simple, Rapid, and Convenient LuminexTM-Compatible Method of Tissue Isolation. *J. Clin. Lab. Anal.* **2008**, *22*, 278–281. [CrossRef] [PubMed]

7. Kruse, C.P.S.; Basu, P.; Luesse, D.R.; Wyatt, S.E. Transcriptome and Proteome Responses in RNAlater Preserved Tissue of Arabidopsis Thaliana. *PLoS ONE* **2017**, *12*, 1–10. [CrossRef] [PubMed]

8. Abbaraju, N.V.; Yang, C.; Bernard, B.R. Protein Recovery and Identification from the Gulf Killifish, Fundulus Grandis: Comparing Snap-Frozen and RNAlater® Preserved Tissues. *Proteomics* **2011**, *11*, 4257–4261. [CrossRef] [PubMed]

9. Belew, J.B.; Brooks, J.C.; McKenna, D.R.; Savell, J.W. Warner–Bratzler Shear Evaluations of 40 Bovine Muscles. *Meat Sci.* **2003**, *64*, 507–512. [CrossRef]

10. Bradford, M.M. A Rapid and Sensitive Method for the Quantitation of Microgram Quantities of Protein Utilizing the Principle of Protein-Dye Binding. *Anal. Biochem.* **1976**, *72*, 248–254. [CrossRef]

11. Laemmli, U.K. Cleavage of Structural Proteins during the Assembly of the Head of Bacteriophage T4. *Nature* **1970**, *227*, 680–685. [CrossRef] [PubMed]

12. Wilm, M.; Shevchenko, A.; Houthaeve, T.; Breit, S.; Schweigerer, L.; Fotsis, T.; Mann, M. Femtomole sequencing of proteins from polyacrylamide gels by nano-electrospray mass spectrometry. *Nature* **1996**, *379*, 466–469. [CrossRef] [PubMed]

13. Buts, B.; Claeys, E.; Demyer, D. Relation between concentration of troponin-T, 30,000-Dalton, and titin on SDS–PAGE and tenderness of bull longissimus dorsi. In Proceedings of the 32nd European Meeting of Meat Research Workers, Ghent, Belgium, 24–29 August 1986.

14. MacBride, M.A.; Parrish, F.C. 30,000-Dalton component of tender bovine longissimus muscle. *J. Food Sci.* **1977**, *42*, 1627–1629. [CrossRef]

15. Ho, C.Y.; Stromer, M.H.; Robson, R.M. Identification of the 30 kDa Polypeptide in Post Mortem Skeletal Muscle as a Degradation Product of Troponin-T. *Biochimie* **1994**, *76*, 369–375. [CrossRef]

16. Olson, D.G.; Parrish, F.C.; Dayton, W.R.; Goll, D.E. Effect of postmortem storage and calcium activated factor on myofibrillar proteins of bovine skeletal muscle. *J. Food Sci.* **1977**, *42*, 117–124. [CrossRef]

Article

Study of the Chronology of Expression of Ten Extracellular Matrix Molecules during the Myogenesis in Cattle to Better Understand Sensory Properties of Meat

Anne Listrat *, Mohammed Gagaoua and Brigitte Picard

Université Clermont Auvergne, INRA, VetAgro Sup, UMR Herbivores, F-63122 Saint-Genès-Champanelle, France; mohammed.gagaoua@inra.fr or gmber2001@yahoo.fr (M.G.); brigitte.picard@inra.fr (B.P.)
* Correspondence: anne.listrat@inra.fr; Tel.: +33-4-73-62-41-05

Received: 30 January 2019; Accepted: 8 March 2019; Published: 13 March 2019

Abstract: The sensory properties of beef are known to depend on muscle fiber and intramuscular connective tissue composition (IMCT). IMCT is composed of collagens, proteoglycans and glycoproteins. The differentiation of muscle fibers has been extensively studied but there is scarcity in the data concerning IMCT differentiation. In order to be able to control muscle differentiation to improve beef quality, it is essential to understand the ontogenesis of IMCT molecules. Therefore, in this study, we investigated the chronology of appearance of 10 IMCT molecules in bovine *Semitendinosus* muscle using immunohistology technique at five key stages of myogenesis. Since 60 days *post-conception* (dpc), the whole molecules were present, but did not have their final location. It seems that they reach it at around 210 dpc. Then, the findings emphasized that since 210 dpc, the stage at which the differentiation of muscle fibers is almost complete, the differentiation of IMCT is almost completed. These data suggested that for the best controlling of the muscular differentiation to improve beef sensory quality, it would be necessary to intervene very early (before the IMCT constituents have acquired their definitive localization and the muscle fibers have finished differentiating), i.e., at the beginning of the first third of gestation.

Keywords: skeletal muscle; fetus; bovine; extracellular matrix; immunohistology

---

1. Introduction

Beef tenderness is one of the most important quality attributes for the consumer. It is often inconsistent and affects consumer satisfaction. The variability in final meat quality is due to several factors such as the differences in muscle characteristics [1], namely to the proportion of different types of muscle fibers [2–4] and the characteristics of intramuscular connective tissue (IMCT) [5,6]. Among the characteristics of IMCT, the total collagen and its solubility are studied more. IMCT is a composite network that can both withstand and transmit forces generated by muscle contraction. This complex network not only ensures the structural integrity of muscle, but also mediates the development and physiological behavior of muscle cells. IMCT maintains the structural integrity of muscle fibers by two layers, the endomysium that surrounds individual skeletal muscle fibers and the perimysium that bundles groups of muscle fiber. IMCT consists of cells and an ExtraCellular Matrix (ECM) that primarily consists of a composite network of fibrillar collagens wrapped in a matrix of proteoglycans (PGs). These later form large complexes by binding to other PGs and to fibrous proteins such as collagens. The PGs consist of a core protein linked by covalent bonds to several glycosaminoglycan chains (GAGs). The chemical structure of these GAGs varies between the different PGs. The PGs of the skeletal muscle are mainly chondroitin sulfate (CS). Among them, decorin, a small PG and versican, a large PG, have been associated with myogenesis [7].

The fibrillary collagens, more commonly known as total collagen (particularly type I that is the most abundant fibrillar collagen of the skeletal muscle) interact with tenascin-X, a glycoprotein, and type XII and XIV collagens, two non-fibrillar minor collagen types [8,9]. Collagen XII and XIV interact with decorin [10–12]). In vitro, the supposed role of tenascin-X and of collagen XII and XIV is to modulate the flexibility of the ECM and consequently, its mechanical properties [13,14]. From these properties and according to Bailey and Light [15], it has been suggested that these molecules could contribute to the meat texture. However, there is a scarcity in studies that addressed this question unless one work from our groups that studied the relation between decorin, tenascin-X, collagen XII and XIV and beef sensory properties [16]. According to our previous study, it seems important to study all these components together, in combination with total and insoluble collagen, as they would have complementary information and an additive role that would play on the final texture quality of meat.

It is worthwhile to note that the differentiation of muscle fiber during fetal life has been extensively studied [17–20]. In the bovine skeletal muscle, myotube-multi-nucleated syncytium formed by the fusion of several myoblasts during myogenesis-formation occurs in three temporally distinct phases. The first generation of embryonic myoblasts proliferates and differentiates in myotubes to the surroundings of 30 days *post-conception* (dpc). They are completely differentiated around 180 dpc (end of the second trimester of gestation). A second and third generation of fetal myoblasts proliferate and differentiate in secondary myotubes between 60 and 90 dpc. At 180 dpc, almost all the myotubes have the appearance of muscle fibers. At this stage, the total number of myofibers is set. Contractile and metabolic maturation occurs during the last trimester. At the end of the gestation (280 dpc), the differentiation of muscle fiber types is nearly complete.

However, there are few datums on the IMCT differentiation of the bovine fetus in vivo [21–24]. So, we hypothesized that the knowledge of the chronology of the differentiation of the different muscle tissues would allow the development of strategies (for example through maternal feeding) to enhance muscle growth and modify both IMCT and muscle fibers characteristics, and consequently their impact on final meat quality. Accordingly, we investigated the expression of ten ECM molecules thought to play an important role in the myogenesis in adults and its potential link to the quality of beef, at key stages of muscle fiber differentiation previously described by our groups [17]. The results of this study emphasized that the molecules studied are present since the beginning of fetal life in bovine and that they acquired the localization they will have in adults in the first two-thirds of fetal life (between 180 and 210 dpc). Furthermore, it appears that the main step of myogenesis occurs during the same period.

2. Materials and Methods

This study was carried out in compliance with the French recommendations and those of the Animal Care and Use Committee of the National Institute for Agricultural Research (INRA, Institut National de la Recherche Agronomique) of Auvergne-Rhône-Alpes, France (under the slaughterhouse and experimental facilities license numbers #63 345 01 and #63 345.17, respectively), for the use of experimental animals including animal welfare, in accordance with the *Use of Vertebrates for Scientific Purposes Act 1985*.

2.1. Muscle Samples

Fifteen fetuses of 60 (n = 3), 110 (n = 3), 180 (n = 3), 210 (n = 3) and 260 (n = 3) days old were obtained by the artificial insemination of Charolais heifers using pure Charolais sperm. These stages have been chosen according to the key stages of muscle fiber differentiation previously highlighted in our laboratory in several studies cited in the review by Picard, et al. [17]. After the slaughter of pregnant heifers, *Semitendinosus* (ST) muscles were carefully dissected out of the two hind limbs from each animal. An approximate of 10 mm slices were taken at the mid-belly of one muscle, at right angles to the direction of the muscle fibers for histology and immunohistology and frozen in isopentane, cooled in liquid nitrogen. For electrophoresis, 3 fetuses per stage were used for 110, 180, 210 and 260 dpc. They were directly frozen in liquid nitrogen. Then all samples were stored at $-80\,^{\circ}$C until analyses.

2.2. Transverse Sections Preparation

All transverse sections (10 µm thick) of ST muscle were realized with a cryotome MICROM HM 500 M at $-25\,^{\circ}$C.

2.3. Azorubine Staining

The muscle cells were stained with azorubine dye that stained the myofibrillar proteins in red. Sections (3 per animal) were fixed for 5 min with a solution of 5.7% formaldehyde and 18 mM $CaCl_2$, washed in water and then dyed with 3% azorubine solution (Azorubine (CI 14410; Serva, Heidelberg, Germany) and 5% acetic acid for 45 min. Sections were washed in water and dehydrated twice for 1 min in acetone (Prolabo, Sion, Switzerland) and then twice for 1 min in Ottix (Microm, Brignais, France). Finally, the sections were mounted with cover-glass with Canada balsam (Prolabo, Sion, Switzerland).

2.4. Antibodies

Primary antibodies (polyclonal rabbit anti-bovine type I collagen (Col I) (catalog number, 20121), monoclonal mouse anti-human type IV collagen (Col IV) (catalog number, 20421), polyclonal rabbit anti-human type VI collagen (Col VI) (catalog number, 20611) (Novotec, Bron, France), monoclonal mouse anti bovine decorin (DCN) (catalog number, DS1) (DSHB, Iowa City, IA, USA), tenascin-X (Tn-X) and collagen XII and XIV (Col XII and Col XIV) (a generous gift from C. Lethias, IBCP, Lyon, France previously described by Berthod, et al. [25] and Elefteriou, et al. [26], chondroitin-4 and -6 sulfate (C4S and C6S) and versican (VCN) (a generous gift from B. Caterson, Cardiff University, Cardiff, UK, previously described by Hayes, et al. [27]) were diluted with 1% bovine serum albumin (BSA) in $1\times$ phosphate buffer saline (PBS) (pH 7.2) to 1:40. Secondary antibody conjugated to Alexa Fluor 488 (Interchim, Montluçon, France) were diluted in 1% BSA in $1\times$ PBS (pH 7.2) to 1:400.

2.5. Immunohistochemistry

The localization of ten ECM molecules (Col I, IV, VI, XII, XIV, DCN, Tn-X, C4S, C6S and VCN) was evidenced by the indirect immunofluorescence on four serial sections per animal according to the procedure of Listrat, Picard and Geay [22]. Natural collagen fluorescence on muscle sections was blocked with 50 mM ammonium chloride in $1\times$ PBS for 10 min for all the used antibodies. For C4S and C6S antibodies, sections were enzymatically pre-treated with 0.5 U/mL chondroitinase ABC (Sigma-Aldrich, France) and 0.5 U/mL keratanase (Sigma-Aldrich, France) in 100 mM Tris acetate buffer (pH 7.4) for 1 hour at 37 $^{\circ}$C to unmask the epitope then rinsed in a solution containing 0.1% Tween20 (Sigma-Aldrich, France) in $1\times$ PBS. For versican, the sections were only incubated in 0.1% Tween20 (Sigma-Aldrich, France) in $1\times$ PBS (without enzymatic pre-treatment) at room temperature. Then all sections were incubated in 1% BSA in $1\times$ PBS for 10 min and they were reacted with the primary antibodies for 1 h at room temperature, washed in $1\times$ PBS three times 5 min, incubated for 40 min at room temperature in the dark with the secondary antibody and washed in $1\times$ PBS three times for 5 min. The sections were rinsed with 0.3% eriochrome-black T in $1\times$ PBS to completely block natural collagen fluorescence and they were then mounted with cover-glass with Fluoromount (Sigma-Aldrich, Saint-Louis, MO, USA). Negative controls were performed by omitting primary antibodies in the same conditions that were previously described.

2.6. Western Blot Analyses of Type XII and XIV Collagens

Total protein extraction was performed with RIPA (Radio-Immunoprecipitation Assay) lysis buffer (150 mM NaCl, 10 mM Tris-HCl, 1 mM EGTA, 1 mM EDTA, adjusted at pH = 7.4 and completed with 100 mM sodium fluoride, 4 mM sodium pyrophosphate and 2 mM orthovanadate, 1% Triton $100\times$,

0.5% Igepal CA-630 and protease inhibitor cocktail (Complete, Roche Diagnostics GmbH, ref. 11 836 145 001)). After extraction, the protein concentration was determined by spectrophotometry (UVIKON 860) following the Bradford assay [28].

Proteins were separated in denaturing conditions (10% sodium dodecyl sulfate-polyacrylamide and β-mercaptoethanol). Samples were loaded on the gel (stacking gels of 4% and separation gels of 6%) at the rate of 50 μg of protein. Gels were run at 80 V for 20 min and then 120 V for 1 h at +4 °C.

Bands were transferred to a polyvinylidene difluoride (PVDF) membrane (ref. IPVH00010, Millipore, Burlington, MA, USA) at 120 mA for 5 h at +4 °C. The unspecific binding of antibodies to the membranes was blocked with 10% skimmed milk (Régilait) in 1× T-TBS (20 mM Tris base; 137 mM NaCl; 0.05 % Tween 20, pH 8) at 37 °C for 20 min. Membranes were washed 3 × 5 min in 1× T-TBS and then they were incubated overnight at +4 °C with the primary antibodies (anti-type XII or anti-type XIV collagen (the same than for immunohistochemistry) diluted to 1/50 in 1% skimmed milk (Régilait) in 1× T-TBS). Membranes were washed 2 × 10 min in 1× T-TBS and hybridized with the secondary antibody (diluted to 1/5000 in 1% skimmed milk (Régilait) in 1× T-TBS) associated with horseradish peroxidase for chemiluminescence detection (IgG sheep anti-mouse, NA931, GE Healthcare Life Sciences, Grenoble, France).

Each collagen type presented three specific bands that were considered and quantified as a single band under Image Quant software (GE Health Care life Science, Grenoble, France). One volume of total protein extract of each of 15 fetuses was mixed and 50 μg of this mixed were deposited on all the gels for normalization. Each collagen type presented two or three specific bands which were considered and quantified as a single band under Image Quant software. The value obtained for the three bands of each sample was divided by the value obtained for the three bands of the mix of samples loaded on each gel. Each sample was measured in triplicate and results were expressed in arbitrary units. The other molecules were not analyzed by Western blot for technical reasons.

2.7. Image Acquisition and Analysis

Histological sections were visualized under an Olympus fluorescence microscope BX 51 using a 10×, 20× or 40× objective and for fluorescence an adequate band pass filter (Alexa 488: excitation filter 460–495, emission filter 510–550, dichromatic mirror 505LP). High-resolution grayscale images were acquired with an Olympus cooled digital camera DP-72 with cell-F software (Olympus Soft Imaging Solutions, Münster, Germany). Azorubine died-sections were analyzed with a VISILOG 6.7 professional software (Noesis, Gif sur Yvette, France) to calculate the mean cross-section area of fibers according to Meunier, et al. [29].

2.8. Statistical Analysis

The differences of relative amounts of Col XII or XIV between stages post-conception were assessed by analysis of variance (ANOVA) using the GLM procedure of SAS 9.2 Software (Statistical Analysis System, Cary, NC, USA). A probability of less than 5% was considered statistically significant. All results were presented as least square means ± Standart Error of the Mean (SEM).

3. Results

3.1. Sixty and 110 dpc

3.1.1. Perimysium

At 60 and 110 dpc, myotubes were grouped in bundles individualized by a wide perimysium divided into a major (around the muscle cell bundles) and minor network (around and inside muscle cell bundles) (Figure 1A). The average area of the myotubes was 84.2 ± 15.9 μm^2 at 60 dpc and 85.5 ± 17.5 μm^2 at 110 dpc. At 60 and 110 dpc, future perimysium was stained by Col I (Figure 2A,B), VI (Figure 2F,G), DCN (Figure 3F,G), C4S (Figure 3K,L), Tn-X (Figure 4A,B) and Col XII (Figure 4F,G).

At 60 dpc, Col XIV was undetectable (Figure 4K). At this stage, C4S (Figure 3K) and TN-X (Figure 4A) labeling were very low. At 110 dpc, the presence of C4S, Tn-X (Figures 3L and 4B), Col XIV (Figure 4L) in the perimysium was confirmed, but only in major networks.

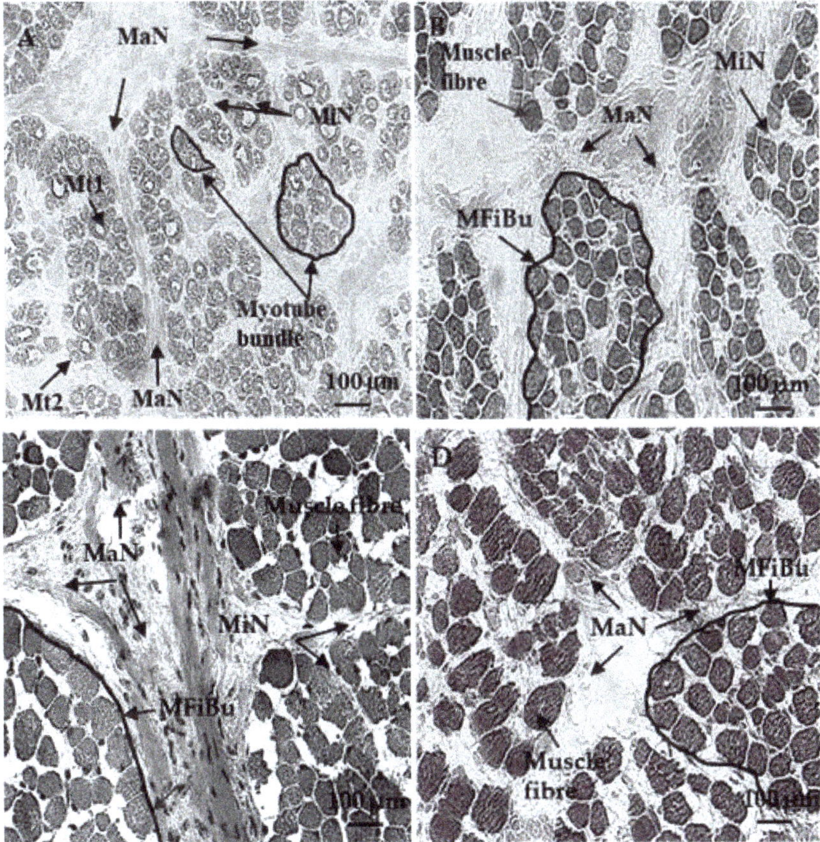

Figure 1. Azorubine staining of transverse sections of foetal Semitendinosus muscle at 110 (**A**), 180 (**B**), 210 (**C**) and 260 (**D**) days post-conception (dpc). (ECM: ExtraCellular Matrix; MaN: perimysium major network, MiN: perimysium minor netork, Mt1: primary myotubes, Mt2: secondary myotube, MFiBu: Muscle Fiber Bundle).

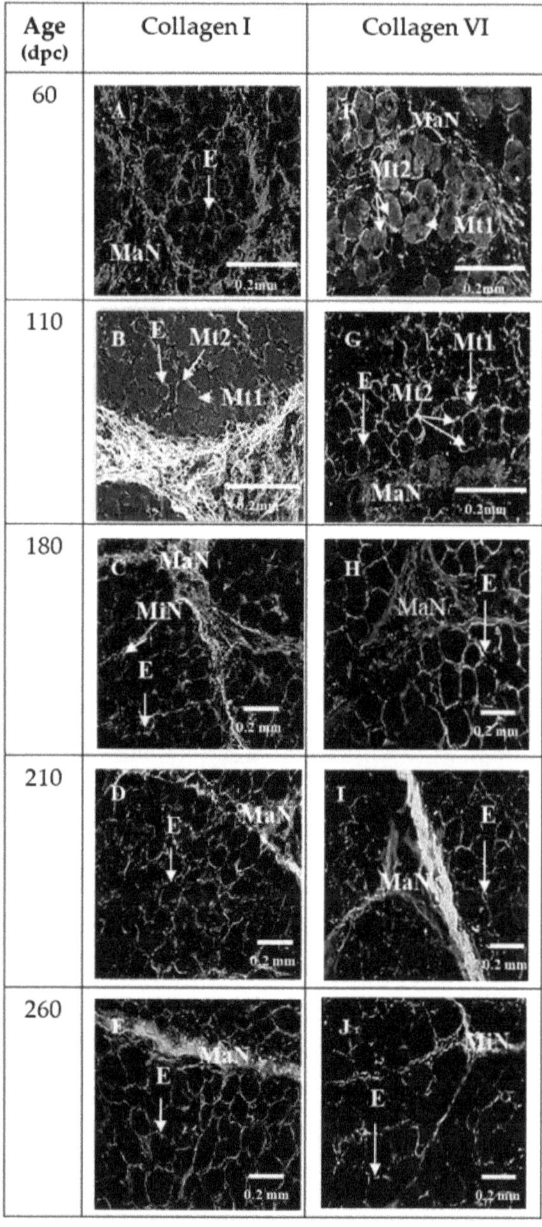

Figure 2. Immunohistochemical labelling with antibodies against type I (**A**, **B**, **C**, **D**, **E**), and type VI (**F**, **G**, **H**, **I**, **J**) collagens of transverse sections of foetal Semitendinosus muscle at 60, 110, 180, 210 and 260 days post-conception (dpc) (Major (MaN) and minor (MiN) networks of perimysium; E: endomysium, Mt1: primary myotubes, Mt2: secondary myotubes).

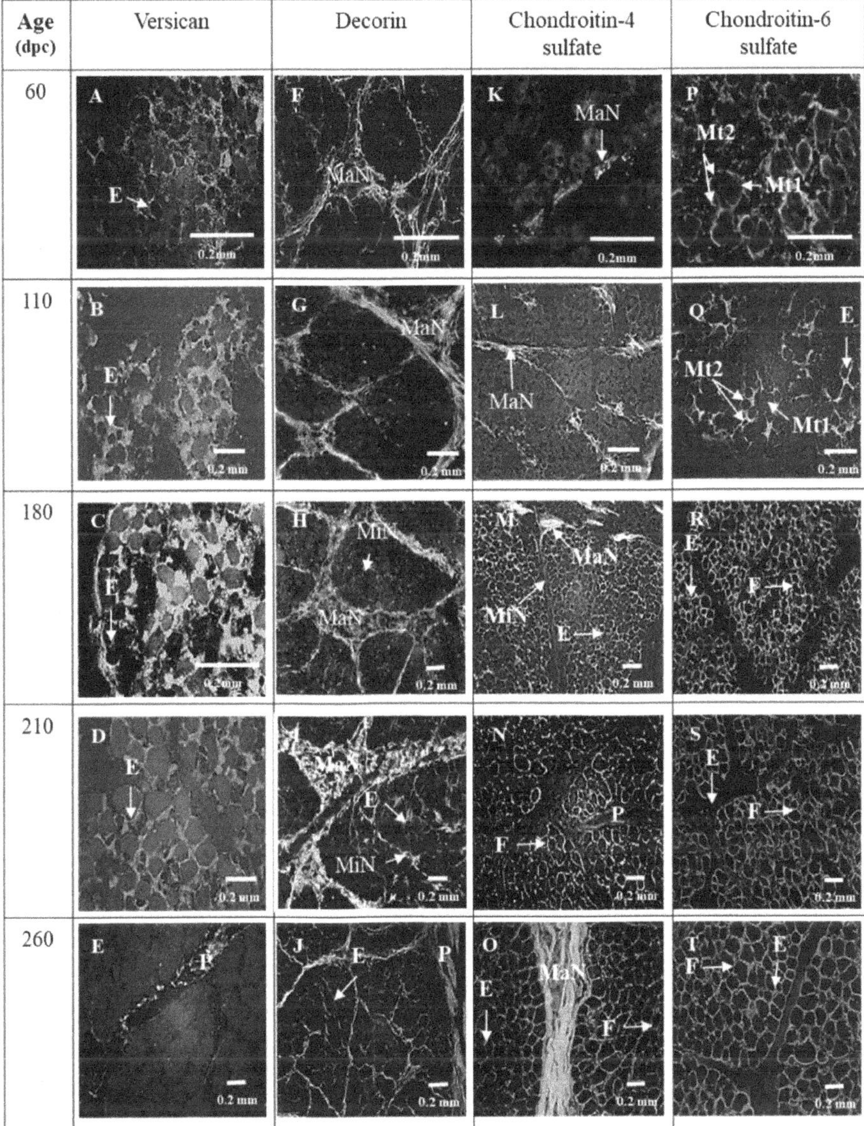

Figure 3. Immunohistochemical labelling with antibodies against versican (**A**, **B**, **C**, **D**, **E**), decorin (**F**, **G**, **H**, **I**, **J**), chondroitin-4-sulfate (**K**, **L**, **M**, **N**, **O**), chondroitin-6-sulfate (**P**, **Q**, **R**, **S**, **T**) of transverse sections of foetal Semitendinosus muscle at 60, 110, 180, 210 and 260 days post-conception (dpc). (Major (MaN) and minor (MiN) networks of perimysium (P), E: endomysium, Mt1: primary myotubes, Mt2: secondary myotubes, F: muscle fiber).

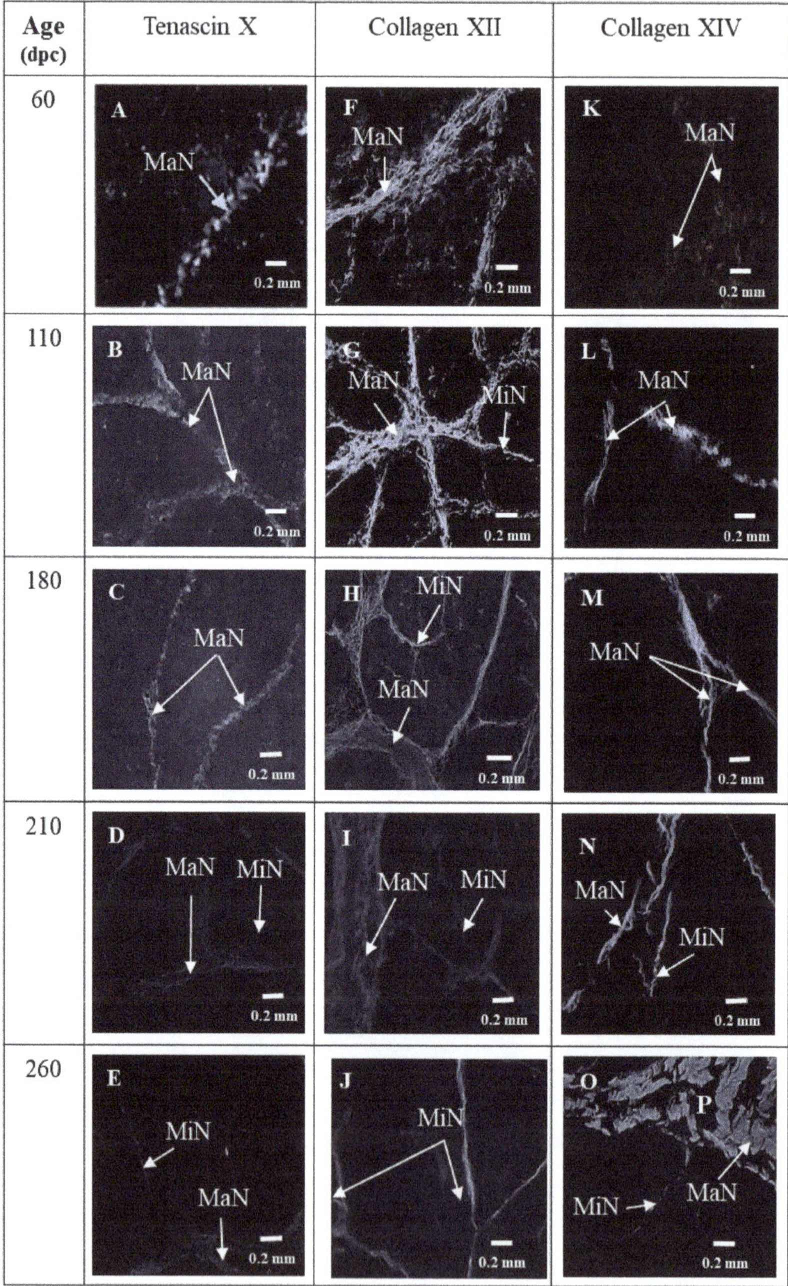

Figure 4. Mmunohistochemical labelling with antibodies against tenascin-X (**A**, **B**, **C**, **D**, **E**), collagen XII (**F**, **G**, **H**, **I**, **J**), collagen XIV (**K**, **L**, **M**, **N**, **O**) of transverse sections of foetal Semitendinosus muscle at 60, 110, 180, 210 and 260 days post-conception (dpc). (Major (MaN) and minor (MiN) networks of perimysium).

At 60 dpc, Col XII was present in major networks of perimysium (Figure 4F) while at 110 dpc (Figure 4G), it was present everywhere in the perimysium.

At all studied stages (110, 180, 210 and 260 dpc), the presence of both Col XII and XIV were observed as three bands of molecular weight between 220 and 290 kDa (Figure 5a,b).

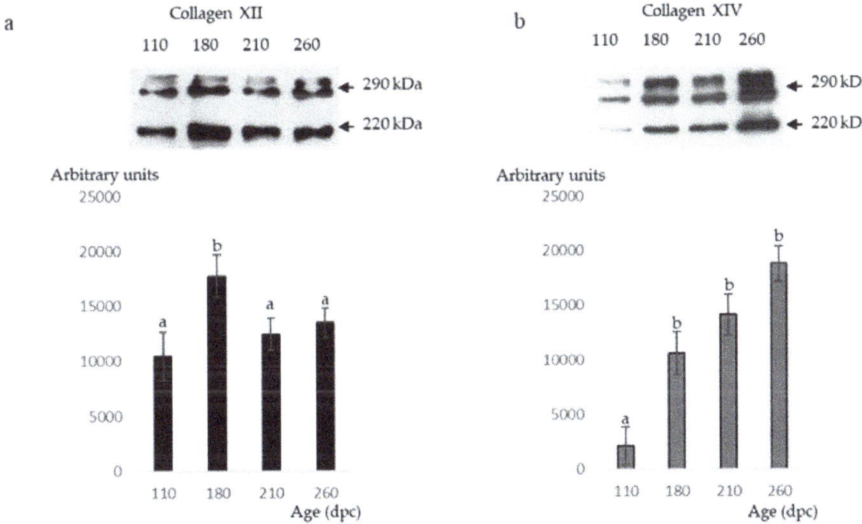

Figure 5. (**a**) Collagen XII and (**b**) Collagen XIV Western blot analysis of foetal Semitendinosus muscle at 110, 180, 210 and 260 days post-conception (dpc). The monoclonal antibodies against collagen XII and XIV, two disulphide bonded polypeptides, recognized three bands on Western blots of bovine muscle extracts, one band at 220 kDa and two others at about 290 kDa. The presence of these three bands was due to the fact that migration conditions were reducing. The relative amounts of collagen XII and XIV (least square means ± standard error of the mean) were expressed in arbitrary units. Different letters on the same graph indicated that relative amounts of collagen XII and XIV differed significantly between foetal stages ($p < 0.05$).

3.1.2. Endomysium

At 60 and 110 dpc, endomysium was distinctly labeled by Col I, VI (Figure 2A,B; Figure 2F,G), IV (Figure 6A,B) and C6S (Figure 3P,Q). At these stages, labeling obtained with an antibody against VCN was diffuse and occupied all the space between myotubes (Figure 3A,B). C4S appeared in endomysium between 60 and 110 dpc (Figure 3K,L).

For VCN, C6S (Figure 3A,B and Figure 3P,Q, respectively), Col VI and IV (Figure 2F,G and Figure 6A,B), the main difference between 60 and 110 dpc was the individualization or non-individualization of primary and secondary myotubes. At 60 dpc, secondary and primary myotubes were not individualized from each other, but seemed wrapped in the same endomysium in bundles of 2 or 3 myotubes while at 110 dpc, some secondary myotubes began to be well individualized from primary myotubes, while others always seemed wrapped in the same endomysium. This result is illustrated in Figure 6 by both labeling with the antibody anti-Col IV (Figure 6a) and with a diagram (Figure 6b).

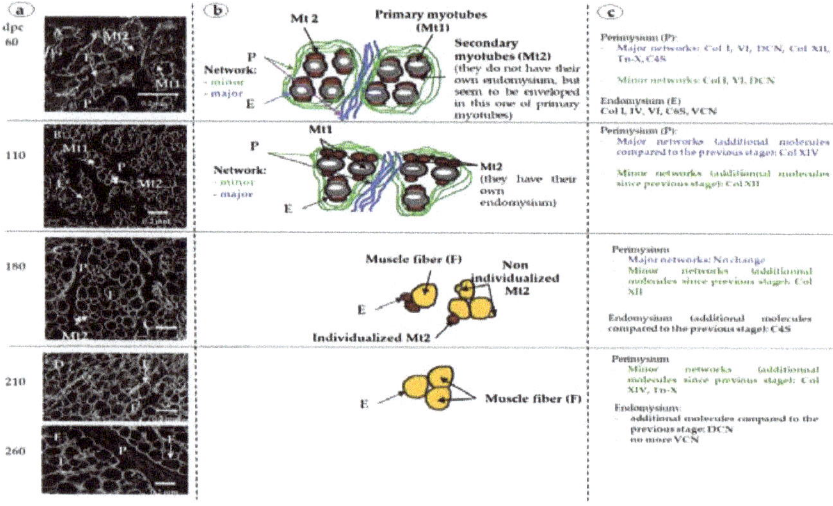

Figure 6. (**a**) The immunohistochemical labeling with antibodies directed against collagen IV (**A**, **B**, **C**, **D**, **E**) of transverse sections of fetal *Semitendinosus* muscle at 60, 110, 180, 210 and 260 days *post-conception* (dpc). (P: perimysium, E: endomysium, Mt1: primary myotubes, Mt2: secondary myotubes, F: muscle fiber). (**b**) Schematic representation of differentiation of muscle fibers at 60, 110, 180, 210 and 260 days post-conception (dpc). At 60 and 110 dpc, muscle fibers are present as primary (cells schematically represented by an oval grey form with a central lumen surrounded by a dark grey line representing endomysium) myotubes (Mt1) and secondary myotubes (Mt2) (cells schematically represented by a brown form without central lumen). At 60 dpc, Mt2 are not individualized; they are wrapped in the same endomysium than Mt1. At 110 dpc, almost all Mt2 are individualized and they have their own endomysium (dark grey line). At 180 and 210 dpc, some Mt2, individualized or not, are still present, but the majority of muscle cells are fibers (schematically represented by a yellow form surrounded by a dark grey line representing endomysium). At 260 dpc, all myotubes are muscle fibers. (**c**) Schematic description of localization of different molecules of ExtraCellular matrix. At each stage, the localization of ECM molecules (Col I, IV, VI, XII and XIV: collagen of type I, IV, VI, XII and XIV; DCN: decorin, TN-X: tenascin-X, C4 and 6S: chondroitin 4 and 6 sulfate; VCN: versican) is indicated in the major and minor networks of perimysium and in the endomysium. From 110 dpc, only the modifications of localization of ECM molecules are indicated.

3.2. 180 dpc to 260 dpc

At 180 dpc, almost all muscle cells had the appearance of muscle fibers and were organized into bundles as in adults (Figure 1B). The average area of muscle fibers (138.2 $\pm$ 34.5 µm^2) was significantly higher ($p < 0.05$) than that at 110 dpc. The fiber cross-section area significantly ($p < 0.05$) and progressively increased (282.53 $\pm$ 74.5 µm^2 at 210 dpc to 391.22 $\pm$ 116.1 µm^2 at 260 dpc) between 180 and 260 days.

3.2.1. ECM Molecules Whose Location Changed

C4S (Figure 3M,N), Tn-X and Col XIV (Figure 4C,D and Figure 4M,N) began to appear in minor networks of perimysium between 180 and 210 dpc. The measures of Col XII and XIV by Western blotting showed that the relative amount of Col XII significantly increased between 110 dpc and 180 dpc then significantly decreased between 180 and 260 dpc (Figure 5a). The relative amounts of Col XIV significantly increased between 110 and 260 dpc (Figure 5b).

DCN appeared in the endomysium and in the minor networks of perimysium between 180 and 210 dpc (Figure 3H,I). C4S appeared in the minor networks of the perimysium between 110 and 180 dpc

(Figure 3L,M). VCN was present in the endomysium until 210 dpc (Figure 3D) then disappeared from the endomysium between 210 and 260 dpc (Figure 3D,E). At 260 dpc, it was only expressed in the perimysium (Figure 3E).

3.2.2. ECM Molecules Whose Location Did not Change

From 180 dpc, the localization of Col I, VI (Figure 2C,H), IV (Figure 6C), and of C6S (Figure 3R) did not change anymore. Col I and VI were present both in the endo- and perimysium, Col IV and C6S only in endomysium. As at 110 dpc, Col XII was present in the entire network until 260 dpc (Figure 4H,I,J). All modifications of the localization of molecules of ECM were summarized in Figure 6c.

4. Discussion

This report described the chronology of differentiation of 10 ECM molecules in the bovine fetus. According to their chronology of differentiation and their localization, the molecules can be classified into several groups. The molecules (i) exclusively located in endomysium (Col IV, C6S), those (ii) exclusively located in perimysium (Col XII and XIV, TN-X), those (iii) first located in perimysium then appearing in endomysium more or less precociously (Col I, VI, DCN and C4S) or (iv) such as VCN first located in endomysium, then (v) that disappeared and appeared in perimysium at the end of the fetal life.

In this study, Col IV was the collagen type detected the earliest in endomysium (since 60 dpc). It remained present in endomysium throughout fetal life as well as in adults. Pöschl, et al. [30] observed that embryos developed up to E9.5 on a null allele of the Col 4a1/2 locus in mice. However, lethality occurred between E10.5–E11.5 (about 30 dpc for the bovine) because of structural deficiencies in the basement membranes. The data of these authors highlighted that Col IV is fundamental for the maintenance of the integrity and function of basement membranes and, consequently, must be present very early in the process of differentiation.

As previously shown, Col I and VI were present since 60 dpc in the perimysium [23,31]. Col VI appeared progressively in the endomysium of primary and secondary myotubes between 60 and 110 dpc; it was colocalized with Col I. At 180 dpc, Col I and VI had the localization that they will have in the adult muscle in the endomysium and the perimysium. The period between 60 and 110 dpc coincides with the differentiation of fetal myoblasts in myotubes. Thus, as previously suggested in vivo by Listrat, Picard and Geay [22] and Listrat, Picard and Geay [23] and in vitro by several other authors [32,33], our results suggest that Col I and VI might be involved in the early differentiation of myotubes. However, recent studies support the idea that these collagens do not play a direct role in differentiation. Their presence would enable the preservation of collagen-binding molecules like proteoglycans, laminins and fibronectin and, consequently, the cytokines and growth factors that bind these molecules and are essential for myoblast differentiation in myotubes [34].

In chicken, VCN is transiently upregulated in myoblasts and newly formed myotubes. It is located in the endomysium. At the same stages and from chicken, decorin is observed in the perimysium, then its distribution gradually spreads to the perimysium and endomysium at the end of ovogenesis [35,36]. These observations are in line with what we observed in this study in bovine fetuses since VCN was present in the endomysium between 60 and 210 dpc, disappeared and were then replaced by DCN between 210 and 260 dpc. In bovine fetuses, the proliferation and differentiation of myoblasts in myotubes are at their highest peak between 60 and 210 dpc [37]; stages of the presence of VCN in endomysium. The role of VSC in the myogenesis is poorly defined, even though it may be an important driver during myogenesis [38]. However, it is known that it contributes to the formation of a hydrated pericellular matrix—part of ECM in close contact with cells—whose remodeling by some proteinases of ADAMTS family (a disintegrin and metalloproteinase with thrombospondin motifs) contributes to the formation of myotubes [38].

At 210 dpc, the total number of muscle fibers is fixed, the contractile differentiation of the first generation of myotubes is completed, and various adult myosin heavy chain (MHC) isoforms begin their expression [37]. We suggest that DCN could have an indirect but crucial role at this stage. Indeed, Miura, et al. [39] proposed that DCN would decrease the activity of myostatin—a growth factor—by its sequestration. A lack of myostatin [40] and an over-expression of DCN [41] would lead to the upregulation of MyoD (myoblast determination protein) and reduce the levels and activity of MEF2 (myocyte enhancer factor-2)—a transcription factor—and calcineurin—a protein phosphatase. MyoD is a transcription factor of the myogenic factors subfamily involved in muscle differentiation which is important for the activation of MHC IIB gene expression [41]. The loss of MEF2 and calcineurin may result in a reduction of slow fibers [42]. Thus, when decorin began to be present in endomysium, it could begin to regulate the fiber-type composition of skeletal muscles by regulating MEF2, calcineurin and MyoD gene expression.

The results of this study showed that, between 110 and 180 dpc, the relative amounts of Col XII increased then, after 180 dpc, it decreased. This decrease could be involved in muscle fiber type transitions that occur at the end of fetal life and, more precisely, in the differentiation of type II fibers as suggested by the work of Zou, et al. [43] in the model of mouse Col12A1$^{-/-}$. In this model, in two muscles, the *Soleus* (slow oxidative) and the *Tibialis* (fast glycolytic), the absence of Col XII results in a more glycolytic metabolism. Nishiyama, McDonough, Bruns and Burgeson [14] have shown that the addition of this collagen type to collagen gels modifies their mechanical properties, suggesting that Col XII might be directly involved in determining the mechanical properties of collagen-rich tissues in vivo and their elasticity. Consequently, it has been proposed that the reduction of relative quantities of Col XII could change the matrix elasticity and thereby facilitate the terminal differentiation of muscle fibers [43].

In our model, the decrease of the relative amounts of collagen XII was associated with an increase of Col XIV. At the same period (between 180 and 210 dpc), Tn-X, which was already present in major perimysium networks, appeared in minor ones. To our knowledge, it is not known whether Tn-X or Col XIV play a role in the regulation of myogenesis. However, when Tn-X [13] and Col XIV [14] are added to collagen gels, they modify their mechanical properties and could act in synergy with Col XII.

5. Conclusions

This study is the first one to investigate the chronology of appearance of 10 ECM molecules of several families (collagens, PGs, . . .) in bovines along the fetal life. The main factor investigated is the age of fetuses at key stages of myogenesis as previously defined in our laboratory [38]. This approach highlighted, for the first time, that since 60 dpc, all the studied ECM molecules were present. They acquired the localization they will have in adults in the first two-thirds of fetal life (between 180 and 210 dpc). The main step of myogenesis terminates during the same period. Around 180 dpc, the number of muscle fibers is fixed and adult MHC forms begin to be present. The findings of this study also suggest that to consider controlling the muscular differentiation, it would be necessary to act very early (before the ECM constituents have acquired their definitive localization and that the muscle fibers have finished differentiating) at the beginning of the first third of gestation. The knowledge of key stages of differentiation of different tissues of muscle is essential in order to be able to control the characteristics in relation to beef quality, for example, through maternal feeding. Indeed, in ruminants, it is known that it is possible to decrease the number of secondary myotubes through the nutrient deficiency of the dam from early-to-mid gestation and during mid-to-late gestation to decrease the number of intramuscular adipocytes and muscle fiber sizes [44]. The role of maternal feeding on IMCT is less known, but Huang, et al. [45] have shown that maternal over-nutrition increased collagen accumulation and cross-linking in skeletal muscle offspring.

Author Contributions: Conceptualization, A.L. and B.P.; writing—original draft preparation, A.L.; results discussion and paper writing, M.G and B.P.

Funding: This research received no external funding.

Acknowledgments: The authors thank the staff of Experimental Unit (UERT) of INRA (Theix) for animal management and slaughtering and the staff of the BIOMARKERS team for sampling, Claire Lethias (LBTI

Laboratoire de Biologie Tissulaire et Ingénierie Thérapeutique, UMR 5305, Unité Mixte CNRS/Université de Lyon, 7 Passage du Vercors, 69367 Lyon, Cedex 07, France) for the generous gift of antibodies anti collagen XII and XIV, Tenascin-X and Bruce Caterson (Cardiff School of Biosciences, The Sir Martin Evans Building, Museum Avenue, Cardiff, CF10 3AX, UK) for the generous gift of antibodies anti chondroitin 4S and 6S and versican. The authors convey special thanks to Salah Eddine Bentebibel and Estelle Jean Dit-Gauthier for their valuable technical help.

Conflicts of Interest: The authors declare no conflict of interest.

References

1. Listrat, A.; Lebret, B.; Louveau, I.; Astruc, T.; Bonnet, M.; Lefaucheur, L.; Picard, B.; Bugeon, J. How muscle structure and composition influence meat and flesh quality. *Sci. World J.* **2016**, *2016*, 3182746. [CrossRef] [PubMed]
2. Chriki, S.; Gardner, G.; Jurie, C.; Picard, B.; Micol, D.; Brun, J.-P.; Journaux, L.; Hocquette, J.-F. Cluster analysis application identifies muscle characteristics of importance for beef tenderness. *BMC Biochem.* **2012**, *13*, 29. [CrossRef] [PubMed]
3. Chriki, S.; Picard, B.; Jurie, C.; Reichstadt, M.; Micol, D.; Brun, J.-P.; Journaux, L.; Hocquette, J.-F. Meta-analysis of the comparison of the metabolic and contractile characteristics of two bovine muscles: *Longissimus thoracis* and *Semitendinosus*. *Meat Sci.* **2012**, *91*, 423–429. [CrossRef] [PubMed]
4. Picard, B.; Gagaoua, M.; Micol, D.; Cassar-Malek, I.; Hocquette, J.F.; Terlouw, C.E. Inverse relationships between biomarkers and beef tenderness according to contractile and metabolic properties of the muscle. *J. Agric. Food Chem.* **2014**, *62*, 9808–9818. [CrossRef] [PubMed]
5. Nishimura, T. The role of intramuscular connective tissue in meat texture. *Anim. Sci. J.* **2010**, *81*, 21–27. [CrossRef]
6. Lepetit, J. A theoretical approach of the relationships between collagen content, collagen cross-links and meat tenderness. *Meat Sci.* **2007**, *76*, 147–159. [CrossRef] [PubMed]
7. Brandan, E.; Gutierrez, J. Role of skeletal muscle proteoglycans during myogenesis. *Matrix Biol.* **2013**, *32*, 289–297. [CrossRef]
8. Kresse, H.; Hausser, H.; Schonherr, E. Small proteoglycans. *Experientia* **1993**, *49*, 403–416. [CrossRef]
9. Lethias, C.; Carisey, A.; Comte, J.; Cluzel, C.; Exposito, J.Y. A model of tenascin-X integration within the collagenous network. *FEBS Lett.* **2006**, *580*, 6281–6285. [CrossRef]
10. Ehnis, T.; Dieterich, W.; Bauer, M.; Kresse, H.; Schuppan, D. Localization of a binding site for the proteoglycan decorin on collagen XIV (undulin). *J. Biol. Chem.* **1997**, *272*, 20414–20419. [CrossRef]
11. Elefteriou, F.; Exposito, J.Y.; Garrone, R.; Lethias, C. Binding of tenascin-X to decorin. *FEBS Lett.* **2001**, *495*, 44–47. [CrossRef]
12. Font, B.; Eichenberger, D.; Rosenberg, L.M.; Van der Rest, M. Characterization of the interactions of type XII collagen with two small proteoglycans from fetal bovine tendon, decorin and fibromodulin. *Matrix Biol.* **1996**, *15*, 341–348. [CrossRef]
13. Margaron, Y.; Bostan, L.; Exposito, J.Y.; Malbouyres, M.; Trunfio-Sfarghiu, A.M.; Berthier, Y.; Lethias, C. Tenascin-X increases the stiffness of collagen gels without affecting fibrillogenesis. *Biophys. Chem.* **2010**, *147*, 87–91. [CrossRef]
14. Nishiyama, T.; McDonough, A.M.; Bruns, R.R.; Burgeson, R.E. Type XII and type XIV collagens mediate interactions between banded collagen fibers in vitro and may modulate extracellular matrix deformability. *J. Biol. Chem.* **1994**, *269*, 28193–28199. [PubMed]
15. Bailey, A.J.; Light, N.D. The effect of ante- and post-mortem factors on meat connective tissue: Growth rate, handling and conditioning. In *Connective Tissue in Meat and Meat Products*; Elsevier Applied Science: London, UK, 1989; pp. 195–224.
16. Dubost, A.; Micol, D.; Lethias, C.; Listrat, A. New insight of some extracellular matrix molecules in beef muscles. Relationships with sensory qualities. *Animal* **2016**, *10*, 821–828. [CrossRef]
17. Picard, B.; Lefaucheur, L.; Berri, C.; Duclos, M. Muscle fibre ontogenesis in farm animal species. *Reprod. Nutr. Dev.* **2002**, *42*, 415–431. [CrossRef]
18. Chal, J.; Pourquié, O. Making muscle: Skeletal myogenesis in vivo and in vitro. *Development* **2017**, *144*, 2104–2122. [CrossRef] [PubMed]
19. Jackson, H.E.; Ingham, P.W. Control of muscle fibre-type diversity during embryonic development: The zebrafish paradigm. *Mech. Dev.* **2013**, *130*, 447–457. [CrossRef]

20. Jiwlawat, N.; Lynch, E.; Jeffrey, J.; Van Dyke, J.M.; Suzuki, M. Current Progress and challenges for skeletal muscle differentiation from human pluripotent stem cells using transgene-free approaches. *Stem Cells Int.* **2018**, *2018*, 6241681. [CrossRef]

21. Listrat, A.; Lethias, C.; Hocquette, J.F.; Renand, G.; Menissier, F.; Geay, Y.; Picard, B. Age-related changes and location of types I, III, XII and XIV collagen during development of skeletal muscles from genetically different animals. *Histochem. J.* **2000**, *32*, 349–356. [CrossRef] [PubMed]

22. Listrat, A.; Picard, B.; Geay, Y. Age-related changes and location of type I, III and IV collagens during skeletal muscle development of double-muscled and normal bovine foetuses. *J. Muscle Res. Cell Motil.* **1998**, *19*, 1–14. [CrossRef]

23. Listrat, A.; Picard, B.; Geay, Y. Age-related changes and location of type I, III, IV, V and VI collagens during development of four foetal skeletal muscles of double-muscled and normal bovine animals. *Tissue Cell* **1999**, *31*, 17–27. [CrossRef] [PubMed]

24. Robelin, J.; Lacourt, A.; Béchet, D.; Ferrara, M.; Briand, Y.; Geay, Y. Muscle differentiation in the bovine fetus: A histological and histochemical approach. *Growth Dev. Aging* **1991**, *55*, 151–160. [PubMed]

25. Berthod, F.; Germain, L.; Guignard, R.; Lethias, C.; Garrone, R.; Damour, O.; Van der Rest, M.; Auger, F.A. Differential expression of collagens XII and XIV in human skin and in reconstructed skin. *J. Investig. Dermatol.* **1997**, *108*, 737–742. [CrossRef] [PubMed]

26. Elefteriou, F.; Exposito, J.Y.; Garrone, R.; Lethias, C. Characterization of the bovine tenascin-X. *J. Biol. Chem.* **1997**, *272*, 22866–22874. [CrossRef]

27. Hayes, A.J.; Hall, A.; Brown, L.; Tubo, R.; Caterson, B. Macromolecular organization and in vitro growth characteristics of scaffold-free neocartilage grafts. *J. Histochem. Cytochem.* **2007**, *55*, 853–866. [CrossRef] [PubMed]

28. Bradford, M.M. A rapid and sensitive method for quantification of microgram quantities of protein utilizing principle of protein-dye binding. *Anal. Biochem.* **1976**, *72*, 248–254. [CrossRef]

29. Meunier, B.; Picard, B.; Astruc, T.; Labas, R. Development of image analysis tool for the classification of muscle fibre type using immunohistochemical staining. *Histochem. Cell Biol.* **2010**, *134*, 307–317. [CrossRef]

30. Pöschl, E.; Schlötzer-Schrehardt, U.; Brachvogel, B.; Saito, K.; Ninomiya, Y.; Mayer, U. Collagen IV is essential for basement membrane stability but dispensable for initiation of its assembly during early development. *Development* **2004**, *131*, 1619–1628. [CrossRef]

31. Cassar-Malek, I.; Ueda, Y.; Bernard, C.; Jurie, C.; Sudre, K.; Listrat, A.; Barnola, I.; Gentes, G.; Leroux, C.; Renand, G. Molecular and biochemical muscle characteristics of Charolais bulls divergently selected for muscle growth. *Indic. Milk Beef Qual.* **2005**, *112*, 371–377.

32. Nusgens, B.; Delain, D.; Senechal, H.; Winand, R.; Lapierre, C.M.; Wahrmann, J.P. Metabolic changes in the extracellular matrix during differentiation of myoblasts of the L6 line and of a Myo- non-fusing mutant. *Exp. Cell Res.* **1986**, *162*, 51–62. [CrossRef]

33. Nandan, D.; Clarke, E.P.; Ball, E.H.; Sanwal, B.D. Ethyl-3,4-dihydroxybenzoate inhibits myoblast differentiation: Evidence for an essential role of collagen. *J. Cell Biol.* **1990**, *110*, 1673–1679. [CrossRef] [PubMed]

34. Chaturvedi, V.; Dye, D.E.; Kinnear, B.F.; van Kuppevelt, T.H.; Grounds, M.D.; Coombe, D.R. Interactions between Skeletal Muscle Myoblasts and their Extracellular Matrix Revealed by a Serum Free Culture System. *PLoS ONE* **2015**, *10*, e0127675. [CrossRef] [PubMed]

35. Carrino, D.A.; Oron, U.; Pechak, D.G.; Caplan, A.I. Reinitiation of chondroitin sulphate proteoglycan synthesis in regenerating skeletal muscle. *Development* **1988**, *103*, 641–656.

36. Carrino, D.A.; Sorrell, J.M.; Caplan, A.I. Dynamic expression of proteoglycans during chicken skeletal muscle development and maturation. *Poult. Sci.* **1999**, *78*, 769–777. [CrossRef] [PubMed]

37. Gagnière, H.; Picard, B.; Geay, Y. Contractile differentiation of foetal cattle muscles: Intermuscular variability. *Reprod. Nutr. Dev.* **1999**, *39*, 637–655. [CrossRef]

38. Stupka, N.; Kintakas, C.; White, J.D.; Fraser, F.W.; Hanciu, M.; Aramaki-Hattori, N.; Martin, S.; Coles, C.; Collier, F.; Ward, A.C.; et al. Versican processing by a disintegrin-like and metalloproteinase domain with thrombospondin-1 repeats proteinases-5 and -15 facilitates myoblast fusion. *J. Biol. Chem.* **2013**, *288*, 1907–1917. [CrossRef]

39. Miura, T.; Kishioka, Y.; Wakamatsu, J.-I.; Hattori, A.; Nishimura, T. Interaction between myostatin and extracellular matrix components. *Anim. Sci. J.* **2010**, *81*, 102–107. [CrossRef]

40. Hennebry, A.; Berry, C.; Siriett, V.; O'Callaghan, P.; Chau, L.; Watson, T.; Sharma, M.; Kambadur, R. Myostatin regulates fiber-type composition of skeletal muscle by regulating MEF2 and MyoD gene expression. *Am. J. Physiol.-Cell Physiol.* **2009**, *296*, C525–C534. [CrossRef]

41. Kishioka, Y.; Thomas, M.; Wakamatsu, J.; Hattori, A.; Sharma, M.; Kambadur, R.; Nishimura, T. Decorin enhances the proliferation and differentiation of myogenic cells through suppressing myostatin activity. *J. Cell Physiol.* **2008**, *215*, 856–867. [CrossRef]

42. Wu, H.; Naya, F.J.; McKinsey, T.A.; Mercer, B.; Shelton, J.M.; Chin, E.R.; Simard, A.R.; Michel, R.N.; Bassel-Duby, R.; Olson, E.N.; et al. MEF2 responds to multiple calcium-regulated signals in the control of skeletal muscle fiber type. *EMBO J.* **2000**, *19*, 1963–1973. [CrossRef] [PubMed]

43. Zou, Y.Q.; Zwolanek, D.; Izu, Y.Y.; Gandhy, S.; Schreiber, G.; Brockmann, K.; Devoto, M.; Tian, Z.Z.; Hu, Y.; Veit, G.; et al. Recessive and dominant mutations in COL12A1 cause a novel EDS/myopathy overlap syndrome in humans and mice. *Hum. Mol. Genet.* **2014**, *23*, 2339–2352. [CrossRef] [PubMed]

44. Du, M.; Tong, J.; Zhao, J.; Underwood, K.R.; Zhu, M.; Ford, S.P.; Nataniellz, P.W. Fetal programming of skeletal muscle development in ruminant animals. *J. Anim. Sci.* **2010**, *88*, E51–E60. [CrossRef] [PubMed]

45. Huang, Y.; Zhao, J.-X.; Yan, X.; Zhu, M.-J.; Long, N.M.; McCormick, R.J.; Ford, S.P.; Nataniellz, P.W.; Du, M. Maternal Obesity Enhances Collagen Accumulation and Cross-Linking in Skeletal Muscle of Ovine Offspring. *PLoS ONE* **2012**, *7*, e31691. [CrossRef] [PubMed]

Article

Acceptability of Dry-Cured Belly (Pancetta) from Entire Males, Immunocastrates or Surgical Castrates: Study with Slovenian Consumers

Marjeta Čandek-Potokar [1,2,*], Maja Prevolnik-Povše [1,2], Martin Škrlep [1], Maria Font-i-Furnols [3], Nina Batorek-Lukač [1], Kevin Kress [4] and Volker Stefanski [4]

[1] Agricultural institute of Slovenia, Hacquetova ul. 17, 1000 Ljubljana, Slovenia; maja.prevolnik@kis.si (M.P.-P.); martin.skrlep@kis.si (M.Š.); nina.batorek@kis.si (N.B.-L.)
[2] University of Maribor, Faculty of Agriculture and Life Sciences, Pivola 10, Hoče, Slovenia
[3] IRTA, Granja Camps i Armet, 17121 Monells, Spain; maria.font@irta.cat
[4] University of Hohenheim, Garbenstr. 17, 70599 Stuttgart, Germany; kress.kevin@uni-hohenheim.de (K.K.); Volker.Stefanski@uni-hohenheim.de (V.S.)
* Correspondence: meta.candek-potokar@kis.si; Tel.: +386-1-2805124

Received: 16 March 2019; Accepted: 11 April 2019; Published: 13 April 2019

Abstract: Abandoning of male piglets castration in the European Union is a challenge for the pork production sector in particular for high-quality dry-cured traditional products. The information on consumer acceptability of dry-cured products from alternatives is limited, so the objective was to test the consumer acceptability of unsmoked traditional dry-cured belly (Kraška panceta) processed from three sex categories, i.e., surgical castrates (SC), entire males (EM) and immunocastrates (IC). Consumers (n = 331) were asked to taste dry-cured bellies from EM, IC and SC and to score the taste appreciation on a 9 cm unstructured scale. After tasting the pancetta of three sex categories, the consumers attributed the lowest acceptability scores to SC, whereas IC and EM received similar scores. Only about a quarter of consumers attributed the lowest score to EM, mainly when boar taint compounds were present. The results of this study indicate that a certain share of consumers was sensitive to taste deficiencies and that the leanness of this product is very important for consumers.

Keywords: consumer; sensory acceptability; dry-cured belly; pancetta; pig; boar taint

---

1. Introduction

Surgical castration of male piglets is a worldwide used practice in pig production with the main goal to prevent so-called boar taint—an unpleasant aroma and taste of pork. Boar taint has been associated with the presence of high levels of androstenone and/or skatole, two lipophilic substances that accumulate in fat tissue of uncastrated male pig [1,2]. Androstenone is a steroid that serves as a pheromone and is produced by testicular Leydig cells, whereas skatole is a product of bacterial degradation of the amino acid tryptophan in the large intestine. Hepatic metabolism of skatole is hindered by steroid hormones (including androstenone), so increased concentrations of androstenone are responsible for higher levels of skatole [3]. The levels above which the consumers can detect androstenone and skatole are considered to be between 0.5–1.0 µg/g fat and 0.20–0.25 µg/g fat for androstenone and skatole, respectively [4]. In practice surgical castration is done early in a piglet's life; the European Union legislation allows this procedure to be done without the use of anesthesia or analgesia within the first week after birth [5]. However there is a growing public dissent concerning this method due to a negative effect on animal welfare. For this reason the European pork chain committed to voluntary stop male piglet castration without pain relieving [6]. However, a sustainable exit from piglet castration only works if unsolved issues are discussed critically and alternative solutions are evaluated. In particular, the industry is facing major challenges when fattening boars to higher age and weight at slaughter, which is problematic for high-quality traditional products, as the highest risks are associated with fat quantity and quality [7]. Boar taint is more easily perceived when fat content is high, no masking ingredients are used and the product is consumed warm [8]. Many processing technologies have been tested in order to mask boar taint; however, a recent review study indicated that in order for consumers to not detect it, processed pork would need to have androstenone levels lower than 0.4 µg per g and to be served at below 23 °C [9]. Boar taint was shown to be perceived in dry-cured products, even if these are not consumed warm [10,11]. A recent review of studies with consumers regarding boar taint and consumer acceptability [8] demonstrated a need for a better understanding of the risks related to the perception of boar taint by the consumer in the case of different product types. Moreover, there is a lack of studies comparing sensory acceptability of meat products made from entire males but also other alternative options like immunocastrated males.

The objective of the present study was to test the sensory acceptability of a traditional Slovenian product, dry-cured belly Kraška panceta with Slovenian consumers. Kraška panceta is a product protected with geographical indication (PGI) according to EU legislation. The origin of raw material is not prescribed which denotes procurement on different EU markets and no control over the sex category of pigs used. With increasing probability of meat from uncastrated male pigs on the market, it is important to verify the acceptability of this product made from meat of different alternatives, i.e., pig sex categories.

2. Materials and Methods

Bellies came from pigs (crosses of German Landrace sows and Pietrain boars raised on the same farm, fed the same standard commercial feed mixture) of three sex categories, i.e., entire males (EM), immunocastrates (IC) and surgical castrates (SC). Pigs derived from one slaughter batch and were of similar age (185.0 ± 3.4 days) and weight (121.1 ± 9.9 kg). IC received two applications of the vaccine IMPROVAC®(Zoetis Belgium SA, Louvain la Neuve, Belgium) according to the manufacturer's recommendations. Fresh bellies were submitted to processing (regular commercial production respecting food safety legislation) according to the rules of traditional Slovenian dry-cured belly product protected with geographical designation (PGI), Kraška panceta. The processing procedure consisted of seven days dry-salting, surface addition of spices (black pepper and garlic), no smoking, and air-drying for 12 weeks. Bacons (*n* = 18, six per sex group with average fresh belly leanness evaluated on a carcass cross section at the last rib being 72.8%, 68.7% and 54.4% for EM, IC and SC, respectively) were selected at the end of the processing for the sensory analysis with consumers. In the case of products from EM, dry-cured bellies were selected so as to cover a large range of boar taint levels (i.e., of androstenone and skatole levels, Table 1), whereas in the case of SC and IC the levels of androstenone and skatole in the fat tissue were below the limit of detection of the analytical method. Surveys with consumers were conducted at the main agricultural fair in Slovenia (AGRA 2018) and comprised of 331 volunteers (210 males, 121 females), visitors of the fair that agreed to taste and evaluate in parallel three slices of pancetta, one from EM, one from IC and one from SC. For that purpose, six sets of EM-IC-SC products were used (Table 1). Each triplet of products was tasted by 50 to 65 consumers. Slices were codified as A, B and C for SC, IC and EM, respectively and all three slices were given to the consumer at the same time. The order of tasting was a decision of the consumer. No personal data were asked and collected from the visitors that tasted the products.

Table 1. Level of boar taint substances in the fat tissue of entire males (EM).

	Set					
	1	**2**	**3**	**4**	**5**	**6**
Boar Taint Substance	**Low Boar Taint**		**High Boar Taint**			**No tainT**
Androstenone [1]	0.55	0.45	19.5	3.07	9.10	b.d.
Skatole [1]	0.03	b.d.	0.19	0.03	0.20	b.d.

[1] µg per g liquid fat; b.d. = below the limit of detection.

Each visitor was asked to evaluate the acceptability of the taste of three pancetta slices (one from EM, one from IC and one from SC) and to provide a note on a 9 cm non-structured scale anchored at two ends (from dislike extremely to like extremely). The distance from the left anchor to mark was measured. Data obtained were submitted to an analysis of variance (ANOVA), using the procedure Mixed of statistical software SAS®(SAS Institute Inc., Cary, NC, USA). The model included the fixed effects of the sex category (EM, IC and SC), set of products (1 to 6) and their interaction and gender of the consumer and a random effect of the consumer. Due to a significant interaction, the slicing by sex category and set of products were used to evaluate the effect of sex category within the set of products and vice versa. The least squares means (LSM) were calculated and compared (using a Tukey test for multiple comparisons). A threshold probability level considered for statistical significance was at $p < 0.05$.

3. Results

The analysis of variance (Table 2) showed that the pig sex category had a significant effect on sensory acceptability of pancetta, and that the consumers attributed the pancetta from EM and IC with the highest average acceptability score (6.3 and 6.2, respectively), whereas SC products received a significantly ($p < 0.0001$) lower average score (5.4). There was no difference in the acceptability score between male and female consumers.

Table 2. Analysis of variance for the pancetta sensory acceptability score.

Residual	Sex Category			Set of Products						Interaction Sex × Set	Consumer Gender	
						p-value						
2.3	<0.0001			0.4308						<0.0001	0.8334	
						LSM						
	EM	**IC**	**SC**	**1**	**2**	**3**	**4**	**5**	**6**		**male**	**female**
	6.3 [b]	6.2 [b]	5.4 [a]	5.9	5.9	6.4	5.9	6.0	6.0		6.0	6.0

EM = entire males, IC = immunocastrates, SC = surgical castrates; 1 to 6 sets of belly with different boar taint levels as defined in Table 1. Least squares mean (LSM) values with a different letter are statistically different at $p < 0.05$.

There was a significant effect ($p < 0.0001$) of the interaction between the set of tested products and pig sex category on the acceptability score, indicating that the result (i.e., acceptability according to the pig sex category) was not the same in all sets of products. Differences in the liking score between EM, IC and SC were significant in the case of high tainted EM products (sets 3 and 4) and the untainted EM product (set 6) but not in low tainted EM products (Table 3).

Pancetta of SC received the lowest average score in four out of six sets (Table 3) and in total and was scored the lowest by 43.5% of consumers. In the absence of boar taint substances in EM (i.e., set 6), EM was liked the most and SC the least with IC taking an intermediate position. When EM had low boar taint (sets 1 and 2), the average acceptability score of EM was similar to IC and SC, whereas when EM had high boar taint (sets 3 and 4), IC had the highest acceptability. Overall, there was 23.3% of consumers who gave the lowest score to pancetta from EM. When androstenone level was about 0.5 ppm (sets 1 and 2), the lowest score was given by 22.8% of consumers, when androstenone level was very high (sets 3, 4 and 5), the lowest score was given by 27.2% of consumers. In the case of EM pancetta with boar taint compounds below detection limit (set 6), there was still 11.3% of consumers which gave the lowest score to EM pancetta. Figure 1 shows a decrease of liking of EM pancetta with an increasing level of boar taint substances, whereas just the opposite trend can be observed for SC and IC (increased liking of IC and SC products with increasing boar taint in EM products).

Table 3. Least squares mean (LSM) for sensory acceptability according to the sex category and set of products.

	Set of Products						
Sex category	**1**	**2**	**3**	**4**	**5**	**6**	***p*-value**
EM	6.4	6.1	6.3 [AB]	5.8 [AB]	6.2	7.2 [C]	0.1028
IC	5.4 [a]	5.8 [ab]	7.1 [B,c]	6.9 [B,bc]	6.3 [ab]	5.9 [B,ab]	0.0014
SC	5.9	5.8	5.7 [A]	4.9 [A]	5.5	4.8 [A]	0.0765
***p*-value**	0.0858	0.6956	0.0074	0.0003	0.1014	<0.0001	

EM = entire males, IC = immunocastrates, SC = surgical castrates. Significant ($p < 0.05$) differences in least squares mean (LSM) values are assigned different letters (between sex categories' uppercase letters and between sets of products' lowercase letters).

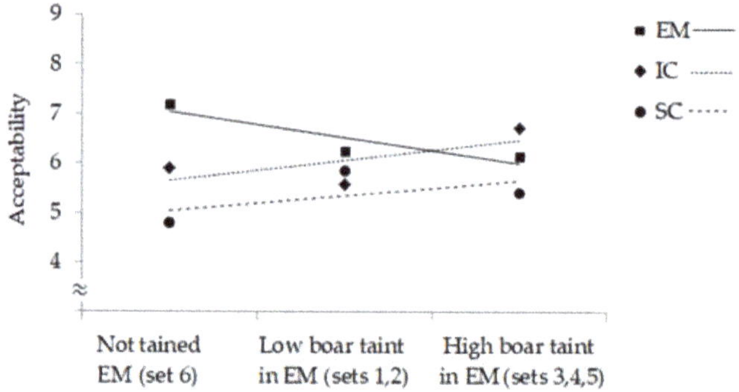

Figure 1. Change of consumer acceptability of pancetta according to the level of boar taint in EM. EM = entire males, IC = immunocastrates and SC = surgical castrates.

4. Discussion

In the present study, the slices of products were coded A, B and C for SC, IC and EM, respectively, which could bias the results as a serial position has been shown to affect preferences (large primacy effect, i.e., advantage of the first) [12]. However, the slices were presented together, the consumers were free to choose the order of their tasting (not recorded) and SC, which were coded A, were the least appreciated. The overall lowest consumer sensory appreciation of dry-cured belly products from SC was unexpected, in particular because boar taint substances were present in most of the EM products and in some of them at very high levels. It can be speculated that the consumers, despite being asked to score the taste, were evaluating the overall acceptability of the product. The context in which the sensory properties are perceived alter the perception [13]. Implicit sensory experience (i.e., the 'taste' of a food) represents only part of the eating experience; visual and haptic perception reveal the strongest correlations with the overall experience [14]. It can be assumed that they were biased by visual appearance, most likely the leanness with EM having the most lean and SC the fattiest bellies with IC in the middle but closer to EM (72.8, 54.4 and 68.7 lean %, respectively). This is also substantiated by the fact, that SC was scored the lowest in majority of the product sets and that IC products were scored the highest (and higher than EM) when boar taint substances were at a high level in the EM product. The maximum difference between scores given to SC, IC and EM were obtained in products of set 6, thus, without taint in EM. However, despite this generally best appreciation observed for EM products, the consumers were sensible to taste deficiencies. Firstly, about a quarter of consumers gave the lowest score to the EM product. Secondly, when EM had low boar taint, it was more similar to IC and SC, probably because consumers were not completely satisfied with the taste of EM although not enough dissatisfied to show a preference for SC or IC. However, when EM had high boar taint, IC had higher acceptability than EM, probably because the consumers did not like the EM taste and choosing between IC and SC, the first were less fatty so were preferred. This is demonstrated in Figure 1, where a downward trend could be observed for EM products, and an upward trend for products of IC and SC, which confirms a reduced appreciation in the case of EM products with boar taint. The review carried out by Font-i-Furnols in 2012 [8] showed that in two studies, the acceptability of EM pork depended on the level of androstenone or boar taint, but in two studies, the bacon from EM was equally accepted as those from other sexes (IC not included). Contrary as in the present study, Spanish consumers scored the loins from EM lower than those from SC and IC, even when the levels of androstenone were low [15]. However, in this study with loins, consumers did not see the fatness of the product, which could influence a lot the appreciation of the meat by consumers [16] and may explain the results of the present study. The literature dealing with the effect of processing technologies on boar tainted meat generally shows that dry-curing does not eliminate its perception in the products [10,11]. It is also considered that boar taint substances are not degraded or lost during the long dry-curing process [17], although in one of our studies we determined some reduction of the concentration of boar taint substances with very long maturing time in dry-cured ham [18]. However, in the case of pancetta, the process is much shorter. It has also been suggested that the curing process leads to a formation of aromatic compounds that are able to mask boar taint [19]. The added spices, garlic and black pepper could also contribute to the masking of boar taint [20]. It is also possible that high leanness of pancetta from EM used in the present study could partly explain why the consumers did not perceive taste deficiencies even more as boar taint is more easily perceived when fat content is high [8]. This investigation shows that boar taint levels cannot solely explain taste acceptability of pancetta, which corroborates with recently published meta-analysis [21] indicating that uncastrated males (EM) apart from boar taint, may present reduced meat quality and are more prone to oxidation [21,22]. Despite some methodological limitations due to the study design, this investigation provides an insight to the sensory acceptability of dry-cured belly by consumers in relation to the sex category as characterized by leanness and boar taint level.

Author Contributions: Conceptualization, M.Č.-P., M.Š.; methodology, M.F.-i.-F., N.B.-L., K.K.; formal analysis, M.Č.-P., M.P.-P; data curation, M.P.-P; writing—original draft preparation, M.Č.-P.; writing—review and editing, M.Š., N.B.-L., K.K., V.S., M.F.-i.-F.; funding acquisition, M.Č.-P., V.S.

Funding: This research was conducted within program SuSAN ERA-NET Cofund, project SuSI (co-financing by the European Commission and Ministry of agriculture, forestry and food of the Republic of Slovenia). Core financing of the Slovenian research agency (grant P4-0133) is also acknowledged. In addition, this article is based upon work from COST Action CA 15215 Innovative approaches in pork production with entire males (IPEMA) supported by COST (European Cooperation in Science and Technology).

Acknowledgments: Authors greatly acknowledge the support of Pršutarna s´Krasa d.o.o. and its manager Tadej Kaltnekar for conduct of the processing.

Conflicts of Interest: The authors declare no conflict of interest. The funders had no role in the design of the study; in the collection, analyses, or interpretation of data; in the writing of the manuscript, or in the decision to publish the results.

References

1. Patterson, R.L.S. 5α-androst-16-ene-3-one: Compound responsible for taint in boar fat. *J. Sci. Food Agric.* **1968**, *19*, 31–38. [CrossRef]
2. Walstra, P.; Maarse, G. *Onderzoek Gestachlengen van Mannelijke Mestvarkens*; IVO-rapport C-147, Rapport 2; Researchgroep voor Vlees en Vleeswaren TNO: Zeist, The Netherlands, 1970; p. 30.
3. Zamaratskaia, G.; Squires, E.J. Biochemical, nutritional and genetic effects on boar taint in entire male pigs. *Animal* **2009**, *3*, 1508–1521. [CrossRef] [PubMed]

4. Walstra, P.; Claudi-Magnussen, C.; Chevillon, P.; von Seth, G.; Diestre, A.; Matthews, K.R.; Homer, D.B.; Bonneau, M. An international study on the importance of androstenone and skatole for boar taint: Levels of androstenone and skatole by country and season. *Livest. Prod. Sci.* **1999**, *62*, 15–28. [CrossRef]

5. Council Directive 2008/120/EC of 18 December 2008 laying down minimum standards for the protection of pigs. *Off. J. Eur. Union* **2009**, *52*, 5–13.

6. European Declaration on Alternatives to Surgical Castration of Pigs. 2010. Available online: http://ec.europa. eu/food/animals/welfare/practice/farm/pigs/castration_alternatives_en (accessed on 18 January 2017).

7. Bonneau, M.; Čandek-Potokar, M.; Škrlep, M.; Font-i-Furnols, M.; Aluwé, M.; The Castrum Network; Fontanesi, L. Potential sensitivity of pork production situations aiming at high-quality products to the use of entire male pigs as an alternative to surgical castrates. *Animal* **2018**, *12*, 1287–1295. [CrossRef] [PubMed]

8. Font-i-Furnols, M. Consumer studies on sensory acceptability of boar taint: A review. *Meat Sci.* **2012**, *92*, 319–329. [CrossRef] [PubMed]

9. Tørngren, M.A.; Claudi-Magnussen, C.; Støier, S.; Kristensen, L. Boar taint reduction in smoked, cooked ham. In Proceedings of the 57th International Congress of Meat Science and Technology, Ghent, Belgium, 7–12 August 2011.

10. Bañón, S.; Gil, M.D.; Garrido, M.D. The effects of castration on the eating quality of dry-cured ham. *Meat Sci.* **2003**, *65*, 1031–1037. [CrossRef]

11. Corral, S.; Salvador, A.; Flores, M. Effect of the use of entire male fat in the production of reduced salt fermented sausages. *Meat Sci.* **2016**, *116*, 140–150. [CrossRef] [PubMed]

12. Mantonakis, A.; Rodero, P.; Lesschaeve, I.; Hastie, R. Order in Choice: Effects of Serial Position on Preferences. *Psychol. Sci.* **2009**, *20*, 1309–1312. [CrossRef] [PubMed]

13. Bell, T.; Meiselman, H.L.; Pierson, B.J.; Reeve, W.G. Effects of adding an Italian theme to a restaurant on the perceived ethnicity, acceptability and selection of foods. *Appetite* **1994**, *22*, 11–24. [CrossRef] [PubMed]

14. Haase, J.; Wiedmann, K.-P.; Labenz, F. Effects of consumer sensory perception on brand performance. *J. Cons. Mark.* **2018**, *35*, 565–576. [CrossRef]

15. Font-i-Furnols, M.; Gispert, M.; Diestre, A.; Oliver, M.A. Acceptability of boar meat by consumers depending on their age, gender, culinary habits, sensitivity and appreciation of androstenone smell. *Meat Sci.* **2003**, *64*, 433–440. [CrossRef]

16. Ngapo, T.M.; Martin, J.-F.; Dransfield, E. International preferences for pork appearance: I. Consumer choices. *Food Qual. Pref.* **2007**, *18*, 26–36. [CrossRef]

17. Haugen, J.E.; Brunius, C.; Zamaratskaia, G. Review of analytical methods to measure boar taint compounds in porcine adipose tissue: The need for harmonised methods. *Meat Sci.* **2012**, *90*, 9–19. [CrossRef] [PubMed]

18. Škrlep, M.; Čandek-Potokar, M.; Batorek Lukač, N.; Prevolnik Povše, M.; Pugliese, C.; Labussière, E.; Flores, M. Comparison of entire male and immunocastrated pigs for dry-cured ham production under two salting regimes. *Meat Sci.* **2016**, *111*, 27–37. [CrossRef] [PubMed]

19. Niderehe, H. Die Verwertung des Fleisches von Schweinen mit Ebergeruch. *Schlachten und Vermarkten* **1977**, *77*, 82–83.

20. Lunde, K.; Egelandsdal, B.; Choinski, J.; Mielnik, M.; Flåtten, A.; Kubberød, E. Marinating as a technology to shift sensory thresholds in ready-to-eat entire male pork meat. *Meat Sci.* **2008**, *80*, 1264–1272. [CrossRef] [PubMed]

21. Pauly, K.; Luginbühl, W.; Ampuero, S.; Bee, G. Expected effects on carcass and pork quality when surgical castration is omitted. *Meat Sci.* **2012**, *92*, 858–862. [CrossRef] [PubMed]

22. Škrlep, M.; Tomažin, U.; Batorek Lukač, N.; Poklukar, K.; Čandek-Potokar, M. Proteomic Profiles of the Longissimus Muscles of Entire Male and Castrated Pigs as Related to Meat Quality. *Animals* **2019**, *9*, 74. [CrossRef] [PubMed]

 *foods*

Article

Lipid Oxidation Inhibition Capacity of 11 Plant Materials and Extracts Evaluated in Highly Oxidised Cooked Meatballs

Stina C. M. Burri [1,*], Kajsa Granheimer [1], Marine Rémy [1], Anders Ekholm [2], Åsa Håkansson [1], Kimmo Rumpunen [2] and Eva Tornberg [1]

[1] Department of Food Technology Engineering and Nutrition, Lund University, Naturvetarvägen 12, 223 62 Lund, Sweden; kajsagranheimer@hotmail.com (K.G.); marine.remy@agrosupdijon.fr (M.R.); asa.hakansson@food.lth.se (Å.H.); eva.tornberg@food.lth.se (E.T.)

[2] Department of Plant Breeding, Swedish University of Agricultural Sciences, Fjälkestadsvägen 459, 291 94 Kristianstad, Sweden; anders.ekholm@slu.se (A.E.); kimmo.rumpunen@slu.se (K.R.)

* Correspondence: stina.burri@food.lth.se; Tel.: +46-462229808

Received: 7 August 2019; Accepted: 9 September 2019; Published: 12 September 2019

Abstract: The underlying mechanism(s) behind the potential carcinogenicity of processed meat is a popular research subject of which the lipid oxidation is a common suspect. Different formulations and cooking parameters of a processed meat product were evaluated for their capacity to induce lipid oxidation. Meatballs made of beef or pork, containing different concentrations of fat (10 or 20 g 100 g^{-1}), salt (2 or 4 g 100 g^{-1}), subjected to differing cooking types (pan or deep frying), and storage times (1, 7, and 14 days), were evaluated using thiobarbituric reactive substances (TBARS). The deep-fried meatball type most susceptible to oxidation was used as the model meat product for testing the lipid oxidation inhibiting capacity of 11 plant materials and extracts, in two concentrations (100 and 200 mg kg^{-1} gallic acid equivalent (GAE)), measured after 14 days of storage using TBARS. Summer savory lyophilized powder was the most efficient plant material, lowering lipid oxidation to 13.8% and 21.8% at the 200 and 100 mg kg^{-1} concentration, respectively, followed by a sea buckthorn leaf extract, lowering lipid oxidation to 22.9% at 100 mg kg^{-1}, compared to the meatball without added antioxidants. The lipid oxidation was thus successfully reduced using these natural antioxidants.

Keywords: natural antioxidant; phenol; malondialdehyde; processed meat; Folin-Ciocalteu

1. Introduction

Since the International Agency for Research on Cancer (IARC) released their monograph stating that consumption of red and processed meats are linked to colorectal cancer (CRC) in 2015, research regarding this topic has increased, particularly since the mechanisms leading to these links remain partly unknown [1]. The most commonly proposed factors underlying the link between consumption of red/processed meat and CRC have been attributed to the following, partly overlapping, mechanisms: (I) An increase in oxidative or N-nitrosation load leading to lipid oxidation and DNA adducts in the intestinal epithelium, respectively; (II) stimulation of proliferation of the epithelium by heme or other food metabolites acting either directly or following conversion, e.g., heterocyclic amines (HCAs) and polycyclic aromatic hydrocarbons (PAHs) through high-temperature cooking; and (III) pro-malignant processes triggered by a higher inflammatory response, e.g., by a process where N-glycolylneuraminic acid (Neu5Gc) is involved in developing xenosialitis, an inflammatory syndrome inducing cancer formation and progression [2,3].

Processed meat refers to products typically made of red meat that have undergone curing, salting, or smoking, often containing high amounts of fatty tissues together with endogenous phospholipids [4].

These factors, together with high-temperature cooking, make them susceptible to oxidative reactions, which may contribute to health hazards [5]. The meatball dish, a minced meat product, is one of the Swedish trademarks in traditional cuisine. The world's largest furniture retailer, IKEA, claims they sell more than one billion meatballs per year [6]. Meatballs are industrially deep-fried, while usually pan-fried when homemade, and are typically made of pork and/or beef meat (fat content ranging between 10 and 20 g 100 g^{-1}), onion, egg, breadcrumbs, and salt (ranging between 2 and 4 g 100 g^{-1}). Meatballs were chosen as the study material due to the large market and the fact that the formulation and cooking process covers most of the main components of processed meat products in general.

The addition of antioxidants to meat, meat products, and meat model systems has been widely studied for oxidation preventing purposes [7]. Synthetic antioxidants, such as butylated hydroxytoluene (BHT), butylated hydroxyanisole (BHA), propyl gallate (PG), and tertiary butylhydroquinone (TBHQ), are commonly used in the food industry for oxidation-inhibiting purposes but are decreasing in use due to their suspected genotoxicity [5,7]. Hence, there is an increased demand of natural antioxidants, which could prevent the oxidation process of different meat products, potentially decreasing the negative health effects of processed meat products, as well as prolonging the shelf-life and promoting sustainability. For an evaluation of the lipid oxidation-inhibiting capacity of antioxidants in meat and meat products, the thiobarbituric reactive substances (TBARS) assay is frequently used [8–10]. Moreover, to obtain reliable results, it is important that the meat model used is appropriate for the product and supplements tested.

The aim of the present study was (I) to develop a relevant oxidized processed meatball model to study the effects of supplemented antioxidants, and (II) investigate lipid oxidation in meatballs without and with a range of plant materials and extracts at different concentrations.

2. Materials and Methods

2.1. Plant Materials and Extracts

Eleven plant materials and extracts were collected from Denmark, Estonia, Finland, Latvia, and Sweden (Table 1), i.e., from partners of the EU-project "Sustainable plant ingredients for healthier meat products-proof of concepts". Finnish phenol-rich extracts were prepared using pressurized hot water extraction (PHWE). Samples were obtained using a Dionex ASE 350 accelerated solvent extractor (Thermo Fischer Scientific Inc., Waltham, MA, USA). Extraction temperatures were 110 °C and 120 °C for sea buckthorn (*Hippophae rhamnoides* L.) leaves and bilberry leaves (*Vaccinium myrtillus* L.), respectively. The static extraction time was 1 min for both samples. After PHWE, the extracts were filtered and lyophilized. Estonian phenol-rich extracts were prepared using a pilot-scale solid-liquid Naviglio extractor (Atlas Filtri, Limena, Italy) with 20 mL 100 mL^{-1} ethanol (aq). Extracts of rhubarb root (*Rheum rhabarbarum* L.) and black currant (*Ribes nigrum* L.) leaves were concentrated to half their volume and were lyophilized to a powder. The ethanol of the Swedish extracts (Table 1) was evaporated before they were diluted to the wanted concentrations. The summer savory (*Satureja hortensis* L.) powder was added non-extracted to the meat batter with the same amount of tap water as the extracts. The extracts were dissolved in MilliQ water prior to dilution in tap water and addition to the meat batter. All samples were analyzed for their total phenols content using Folin-Ciocalteu reagent [11].

Table 1. Samples of antioxidant plant materials and extraction methods used.

Plant Material	Latin Name	Cultivar	Country	Extraction Solution	Abbreviation	Extraction Method	GAE mg mL^{-1} Extract
Lyophilized Sea buckthorn leaves	*Hippophae rhamnoides* L.	'Botnia Guldklimp'	Finland	Pressurized hot water	SBTPHWE	Section 2.1	7.0
Lyophilized Bilberry leaves	*Vaccinium myrtillus* L.	Native stands	Finland	Pressurized hot water	BBPHWE	Section 2.1	11.6
Sea buckthorn leaves and sprouts	*Hippophae rhamnoides* L.	Mix of 'Botaničeskaja Ļubiteļskaja' and 'Prozračnaja'	Latvia	80% Ethanol	SBT80	Gornas et al. [12]	13.2
Sea buckthorn leaves and sprouts	*Hippophae rhamnoides* L.	Mix of 'Botaničeskaja Ļubiteļskaja' and 'Prozračnaja'	Latvia	H_2O	SBTH$_2$O	Modified from Gornas et al. [12]	9.2
Summer savory leaves	*Satureja hortensis* L.	Seed origin: Hild Samen	Denmark	Non-extracted	SS	Non-extracted	12.0
Sea buckthorn leaves	*Hippophae rhamnoides* L.	'Finskaja'	Sweden	50% ethanol	SBT	Burri et al. [11]	8.8
Olive Polyphenols-Phenoliv	*Olea europaea* L.	Phenoliv™	Sweden	50% ethanol	OS	Burri et al. [11]	3.8
Onion skin	*Allium cepa* L.	'Donna'	Sweden	50% ethanol	OPP	Burri et al. [11]	3.0
Beetroot leaves	*Beta vulgaris* subsp. *vulgaris*	'Action'	Sweden	50% ethanol	BR	Burri et al. [11]	1.0
Lyophilized rhubarb root	*Rheum rhabarbarum* L.	'Victoria'	Estonia	20% ethanol	LRR	Section 2.1	18.1
Lyophilized black currant leaves	*Ribes nigrum* L.	'Pamjat Vavilova'	Estonia	20% ethanol	LBC	Section 2.1	10.1

2.2. Chemicals

2-Thiobarbituric acid ≥98 g 100 g^{-1} (TBA), 1,1,3,3-tetramethoxypropane 99 mL 100 mL^{-1} (TMP), trichloroacetic acid ≥99.0 g 100 g^{-1} (TCA), and ethanol 96 mL 100 mL^{-1} were obtained from Sigma-Aldrich Inc., St. Louis, MO, USA. Hydrochloric acid (40 mM L^{-1}) and 85 mL 100 mL^{-1} ortho-phosphoric acid (H$_3$PO$_4$) were obtained from Merck KGaA, Darmstadt, Germany.

2.3. Preliminary Trial for Meatball Model Selection

Meatballs were produced in triplicates with two concentrations of NaCl (2 and 4 g 100 g^{-1}), fat (10 and 20 g 100 g^{-1}) from two types of meat (pork and beef obtained from Atria Sverige AB) and were either pan-fried or deep-fried, and stored for 1, 7, and 14 days (Figure 1). Lean meat from pork shoulder (*Musculus trapezius*) (fat content 4.3 g 100 g^{-1}) was adjusted to fat contents of 10 and 20 g 100 g^{-1}, respectively, during mincing using pork belly (*M. external abdominal oblique*) (fat content 28.2 g 100 g^{-1}). Lean meat from boneless beef knuckle (*M. semitendinosus*) (fat content 5.5 g 100 g^{-1}) was adjusted to the same fat contents using a mixture of cuts from beef chuck (*M. deltoideus*) and clod (*M. latissimus dorsi, M. trapezius, M. serratus ventralis*) (fat content 27.5 g 100 g^{-1}). After fat adjustment, NaCl was added and blended to the mince before vacuum packing and freezing at −18 °C until trials begun. The mince was thawed in a refrigerator (4 °C) overnight before meatballs were manufactured for the trial. Meatballs were either pan-fried or deep-fried in randomized order in refined rapeseed oil (Zeta, Stockholm, Sweden), containing 7.5 g 100 g^{-1} saturated fatty acids, 62.5 g 100 g^{-1} mono-unsaturated fatty acids, and 30 g 100 g^{-1} poly-unsaturated fatty acids, where the temperature in the pan was kept stable at 175 °C ± 1 °C controlled by a laser thermometer (IR-termometer Basetech IRT-350, Plano, Texas, USA) and the deep-frying temperature was kept stable at 160 °C ± 0.5 °C. When the meatballs had reached a 72 °C inner temperature, they were removed from heating and rested on paper towels until room temperature was reached. The meatballs were then stored in a refrigerator (4 °C) in sealed polyethylene bags for 1, 7, and 14 days before the level of lipid oxidation was measured using thiobarbituric reactive substances (TBARS).

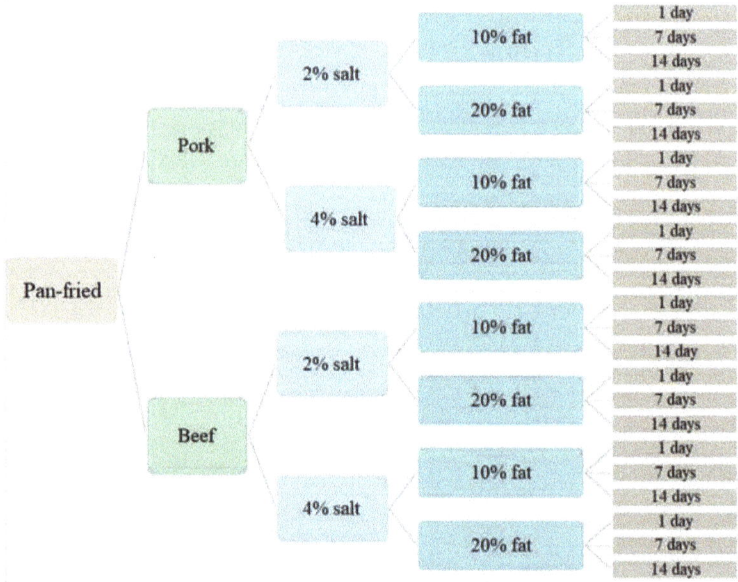

Figure 1. Scheme of the pan-fried meatball study set up as modified by Granheimer [13]. The same set up was used for the deep-fried meatball study.

2.4. Test of Antioxidants in Most Oxidized Pork Meatball

The 11 antioxidant powders and extracts were added in 100 and 200 mg kg^{-1} of gallic acid equivalents (GAE), respectively, based on their total phenols content (Table 1) to the most oxidized type of meatball from the above presented study, namely, pork with 2 g 100 g^{-1} NaCl and 20 g 100 g^{-1} fat (weight by weight, which was deep-fried. In order to incorporate the antioxidants into the meat batters, 90 g of mince was mixed by hand with 9 mL of antioxidant solution until all fluid had been absorbed by the mince. Meatballs were hand-rolled and weighed in triplicates to a standardized weight of 16.5 g $\pm$ 0.1 g before they were deep-fried as mentioned above at 160 °C until the inner temperature reached 72 °C. The meatballs were then let to rest on paper towels until room temperature was reached. After 14 days in the refrigerator (4 °C), the level of lipid oxidation was measured using thiobarbituric reactive substances (TBARS).

2.5. Thiobarbituric Reactive Substances (TBARS)

The lipid oxidation was measured in triplicates for each meatball using TBARS based on the method of Buege and Aust [14], where the prepared TBA reagent consisted of 15 g 100 mL^{-1} TCA and 0.75 g 100 mL^{-1} TBA in 0.25 M HCl. Samples from meatballs were weighed to 6 $\pm$ 0.1 g and crushed by mortar and pestle prior to being added to a 50-mL falcon tube containing 22.5 mL Milli-Q H$_2$O together with 1.5 mL of 10 g 100 mL^{-1} TCA solution. The samples were then vortexed for 60 s prior to heating at 40 °C for 5 min to precipitate proteins, before 2 mL 96 mL 100 mL^{-1} ethanol was added to solubilize fats. The mixtures were then filtered through a filter paper (Munktell, grade 1F) to obtain a clear filtrate for TBARS analysis. In total, 2.5 mL of the TBA reagent was added to 0.5 mL of the filtrates, which were then heated to 90 °C for 10 min before being cooled in tap water to end the reaction. All samples were centrifuged at 3600 g for 20 min (Allegra® X-15R, Beckman Coulter, Brea, CA, USA) before the absorbance of the supernatants was measured at 534 nm and 600 nm, respectively (Varian Cary® 50 UV-Vis Spectrophotometer, Agilent, Santa Clara, CA, USA). If the coefficient of variation (cv%) between the sample replicates exceeded 10%, the analyses were re-run. Results are reported as μM malondialdehyde (MDA) g^{-1} meatball for meatball parameter evaluations, and as a percentage oxidation of the blank sample (without added antioxidants) for antioxidant evaluation.

2.6. Statistical Analyses

Statistical analyses were carried out using SPSS Statistics 25.0 (IBM Corp. Released 2017. IBM SPSS Statistics for Windows, Version 25.0. Armonk, NY, USA) and R for Windows GUI front-end version 3.5.3 (R version 3.5.2 (2018-12-20)—"Eggshell Igloo" Copyright © 2018 The R Foundation for Statistical Computing Platform: x86_64-w64-mingw32/x64 (64-bit). Box-Cox transformations were used to achieve normal distribution of data. A univariate general linear model (GLM) was performed on logarithmic values of oxidation data in both the preliminary trial and antioxidant study in order to ensure normal distribution of samples (skewness and kurtosis with maximum values of ±1.96 [15]). In the preliminary trial, meat type, salt content, fat content, cooking type, and storage time were all fixed factors tested in a full-factorial GLM model on the dependent variable TBAR (μM MDA g^{-1} meatball). In the antioxidant study, species and concentrations of antioxidants were fixed factors in the full-factorial model of the dependent variable percent lipid oxidation of meatballs compared to meatballs without antioxidants. In both trials, meatball samples were made in triplicates of which TBARS was measured in 3 technical replicates; statistical analyses were conducted on an average of the results of the meatball triplicates. Post-hoc tests were performed using the Scheffe method. A Pearson correlation analysis was conducted using SPSS on previous antioxidant activity data for the samples included in this study together with the lipid oxidation inhibiting capacity to study if, and if so, which antioxidant mode of action gave rise to the lipid oxidation inhibiting capacity.

3. Results

3.1. Preliminary Trial for Meatball Model Selection

Lipid oxidation levels of meatballs differing in composition, cooking method, and time of storage are presented in Figure 2. Statistical results indicate that meat type, storage time, cooking type, and salt content all had significant effects on the lipid oxidation ($p < 0.001$) as did the fat content ($p < 0.01$). Nearly all interactions had significant effects on lipid oxidation ($p < 0.05$) except for interactions between: Salt × cooking, meat × salt × storage, meat × salt × fat × cooking, and meat × salt × fat × storage (Table A1). However, the partial Eta squared (η^2) showed the effect sizes of most interactions were minor. The factors affecting lipid oxidation the most were meat type and storage time ($p < 0.001$) (Table A1), where pork oxidized more than beef, and where oxidation increased with longer storage times for both meat types, as is shown in the interaction between meat type and storage time (Figure 3a). There was an interaction between meat type and fat content (Figure 3b), where beef with 20 g 100 g^{-1} fat oxidized less than with 10 g 100 g^{-1} fat, and the opposite was shown in pork meat. The combination of parameters that oxidized the most contained pork meat, 20 g 100 g^{-1} fat, and 2 g 100 g^{-1} salt; was deep-fried; and was stored for 14 days (Figure 2) (Table A2).

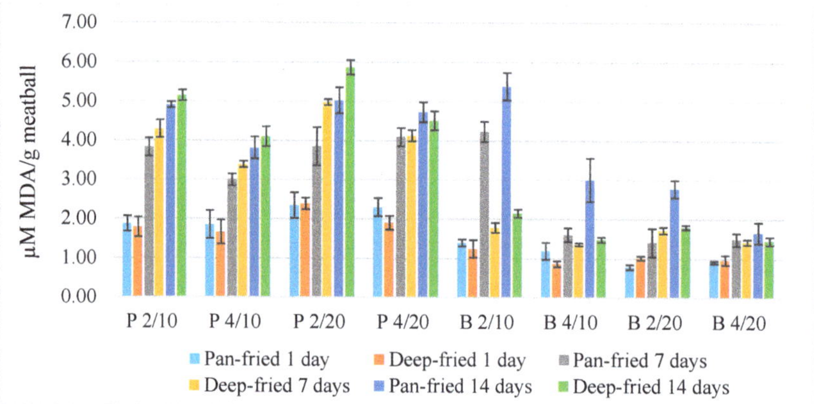

Figure 2. Lipid oxidation in model meatballs with differing parameters shown in µM malondialdehyde (MDA) g^{-1} meatball, where P = pork and B = beef meat. The numbers 2 and 4 correspond to the salt content in % and the numbers 10 and 20 correspond to the fat content in %. The standard deviation is shown by the error bars ($n = 3$).

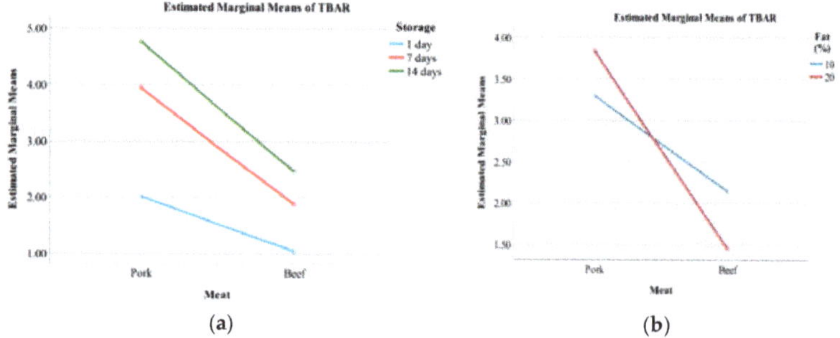

Figure 3. Main effect plots (estimated marginal means) of TBARS (µM malondialdehyde g^{-1} meatball) of (a) the interaction between meat type and storage times and (b) the interaction between meat type and fat contents (%).

3.2. Inclusion of Antioxidants in Most Oxidized Meatball

All meatball samples were evaluated for their TBAR substances (malondialdehyde (MDA) μM g^{-1}) in triplicates after 14 days of storage where the results were calculated as a percentage of oxidation compared to samples without added antioxidants ((μM MDA g^{-1} meatball with plant material/μM MDA g^{-1} meatball without antioxidant) * 100) (Figure 4). Statistical results indicate that antioxidant species, concentrations, and the interaction of both significantly affected the level of lipid oxidation compared to the meatball without added antioxidants ($p < 0.001$) (Figure 4) of which the antioxidant species was shown to have the largest effect size on the level of lipid oxidation (Table A3). The summer savory powder (*Satureja hortensis* L., SS) at 200 mg kg^{-1} and 100 mg kg^{-1}, water extracted sea buckthorn (*Hippophae rhamnoides* L., SBTH2O) at 100 mg kg^{-1}, and olive polyphenols (*Olea europaea* L., OPP) at 200 mg kg^{-1} were the most efficient antioxidants, lowering lipid oxidation to 13.8%, 21.8%, 22.9%, and 26.1%, respectively (Table A4), compared to the meatball with no added antioxidants. There were no significant correlations (Pearson) between the total phenols content (GAE mg mL^{-1}) and the lipid oxidation inhibition capacity of the antioxidants (Table A5).

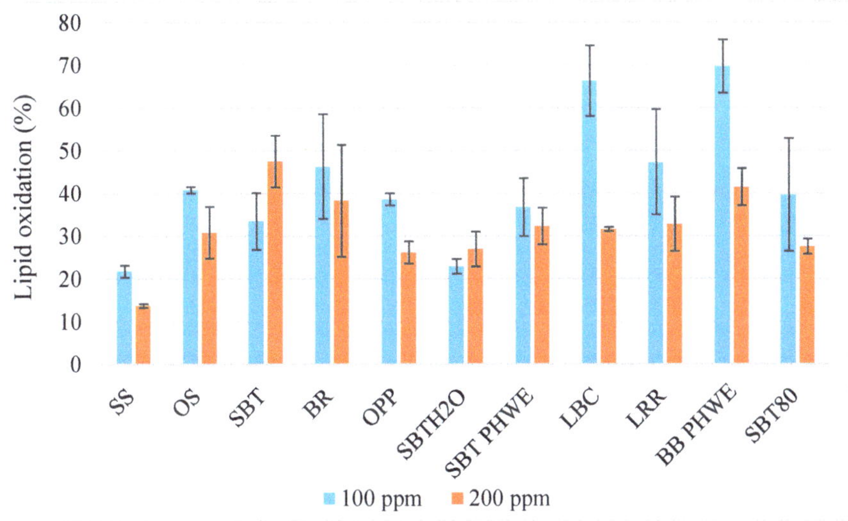

Figure 4. Lipid oxidation in the meatball type most prone to oxidize (pork with 2% NaCl, 20% fat (w/w), deep-fried), with different antioxidants shown as a percentage of oxidation compared to the meatball without added antioxidants at two concentrations. SS = Summer Savory, OS = Onion skin, SBT = Sea buckthorn leaves, BR = Beetroot leaves, OPP = Olive polyphenols, SBTH2O = water extracted sea buckthorn leaves and sprouts, SBT PHWE = Sea buckthorn leaves and sprouts-pressurized hot water extraction, LBC = Lyophilized black currant leaves, LRR = Lyophilized rhubarb root, BB PHWE = Bilberry leaves-pressurized hot water extraction, and SBT80 = ethanol (80%) extracted sea buckthorn leaves and sprouts. The standard deviation is shown by the error bars ($n = 3$).

4. Discussion

This study was divided into two parts, where the aim of the first part was to select a meatball model prone to oxidize. The aim of the second part of this work was to include and test various types of natural antioxidants for lipid oxidation inhibiting purposes. Both parts of the study will be discussed separately hereunder.

4.1. Preliminary Trial for Meatball Model Selection

We found that meat type was the most important factor affecting lipid oxidation (Table A1) where the pork meatballs overall were significantly ($p < 0.001$) more prone to oxidize than those of beef meat (Figure 3). This was not according to our expectations since pork meat contains less myoglobin (2 mg g^{-1}) than beef meat (8 mg g^{-1}) [16], and since the heme-iron content in meat is commonly hypothesized to be one of the significant substances in red and processed meat inducing carcinogenesis due to its involvement in mutagenic nitroso-compounds (NOCs) and the production of reactive oxygen species (ROS), producing lipid oxidation secondary products, such as malondialdehyde (MDA) [3,17,18]. However, a plausible explanation for the obtained results of the meatballs made of beef or pork meat could be differences in the levels of polyunsaturated fatty acids (PUFA) frequently involved in lipid oxidation. Pork meat has considerably higher amounts of linoleic acid (18:2) in both adipose tissue (14.3 g 100 g^{-1}) and muscle tissue (14.2 g 100 g^{-1}) than beef adipose tissue (1.1 g 100 g^{-1}) and muscle tissue (2.4 g 100 g^{-1}) [19]. This difference in linoleic acid (18:2) content could be of importance, since the oxidation in linoleic acid occurs 10 times faster than in oleic acid (18:1), and 20 to 30 times faster than linolenic acid (18:3) [17].

The meatball that oxidized the most contained pork, 20 g 100 g^{-1} fat, and 2 g 100 g^{-1} salt, and had been deep-fried and stored for 14 days (Figure 2). This meatball type was then chosen for further analysis, although it did not statistically differ from other meatballs (Table A2), since the aim remained to find the meatball type with the most oxidizing combinations of parameters. Interestingly, the beef meatballs that contained 10 g 100 g^{-1} fat and 2 g 100 g^{-1} salt that had been pan-fried and stored for one and 14 days oxidized more than the other combinations of beef meatballs (Figure 2), which could be an explanation for the antagonistic interaction between fat content and meat type (Figure 3b). Initially, we hypothesized that the higher the fat and salt contents in the meatballs, the higher the level of lipid oxidation, but this was not shown to be the case in our study. Salt (NaCl) is generally known to be a pro-oxidant in meat and meat products [5]; however, salt contents over 3 g 100 g^{-1} have shown little to no pro-oxidant effect [17], which is in accordance with the findings in the present study (Figure 3).

Although we found significant effects of cooking types on the level of lipid oxidation (Table A1), both deep-frying and pan-frying are considered to belong to the dry heat cooking category [20]. Hence, the heat transfer could be considered to be equal between the cooking types in matters of changes in physiochemical properties. When meatballs were analyzed for fat loss during pan-frying, Granheimer [13] found that the beef meatballs had higher fat loss than those of pork. However, in our study, pan-fried beef meatballs with 10 g 100 g^{-1} fat and 2 g 100 g^{-1} salt, which had higher TBARS levels (Figure 3), instead gained fat. In the same study, beef meatballs were shown to gain more fat during deep-frying than pork meatballs, in which fat was lost [13]. Haak et al. [21] found that the fatty acid composition of pan-fried meats became similar to that of the culinary fat due to its fat uptake. This could then explain why the beef meatball oxidized more than the others since rapeseed oil contains 20.9 g 100 g^{-1} PUFA [22] and thus increases the susceptibility to oxidation compared to the other beef meatballs. That all meatballs oxidized more over time ($p < 0.001$) (Figure 3) was expected due to the chain reaction nature of the lipid oxidation reaction.

4.2. Test of Antioxidants in the Most Oxidized Pork Meatball

Synthetic antioxidants have typically been included into meat products to increase, e.g., shelf-life and nutritional value, and now the industry is demanding new natural sources of antioxidant ingredients [5]. Numerous trials and experiments have successfully been carried out in screening the lipid oxidation inhibiting capacity of extracts from plant materials in various meat products [7]. For instance, olive leaf extracts have previously shown to be a potent antioxidant in bovine and porcine muscle model systems [9], oregano and sage essential oils have significantly reduced levels of lipid oxidation [8], and summer savory has previously been shown to reduce lipid oxidation in pork meatballs significantly [23]. All samples tested in our study effectively inhibited lipid oxidation at both concentrations ($p < 0.001$) (Figure 4), where the summer savory (SS) powder was the most efficient at both concentrations, and statistically differed from all other samples at 200 mg kg^{-1} (Table A4). The statistical results showed that antioxidant species, concentration, and the interaction of both had significant effects on lipid oxidation ($p < 0.001$), where species had the largest effect size (Table A5). Overall, samples inhibited lipid oxidation more efficiently at 200 mg kg^{-1} than at 100 mg kg^{-1} (Figure 4) except for the sea buckthorn sample extracted with 50 mL 100 mL^{-1} ethanol (SBT), which was significantly more effective at 100 mg kg^{-1}, and the sea buckthorn sample extracted with H$_2$O (SBTH2O), which showed tendencies to be more efficient at 100 mg kg^{-1}. Radenkovs et al. [24] previously attributed this phenomenon to the phenol composition of each antioxidant, where reaction speed predominantly depends on each phenol's chemical structure rather than its concentration. This explanation is further reinforced when interpreting the non-significant correlation analysis (Table A5) between the total phenols content and inhibition of lipid oxidation capacity. In further research, meatballs without and with antioxidants will be tested in vivo to evaluate potential changes in intestinal inflammatory reactions following a diet consisting of 20 g 100 g^{-1} of these meatballs.

5. Conclusions

The aim of the current study was to evaluate the lipid oxidation inhibiting capacity of natural antioxidants in a readily oxidized meat product. Various meatball properties were studied in order to find the combination that gave rise to the most lipid oxidation. The meatball most prone to oxidize was deep-fried, made of pork, contained 20 g 100 g^{-1} fat and 2 g 100 g^{-1} salt, and had been stored for 14 days. This meatball type was then manufactured without and with 11 different plant materials and extracts at two concentrations, 100 mg kg^{-1} and 200 mg kg^{-1} GAE, and was stored for 14 days. All samples inhibited lipid oxidation effectively in both tested concentrations, where the summer savory powder was the most efficient in both the 100 mg kg^{-1} and 200 mg kg^{-1} concentration, lowering lipid oxidation to 21.8% and 13.8%, respectively, compared to meatballs with no added antioxidants. Thus, antioxidant rich plant materials and extracts could efficiently prevent lipid oxidation in processed meat products, such as meatballs.

Author Contributions: Conceptualization, E.T. and S.C.M.B.; methodology, E.T.; software, S.C.M.B.; validation, E.T., K.R. and A.E.; formal analysis, S.C.M.B. and K.R.; investigation, S.C.M.B., K.G., A.E. and M.R.; resources, K.R.; data curation, S.C.M.B.; writing—original draft preparation, S.C.M.B.; writing—review and editing, K.R., K.G., Å.H.; visualization, S.C.M.B.; supervision, K.R., Å.H.; project administration, K.R.; funding acquisition, K.R., Å.H., E.T.

Funding: This work was supported by the Swedish Research Council for Environment, Agricultural Sciences and Spatial Planning (Grant 222-2014-1923) within the ERA-NET SUSFOOD transnational cooperation programme (including Denmark, Estonia, Finland, Latvia and Sweden) for the project SUSMEATPRO, "Sustainable plant ingredients for healthier meat products—proof of concepts".

Acknowledgments: We thank Sofia Stüffe and Hans Mathiason at Atria Supply AB for their help in accomplishing this work and for their contribution of meat. We also thank Uko Bleive and Tõnu Püssa from the Estonian University of Life Sciences, Estonia, Martin Jensen from Aarhus University, Denmark, Jarkko Hellström, Sari Mäkinen, Risto Korpinen and Pirjo Mattila from the Natural Resources Institute, Finland, Vitalijs Radenkovs and Dalija Seglina from the Institute of Horticulture, Latvia, for their help in acquiring and extracting plant materials for this study.

Appendix A

Table A1. General Linear model (GLM) table of the lipid oxidation affecting parameters on the variable TBAR, where * = $p < 0.05$, ** = $p < 0.01$, and *** = $p < 0.001$. Df = Degrees of freedom. The parameters with the largest effect sizes are shown in bold.

	Df	Sum of Squares	Mean Square	F Value	*p* Value	Pr (>F)	Partial Eta Squared
Meat	1	**18.12**	18.12	2142.5	<0.001	***	0.957
Salt	1	1.94	1.94	229.3	<0.001	***	0.705
Fat	1	0.06	0.06	7.6	0.007	**	0.074
Cooking type	1	0.40	0.40	47.5	<0.001	***	0.331
Storage time	2	**16.49**	8.24	975.1	<0.001	***	0.953
Meat × Salt	1	0.07	0.07	8.7	0.004	**	0.083
Meat × Fat	1	1.98	1.98	234.2	<0.001	***	0.709
Salt × Fat	1	0.36	0.36	42.4	<0.001	***	0.306
Meat × Cooking	1	1.01	1.01	119.2	<0.001	***	0.554
Salt × Cooking	1	0.01	0.01	0.7	0.405		0.007
Fat × Cooking	1	0.54	0.54	63.5	<0.001	***	0.398
Meat × Storage	2	0.96	0.48	56.9	<0.001	***	0.542
Salt × Storage	2	0.52	0.26	31.0	<0.001	***	0.392
Fat × Storage	2	0.12	0.06	7.1	0.001	**	0.128
Cooking × Storage	2	0.32	0.16	19.1	<0.001	***	0.284
Meat × Salt × Fat	1	0.14	0.14	16.4	0.000	***	0.146
Meat × Salt × Cooking	1	0.27	0.27	31.6	<0.001	***	0.247
Meat × Fat × Cooking	1	0.49	0.49	58.3	<0.001	***	0.378
Salt × Fat × Cooking	1	0.25	0.25	29.5	<0.001	***	0.235
Meat × Salt × Storage	2	0.04	0.02	2.4	0.095		0.048
Meat × Fat × Storage	2	0.08	0.04	5.0	0.009	**	0.094
Salt × Fat × Storage	2	0.17	0.08	10.1	0.000	***	0.173
Meat × Cooking × Storage	2	0.75	0.37	44.2	<0.001	***	0.479
Salt × Cooking × Storage	2	0.09	0.05	5.4	0.006	**	0.100
Fat × Cooking × Storage	2	0.09	0.04	5.2	0.007	**	0.097
Meat × Salt × Fat × Cooking	1	0.01	0.01	0.9	0.336		0.010
Meat × Salt × Fat × Storage	2	0.02	0.01	1.1	0.355		0.021
Meat × Salt × Cooking × Storage	2	0.13	0.06	7.5	0.001	***	0.134
Meat × Fat × Cooking × Storage	2	0.07	0.03	3.8	0.025	*	0.074
Salt × Fat × Cooking × Storage	2	0.14	0.07	8.1	0.001	***	0.145
Meat × Salt × Fat × Cooking × Storage	2	0.08	0.04	4.8	0.011	*	0.090
Residuals	96	0.81	0.01				

Table A2. Combinations of parameters promoting lipid oxidation ranked from highest to lowest value, where DF = Deep-frying, PF = Pan-frying, CI = Confidence interval, and EMM = estimated marginal means. The standard error was 0.053 and the degrees of freedom were 96 for all samples. All samples were compared pairwise and were assigned one or more group letters to present their respective significant difference or lack thereof.

Meat	Salt (%)	Fat (%)	Cooking Type	Storage Time (Days)	TBAR EMM	Lower CI Limit	Upper CI Limit	Group
Pork	2	20	DF	14	2.819	2.714	2.925	a
Beef	2	10	PF	14	2.681	2.576	2.787	ab
Pork	2	10	DF	14	2.610	2.505	2.716	abc
Pork	2	20	PF	14	2.574	2.468	2.679	abcd
Pork	2	20	DF	7	2.560	2.455	2.666	abcd
Pork	2	10	PF	14	2.539	2.433	2.644	abcd
Pork	4	20	PF	14	2.485	2.379	2.590	bcde
Pork	4	20	DF	14	2.417	2.311	2.522	bcde
Pork	2	10	DF	7	2.347	2.242	2.453	cdef
Beef	2	10	PF	7	2.330	2.224	2.435	cdef
Pork	4	20	DF	7	2.296	2.191	2.402	def
Pork	4	10	DF	14	2.286	2.180	2.391	def
Pork	4	20	PF	7	2.284	2.178	2.389	def
Pork	2	20	PF	7	2.199	2.094	2.305	efg
Pork	2	10	PF	7	2.195	2.089	2.300	efg
Pork	4	10	PF	14	2.188	2.083	2.294	efg
Pork	4	10	DF	7	2.043	1.938	2.149	fgh
Beef	4	10	PF	14	1.899	1.793	2.004	ghi
Pork	4	10	PF	7	1.898	1.793	2.004	ghi
Beef	2	20	PF	14	1.816	1.710	1.921	hij
Pork	2	20	DF	1	1.664	1.559	1.769	ijk
Pork	2	20	PF	1	1.645	1.540	1.750	ijkl
Pork	4	20	PF	1	1.627	1.522	1.732	ijkl
Beef	2	10	DF	14	1.565	1.459	1.670	jklm
Pork	4	20	DF	1	1.458	1.353	1.563	klmn
Pork	2	10	PF	1	1.442	1.336	1.547	klmn
Pork	4	10	PF	1	1.430	1.324	1.535	klmno
Beef	2	20	DF	14	1.412	1.306	1.517	klmnop
Beef	2	10	DF	7	1.403	1.298	1.509	klmnop
Pork	2	10	DF	1	1.402	1.296	1.507	klmnop
Beef	2	20	DF	7	1.374	1.268	1.479	klmnop
Beef	4	20	PF	14	1.342	1.237	1.448	lmnop
Pork	4	10	DF	1	1.342	1.237	1.447	lmnop
Beef	4	10	PF	7	1.317	1.212	1.423	mnopq
Beef	4	20	PF	7	1.260	1.155	1.366	mnopqr
Beef	4	10	DF	14	1.257	1.152	1.363	mnopqrs
Beef	4	20	DF	14	1.244	1.139	1.350	nopqrs
Beef	4	20	DF	7	1.230	1.125	1.336	nopqrs
Beef	2	10	PF	1	1.217	1.112	1.322	nopqrst
Beef	2	20	PF	7	1.217	1.111	1.322	nopqrst
Beef	4	10	DF	7	1.198	1.093	1.304	nopqrst
Beef	2	10	DF	1	1.133	1.027	1.238	opqrstu
Beef	4	10	PF	1	1.104	0.999	1.209	pqrstu
Beef	2	20	DF	1	1.011	0.905	1.116	qrstu
Beef	4	20	DF	1	0.978	0.873	1.084	rstu
Beef	4	20	PF	1	0.949	0.844	1.055	stu
Beef	4	10	DF	1	0.919	0.814	1.025	tu
Beef	2	20	PF	1	0.868	0.762	0.973	u

Table A3. GLM table of the lipid oxidation affecting parameters on the variable TBAR of meat products with different supplemented antioxidants, where *** = $p < 0.001$, and Df = Degrees of freedom. The parameter with the largest effect size is shown in bold.

	Df	Sum of Squares	Mean Square	F Value	p Value	Pr (>F)	Partial Eta Squared
Species	10	**0.024**	0.002	37.6	<0.001	***	0.895
Concentration	1	0.005	0.005	72.7	<0.001	***	0.623
Species × Concentration	10	0.005	0.001	8.1	<0.001	***	0.647
Residuals	44	0.003	0.000				

Table A4. Combinations of species and concentrations of antioxidants resulting in the lowest lipid oxidation level, where SS = Summer Savory, OS = Onion skin, SBT = Sea buckthorn leaves, BR = Beetroot leaves, OPP = Olive polyphenols, SBTH2O = water extracted sea buckthorn leaves and sprouts, SBT PHWE = Sea buckthorn leaves and sprouts-pressurized hot water extraction, LBC = Lyophilized black currant leaves, LRR = Lyophilized rhubarb root, BB PHWE = Bilberry leaves – pressurized hot water extraction, SBT80 = ethanol (80%) extracted sea buckthorn leaves and sprouts, CI = Confidence interval, and EMM = estimated marginal means. The standard error was 0.005 and the degrees of freedom were 44 for all samples. Note that the higher the value, the lower the level of oxidation due to a boxcox-transformation of the dependent variable with ˆ −0.7. All samples were compared pairwise and were assigned one or more group letters to present their respective significant difference or lack thereof.

Concentration (ppm)	Species	TBAR EMM	Lower CI Limit	Upper CI Limit	Group
200	SS	0.157	0.148	0.166	a
100	SS	0.113	0.104	0.123	b
100	SBTH2O	0.110	0.100	0.119	bc
200	OPP	0.100	0.091	0.109	bcd
200	SBTH2O	0.098	0.089	0.108	bcd
200	SBT80	0.096	0.087	0.105	bcde
200	OS	0.090	0.081	0.099	bcdef
200	BR	0.088	0.079	0.097	cdef
200	LBC	0.087	0.078	0.096	cdef
200	SBT PHWE	0.086	0.077	0.096	cdef
200	LRR	0.086	0.077	0.095	cdef
100	SBT	0.085	0.075	0.094	cdefg
100	SBT PHWE	0.080	0.070	0.089	defg
100	SBT80	0.077	0.068	0.087	defg
100	OPP	0.076	0.066	0.085	defgh
100	OS	0.073	0.063	0.082	efghi
200	BB PHWE	0.072	0.063	0.082	efghi
100	LRR	0.067	0.058	0.076	fghi
200	SBT	0.066	0.056	0.075	fghi
100	BR	0.060	0.051	0.069	ghi
100	LBC	0.052	0.043	0.061	hi
100	BB PHWE	0.050	0.041	0.059	i

Table A5. Pearson correlation between total phenols content (GAE mg mL^{-1}) and TBARS values at the 100 mg kg^{-1} and 200 mg kg^{-1} concentration where $n = 11$.

		Total Phenols	TBARS 100	TBARS 200	Species
Total phenols	Correlation	1	0.100	−0.1333	0.543
	p value		0.768	0.695	0.084
TBARS 100	Correlation	0.100	1	0.482	0.566
	p value	0.768		0.133	0.069
TBARS 200	Correlation	−0.133	0.482	1	0.204
	p value	0.696	0.133		0.547
Species	Correlation	0.543	0.566	0.204	1
	p value	0.084	0.069	0.547	

References

1. International Agency for Research on Cancer (IARC). *IARC Monographs Evaluate Consumption of Red Meat and Processed Meat*; WHO: Lyon, France, 2015.
2. Hammerling, U.; Bergman Laurila, J.; Grafström, R.; Ilbäck, N.-G. Consumption of Red/Processed Meat and Colorectal Carcinoma: Possible Mechanisms Underlying the Significant Association. *Crit. Rev. Food Sci. Nutr.* **2016**, *56*, 20. [CrossRef] [PubMed]
3. Cascella, M.; Bimonte, S.; Barbieri, A.; Del Vecchio, V.; Caliendo, D.; Schiavone, V.; Fusco, R.; Granata, V.; Arra, C.; Cuomo, A. Dissecting the mechanisms and molecules underlying the potential carcinogenicity of red and processed meat in colorectal cancer (CRC): An overview on the current state of knowledge. *Infect. Agent Cancer* **2018**, *13*, 3. [CrossRef] [PubMed]
4. Domingo, J.L.; Nadal, M. Carcinogenicity of consumption of red meat and processed meat: A review of scientific news since the IARC decision. *Food Chem. Toxicol.* **2017**, *105*, 256–261. [CrossRef] [PubMed]
5. Jiang, J.; Xiong, Y.L. Natural antioxidants as food and feed additives to promote health benefits and quality of meat products: A review. *Meat Sci.* **2016**, *120*, 107–117. [CrossRef] [PubMed]
6. Israelsson, L. Ikeas köttbullar inte helt svenska—Kött från Irland. *Expressen* **2019**.
7. Kumar, Y.; Yadav, D.N.; Ahmad, T.; Narsaiah, K. Recent trends in the use of natural antioxidants for meat and meat products. *Compr. Rev. Food Sci. Food Saf.* **2015**, *14*, 796–812. [CrossRef]
8. Fasseas, M.K.; Mountzouris, K.C.; Tarantilis, P.A.; Polissiou, M.; Zervas, G. Antioxidant activity in meat treated with oregano and sage essential oils. *Food Chem.* **2008**, *106*, 1188–1194. [CrossRef]
9. Hayes, J.E.; Stepanyan, V.; Allen, P.; O'Grady, M.N.; O'Brien, N.M.; Kerry, J.P. The effect of lutein, sesamol, ellagic acid and olive leaf extract on lipid oxidation and oxymyoglobin oxidation in bovine and porcine muscle model systems. *Meat Sci.* **2009**, *83*, 201–208. [CrossRef] [PubMed]
10. Lee, S.K.; Han, J.H.; Decker, E.A. Antioxidant activity of phosvitin in phosphatidylcholine liposomes and meat model systems. *Food Chem. Toxicicol.* **2002**, *67*, 37–41. [CrossRef]
11. Burri, C.M.S.; Ekholm, A.; Håkansson, Å.; Tornberg, E.; Rumpunen, K. Antioxidant capacity and major phenol compounds of horticultural plant materials not usually used. *J. Funct. Foods* **2017**, *38*, 119–127. [CrossRef] [PubMed]
12. Gornas, P.; Sne, E.; Siger, A.; Seglina, D. Sea buckthorn (*Hippophae rhamnoides* L.) leaves as valuable source of lipophilic antioxidants: The effect of harvest time, sex, drying and extraction methods. *Ind. Crops Prod.* **2014**, *60*, 1–7. [CrossRef]
13. Granheimer, K. Different Parameters Affecting Lipid Oxidation in Meatballs. Master's Thesis, Lund University, Lund, Sweden, 2017.
14. Buege, A.J.; Aust, S.D. Microsomal Lipid Peroxidation. *Methods Enzym.* **1978**, *52*, 302–310.
15. Kim, H.Y. Statistical notes for clinical researchers: Assessing normal distribution (2) using skewness and kurtosis. *Restor. Dent. Endod.* **2013**, *38*, 52–54. [CrossRef] [PubMed]
16. Hedrick, H.B.; Aberle, E.D.; Forrest, J.C.; Judge, M.D. *Principles of Meat Science*, 3rd ed.; Kendall Hunt: Dubuque, IA, USA, 2013.
17. Amaral, A.B.; da Silva, M.V.; Lannes, S.C.D.S. Lipid oxidation in meat: Mechanisms and protective factors—A review. *Food Sci. Technol.* **2018**, *38*, 15. [CrossRef]
18. Joosen, A.M.; Kuhnle, G.G.; Aspinall, S.M.; Barrow, T.M.; Lecommandeur, E.; Azqueta, A.; Collins, A.R.; Bingham, S.A. Effect of processed and red meat on endogenous nitrosation and DNA damage. *Carcinogenesis* **2009**, *30*, 1402–1407. [CrossRef] [PubMed]
19. Pereira, F.A.L.; Abreu, G.V.K. Lipid Peroxidation in Meat and Meat Products. In *Lipid Peroxidation, Edited Volume*; Catala, A., Mansour, M.A., Eds.; IntechOpen: London, UK, 2018.
20. Pearson, A.M.; Gillett, T.A. *Processed Meats*, 3rd ed.; Oregon State University: Corvallis, OR, USA, 2012.
21. Haak, L.; Sioen, I.; Raes, K.; Camp, J.; Smet, S. Effect of pan-frying in different culinary fats on the fatty acid profile of pork. *Food Chem.* **2007**, *102*, 857–864. [CrossRef]
22. Orsavova, J.; Misurcova, L.; Ambrozova, J.V.; Vicha, R.; Mlcek, J. Fatty Acids Composition of Vegetable Oils and Its Contribution to Dietary Energy Intake and Dependence of Cardiovascular Mortality on Dietary Intake of Fatty Acids. *Int. J. Mol. Sci.* **2015**, *16*, 12871–12890. [CrossRef] [PubMed]

23. Lindberg-Madsen, H.; Andersen, L.; Christiansen, L.; Brockhoff, P.; Bertelsen, G. Antioxidative activity of summer savory (*Satureja hortensis* L.) and rosemary (*Rosmarinus officinalis* L.) in minced, cooked pork meat. *Z. Lebensm. Unters. Forsch.* **1996**, *203*, 333–338. [CrossRef]
24. Radenkovs, V.; Püssa, T.; Juhnevica-Radenkova, K.; Anton, D.; Seglina, D. Phytochemical characterization and antimicrobial evaluation of young leaf/shoot and press cake extracts from Hippophae rhamnoides L. *Food Biosci.* **2018**, *24*, 10. [CrossRef]

MDPI

St. Alban-Anlage 66

4052 Basel

Switzerland

Tel. +41 61 683 77 34

Fax +41 61 302 89 18

www.mdpi.com

Foods Editorial Office

E-mail: foods@mdpi.com

www.mdpi.com/journal/foods